PARASITOLOGY
FOR THE 21ST CENTURY

Parasitology
for the 21st Century

Keynote Papers from the VIII International Congress of Parasitology

Edited by

M. Ali Özcel

President of the Organizing Committee
VIII International Congress of Parasitology
PK 81 35042
Bornova
Izmir, Turkey

M. Ziya Alkan

Secretariat
VIII International Congress of Parasitology
PK 81 35042
Bornova
Izmir, Turkey

CAB INTERNATIONAL

CAB INTERNATIONAL	Tel: +44 (0)1491 832111
Wallingford	Telex: 847964 (COMAGG G)
Oxon OX10 8DE	E-mail: cabi@cabi.org
UK	Fax: +44 (0)1491 833508

A catalogue record for this book is available from the British Library.

ISBN 0 85198 977 2

Typeset in 10/12 pt Mallard by Colset Pte Ltd, Singapore
Printed and bound in the UK by Biddles Ltd, Guildford

Contents

Contributors

H. Akuffo
Microbiology and Tumor Biology Center,
Karolinska Institutet and Smittskyddsinstitutet,
Box 280,
S171 77 Stockholm,
Sweden

S.B. Aley
University of California,
San Diego,
CA 92103–8416,
USA

E.M. Aliouat
INSERM (U.42),
369 rue J. Guesde,
BP 39,
59650 Villeneuve d'Ascq,
France

P. Arhets
Unité de Pathogénie Microbienne Moléculaire,
INSERM U389,
Institut Pasteur,
28 rue de Dr Roux,
75724 Paris Cédex 15,
France

R. Armijos
Laboratory of Immunology,
Central University Medical School,
Quito,
Ecuador

H. Aviles
Laboratory of Immunology,
Central University Medical School,
Quito,
Ecuador

P. Ayala
Departments of Experimental Pathology, Genetics and Molecular Biology, Centro de Investigacion y de Estudios Avanzados del I.P.N.,
México City,
México

S. Britton
Department of Immunology, Microbiology, Pathology and Infectious Diseases,
Division of Infectious Diseases,
Karolinska Institute at Huddinge Hospital,
141 86 Huddinge,
Sweden

Z.I. Cabantchik
Department of Biological Chemistry, The Hebrew University, Jerusalem, Israel

J.M. Calvo-Calle
Department of Medical and Molecular Parasitology, New York University School of Medicine, 341 East 25th Street, New York, NY 10016, USA

D. Camus
INSERM (U.42), 369 rue J. Guesde, BP 39, 59650 Villeneuve d'Ascq, France, and Laboratory of Parasitology-Mycology, Faculty of Medicine and Hospital University Centre, 1 Place Verdun, 59045 Lille, France

E.M. Carvalho
Serviço de Imunologia, Universitaria Prof. Edgar Santos (HUPES), UFBA, Lab Immunologia 30 Rua and Rua Joao das Botas, a/n Canela 40.110.160, Salvador-Bahia, Brazil

K.-P. Chang
Chicago Medical School, North Chicago, IL 60064, USA

M. Chevion
Department of Cellular Biochemistry, Hadassah Medical School, The Hebrew University, Jerusalem, Israel

P. Clavijo
Department of Medical and Molecular Parasitology, New York University School of Medicine, 341 East 25th Street, New York, NY 10016, USA

E. Dei-Cas
INSERM (U.42), 369 rue J. Guesde, BP 39, 59650 Villeneuve d'Ascq, France, and Laboratory of Parasitology-Mycology, Faculty of Medicine and Hospital University Centre, 1 Place Verdun, 59045 Lille, France

M. del Consuelo Gómez
Program of Molecular Biomedicine and Department of Experimental Pathology, Centro de Investigacion y de Estudios Avanzados del I.P.N., México City, México

C.M. Denis
INSERM (U.42), 369 rue J. Guesde, BP 39, 59650 Villeneuve d'Ascq, France

J.B. Dixon
Department of Veterinary Pathology,
University of Liverpool,
P.O. Box 147,
Liverpool L69 3BX,
UK

I. Durand
INSERM (U.42),
369 rue J. Guesde,
BP 39,
59650 Villeneuve d'Ascq,
France

L. Eckmann
UCSD Medical Center,
200 W Arbor Drive,
San Diego,
CA 92103-8426,
USA

Gh.H. Edrissian
School of Public Health and
Institute of Public Health
Researches,
Tehran University of Medical
Sciences,
P.O. Box 6446-14155,
Tehran,
Iran

R. Farfan
Department of Medical and
Molecular Parasitology,
New York University School of
Medicine,
341 East 25th Street,
New York,
NY 10016,
USA

O. Gaye
Faculty of Medicine,
Dakar-FANN,
Dakar,
Senegal

I. Gigli
UCSD Medical Center,
200 W Arbor Drive,
San Diego,
CA 92103-8426,
USA

F.D. Gillin
University of California,
San Diego,
CA 92103-8416,
USA

T. Godal
UNDP/World Bank/WHO,
Special Programme for Research
and Training in Tropical Diseases,
World Health Organization,
CH-1211,
Geneva,
Switzerland

J. Golenser
Department of Parasitology,
Hadassah Medical School,
The Hebrew University,
Jerusalem,
Israel

V.R. Gordon
Department of Veterinary
Pathobiology and Natural History
Survey,
University of Illinois,
Urbana,
IL 61801,
USA

B. Gottstein
Institute of Parasitology,
University of Bern,
P.O. Box 8466,
Bern,
Switzerland

P. Gounon
Station Centrale de Microscopie
Electronique,

Institut Pasteur,
28 rue de Dr Roux,
75724 Paris Cédex 15,
France

N. Guillén
Unité de Pathogénie Microbienne Moléculaire,
INSERM U389,
Institut Pasteur,
28 rue de Dr Roux,
75724 Paris Cédex 15,
France

R.D. Hansen
Department of Veterinary Pathobiology and Natural History Survey,
University of Illinois,
Urbana,
IL 61801,
USA

C. Harf
Laboratoire d'Hydrobiologie de l'Environnement,
Faculté de Pharmacie,
Université Louis Pasteur Strasbourg,
B.P. 24-67401,
Illkirch Cedex,
France

M.L. Hetsko
University of California,
San Diego,
CA 92103-8416,
USA

R. Houin
Faculté de Médecine,
Laboratoire de Parasitologie-Mycologie,
8 rue du Général Sarrail,
94010 Créteil Cedex,
France

L. Hviid
Centre for Medical Parasitology,
Department of Infectious Diseases M 7641,
National University Hospital (Rigshospitalet) and
Institute for Medical Microbiology and Immunology,
University of Copenhagen,
Copenhagen,
Denmark

C.L. Jaffe
Department of Parasitology,
Hebrew University-Hadassah Medical School,
P.O. Box 12272,
Jerusalem 91120,
Israel

M. James
Ochsner Medical Foundation and Department of Tropical Medicine,
Tulane University,
New Orleans,
LA 70112,
USA

M. Kagnoff
UCSD Medical Center,
200 W Arbor Drive,
San Diego,
CA 92103-8426,
USA

I. Kakoma
Department of Veterinary Pathobiology and Natural History Survey,
University of Illinois,
Urbana,
IL 61801,
USA

N.R. Lynch
Instituto de Biomedicina,
Aptdo 4043,
Carmelitas,
1010A Caracas,
Venezuela

K. Maasho
Department of Immunology, Microbiology, Pathology and Infectious Diseases, Division of Infectious Diseases, Karolinska Institute at Huddinge Hospital, 141 86 Huddinge, Sweden

R.C. Mahajan
Department of Parasitology, Postgraduate Institute of Medical Education and Research, Chandigarh - 160 012, India

E. Mazars
INSERM (U.42), 369 rue J. Guesde, BP 39, 59650 Villeneuve d'Ascq, France

J.M. McCaffery
University of California, San Diego, CA 92103-8416, USA

J.H. McKerrow
VA Medical Center, San Francisco, CA 94143, USA

G.L. McLaughlin
Department of Pathology and Laboratory Medicine, Indiana University Medical School, Indianapolis, IN 46202, USA

D.P. McManus
Molecular Parasitology Unit, Tropical Health Program, The Queensland Institute of Medical Research, 300 Herston Road, Brisbane 4029, Oueensland, Australia

R. Mdachi
Kenya Trypanosomiasis Research Institute, Muguga, Kenya

T.-C. Meng
University of California, San Diego, CA 92103-8416, USA

K. Mohan
Department of Parasitology, Postgraduate Institute of Medical Education and Research, Chandigarh - 160 012, India

S. Montenegro-James
Ochsner Medical Foundation and Department of Tropical Medicine, Tulane University, New Orleans, LA 70112, USA

A. Moreno
Instituto de Inmunologia, Hospital San Juan de Dios, Bogota, Colombia

P. Morera
School of Medicine and Institute for Health Research, University of Costa Rica, Pathology Department, Hospital San Juan de Dios, P.O. Box 2177-1000, San José, Costa Rica

K.E. Mott
Schistosomiasis Control Unit,
Division of Control of Tropical Diseases,
World Health Organization,
Geneva 27,
Switzerland

Y. Munesinghe
Department of Medical and Molecular Parasitology,
New York University School of Medicine,
341 East 25th Street,
New York,
NY 10016,
USA

E. Nanteza
Department of Microbiology and Parasitology,
Makerere University,
Kampala,
Uganda

E.H. Nardin
Department of Medical and Molecular Parasitology,
New York University School of Medicine,
341 East 25th Street,
New York, NY 10016,
USA

R.J. Novak
Department of Veterinary Pathobiology and Natural History Survey,
University of Illinois,
Urbana,
IL 61801,
USA

R. Nussenzweig
Department of Medical and Molecular Parasitology,
New York University School of Medicine,
341 East 25th Street,
New York,
NY 10016,
USA

C. Ödberg Ferragut
INSERM (U.42),
369 rue J. Guesde,
BP 39,
59650 Villeneuve d'Ascq,
France

G.A. Oliviera
Department of Medical and Molecular Parasitology,
New York University School of Medicine,
341 East 25th Street,
New York,
NY 10016,
USA

E. Orozco
Program of Molecular Biomedicine and Department of Experimental Pathology,
Centro de Investigacion y de Estudios Avanzados del I.P.N.,
México City,
México

D.G. Pérez
Program of Molecular Biomedicine and Department of Experimental Pathology,
Centro de Investigacion y de Estudios Avanzados del I.P.N.,
México City,
México

N. Rachamim
Department of Membrane Research and Biophysics,
Weizmann Institute of Science,
Rehovot,
Israel

S.L. Reed
UCSD Medical Center,

200 W Arbor Drive,
San Diego,
CA 92103–8426,
USA

D.S. Reiner
University of California,
San Diego,
CA 92103–8416,
USA

R. Rodriguez
Instituto de Inmunologia,
Hospital San Juan de Dios,
Bogota,
Colombia

Rubaire-Akiiki
Department of Microbiology and Parasitology,
Makerere University,
Kampala,
Uganda

J. Samuelson
Department of Tropical Public Health,
Harvard School of Public Health,
Boston,
MA 02115,
USA

P. Sansonetti
Unité de Pathogénie Microbienne Moléculaire,
INSERM U389,
Institut Pasteur,
28 rue de Dr Roux,
75724 Paris Cédex 15,
France

C. Santrich
Biochemistry Department,
Purdue,
West Lafayette,
IN 47906,
USA

N. Saravia
CIDEIM,
Cali,
Colombia, and
University of North Carolina,
Chapel Hill,
NC,
USA

R. Sarfstein
Department of Parasitology,
Hebrew University-Hadassah Medical School,
P.O. Box 12272,
Jerusalem 91120,
Israel

S.Z. Shapiro
Retrovirus Research Laboratory,
CBER/FDA,
Bethesda,
MD 20892,
USA

K.C. Shekhar
Department of Parasitology,
Faculty of Medicine,
University of Malaya,
Kuala Lumpur,
Malaysia

S.S. Ssenyonga
Department of Microbiology and Parasitology,
Makerere University,
Kampala,
Uganda

N.N. Taneli
Dokuz Eylül University,
Faculty of Medicine,
Department of Pediatrics,
Izmir,
Turkey

M.H. Vodkin
Department of Veterinary Pathobiology and Natural History

Survey,
University of Illinois,
Urbana,
IL 61801,
USA

O. Wafula
Department of Microbiology and Parasitology,
Makerere University,
Kampala,
Uganda

K. Weigle
CIDEIM,
Cali,
Colombia, and
University of North Carolina,
Chapel Hill,
NC,
USA

E. Wozniak
School of Veterinary Medicine,
University of California at Davis,
Davis,
CA,
USA

Preface

I am pleased to introduce the *Proceedings of the Eighth International Congress of Parasitology* (ICOPA VIII) held in Izmir, Turkey, on 10–14 October 1994. All chapters were presented at the congress as lectures or as part of a symposium or round table discussion.

The general concept of ICOPA VIII was 'New Dimensions in Parasitology for the 21st Century'. This congress comprised 21 lectures, 32 round table sessions and 26 symposia. There were more than 700 oral presentations (free communications) and 900 poster presentations. These presentations brought together new dimensions, tools, findings and techniques and introduced ideas for future research and control programmes of parasitic diseases all over the world.

Twenty-five presentations, mostly lectures, were selected for publication by the Scientific Commitee of ICOPA VIII, according to the agreement with the publisher CAB INTERNATIONAL.

Among the 25 presentations included in this book, five chapters present the latest information on immunology, serology and epidemiology of visceral leishmaniasis. Three chapters contain recent advances in cell-mediated immunity, multiple antigen peptides with epitopes in *Plasmodium falciparum* and one chapter is on oxidative stress and malaria. There are also chapters on *Entamoeba histolytica*, including drug-resistant strains, molecular and cell biology of invasion, cytoskeleton activities in pathogenic *E. histolytica* and the relationship between amoebae and bacteria.

Other chapters on different aspects of parasitology include: metabolic disturbances in giardiasis; infection sources, reservoirs and transmission of *Pneumocystis*; genetic variability in parasitic helminths; allergic reactivity and helminthic infections; protective immune mechanisms against *Echinococcus multilocularis*; the role of macrophages in host–metacestode interplay; epidemiological research with the tick *Rhinocephalus bursa*; abdominal angiostrongyloides; and polymerase chain reaction (PCR)-based detection and typing of parasites.

The topics discussed in this book will be an excellent source for research proposals in the future. Biological characteristics of parasites, molecular biological investigations, DNA technology, PCR-based detection and typing of parasites are all new dimensions in parasitology for the 21st century.

I would like to express my thanks to the distinguished scientists who contributed by providing the full papers of their presentations. Their enthusiasm and kind efforts greatly helped to make this congress successful and memorable. I want to express our appreciation to CAB INTERNATIONAL for publishing these proceedings and making them available to scientists in the field of parasitology. I also want to thank my co-editor Dr M. Ziya Alkan for his assistance in preparing the manuscripts for publication.

Professor Dr M. Ali Özcel
President, World Federation of Parasitologists
Head, Department of Parasitology,
Ege University Medical Faculty
President, Turkish Society for Parasitology

Foreword

One attends a congress like the Eighth International Congress of Parasitology for many reasons: to renew old friendships and to establish new friendships and collaborations with colleagues from all over the world; an opportunity to experience the history, culture and pleasures of an area like the Aegean cost of Turkey, and to experience the hospitality of its citizens, particularly our hosts; and finally to learn of the remarkable advances in one's field of scientific endeavour. In the case of the field of parasitology, there is also the opportunity to learn of the serious impact that parasites continue to have on the health and wellbeing of human and animal populations throughout the world.

Having experienced this Congress I can certainly say that from my perspective it was a remarkable success. The opportunity to interact with so many of my colleagues and friends was wonderful and meaningful both on a personal and scientific level. I am not certain that I have ever experienced the warmth and hospitality offered by Professor Özcel and his colleagues who organized the meeting.

In my role as scientific chairman I was delighted at the breadth and sophistication of the scientific programme, and the participation of so many outstanding scientists from throughout the world. From the point of malaria, I am certain that there has never been a national or international meeting with a programme that brought together so many of the principal scientists in the world working on malaria vaccine development, with such a diverse and intense series of plenaries, round tables and symposia. However, malaria vaccine development was not alone. Numerous participants told me that the same held true of the sessions on leishmaniasis and other parasitic diseases.

The theme of this congress was 'New Dimensions in Parasitology for the 21st Century'. Professor Özcel and his colleagues did a remarkable job in providing a setting where we could fulfil these objectives. In doing so they set a standard of excellence that should be a model for future congresses.

Stephen L. Hoffman, MD, DTMH
Scientific Chairman, ICOPA VIII

1

New Dimensions for Parasitology in the 21st Century

T. GODAL

UNDP/World Bank/WHO Special Programme for Research and Training in Tropical Diseases, World Health Organization, CH-1211, Geneva, Switzerland

Parasitology is an enormous subject. The very size of the audience at this Congress demonstrates this fact. No one person can possibly address the entire subject. This paper therefore confines itself to those parasites I know best, those that cause the major human tropical diseases. Tropical diseases will still provide an enormous challenge to parasitology in the 21st century: to work against the great misery they cause worldwide and to provide hope. I will discuss here how this may be achieved, and against what forces we shall have to work.

At least 500 million people on Earth – one living person in ten – suffer from one or more of the eight major tropical diseases: malaria, schistosomiasis (bilharzia), lymphatic filariasis (leading to elephantiasis), onchocerciasis (river blindness), leprosy, African trypanosomiasis (sleeping sickness), Chagas disease (American trypanosomiasis) and leishmaniasis (including kala azar). Some of these tropical diseases, in particular malaria, schistosomiasis and Chagas disease, are responsible for enormous losses of global disability adjusted life-years (DALYs), and others, while less prevalent, are fearsome or gruesome diseases, creating great personal social stigma or fear (such as leprosy, river blindness, elephantiasis, sleeping sickness, kala azar and mucocutaneous leishmaniasis). The World Health Organization (WHO) estimates that there are 300–500 million clinical cases of malaria per year, 90% of them in Africa, and 1.5–2.7 million deaths, at least one million of them African children under five years old. Of the 10% of estimated clinical cases outside Africa, some 70% come from just seven countries: in decreasing order of incidence these are: India, Brazil, Sri Lanka, Afghanistan, Thailand, Vietnam, and Colombia.

Some 200 million people are infected with schistosomiasis, and tens of millions suffer with debilitating chronic morbidity. Around 600 million people are at risk in 74 countries, particularly in urban areas, refugee camps, and where water resources are being developed.

Lymphatic filariasis affects 76 countries. Almost a billion people are at risk, and 100 million people are infected (including 45 million in India and 32 million in Africa); one million people are severely disabled by the disease.

River blindness affects 34 countries, mostly in West Africa. Some 90 million people at risk, and 18 million are infected. Over 300,000 people are blind from the disease, reducing their life expectancy by two-thirds; and hundreds of thousands more suffer with severe skin problems.

Leprosy affects 80 countries; and two billion people live in countries seriously affected. India accounts for 64% of the registered cases in the world. The six most endemic countries (Bangladesh, Brazil, India, Indonesia, Myanmar and Nigeria) account for 84%, and 25 countries account for 93% of cases.

African trypanosomiasis or sleeping sickness affects 36 countries. Prevalence rates are usually 1–2%, but can explode to 30% or higher in epidemics. Some 15–20,000 cases per year are reported, but only three to four million of the 55 million people at risk are under surveillance. The disease is fatal if untreated, and there is a constant risk of epidemics.

Chagas disease affects 21 countries in South America, putting 90 million people at risk. Some 12–16 million people are infected with the disease. There are one million new cases of chronic disease and some 45,000 deaths annually.

Leishmaniasis, a complex of several related parasitic diseases, affects 80 countries, and puts 350 million people at risk. Some 12 million people are infected, with 1–1.5 million new cases per year of cutaneous leishmaniasis, which leaves year-long scarring sores. There are 500,000 cases, and 75–80,000 deaths per year from fatal visceral leishmaniasis (kala azar) in continuing epidemics in India (where there are now 350,000 cases per year, particularly in Bihar state) and Sudan. The incidence of human immunodeficiency virus (HIV)/leishmania co-infections is also rising, especially in southern Europe.

Parasitologists can take some comfort in looking back to the last two decades of this century which have seen an extraordinary period of progress. Tools have been developed which have enabled WHO to take a series of strategic control decisions (Box 1.1).

Until the mid-1970s, leprosy was a disease more or less beyond hope. The parasite was developing resistance to dapsone monotherapy – which had been a lifelong treatment – and the many non-governmental organizations concerned with the disease were devoted mostly to

Box 1.1. Landmark WHO decisions on tropical disease control since 1975, following the development of new tools.

1981 – Leprosy (MDT)
1989 – Onchocerciasis (ivermectin)
1992 – Chagas disease (vector control – 'southern cone' countries)
1993 – Schistosomiasis in selected areas (praziquantel + vector control)
1996(?) – Lymphatic filariasis (single-dose ivermectin + diethyl carbamazine (DEC); DEC alone; DEC-impregnated salt)

caring for the disabled. Some 10–12 million people were estimated to be infected. Then in 1981 WHO, encouraged by the UNDP/World Bank/WHO Special Programme for Research and Training in Tropical Diseases (TDR), took the bold decision of recommending a simple three-drug regimen of therapy to combat resistance – multidrug therapy (MDT). TDR mapped and quantified the main areas of dapsone resistance, and established large-scale 'post-marketing surveillance' of MDT. By the end of 1993 the number infected was estimated to be 2.4 million – a fourfold decrease – with 600,000 new cases being detected annually. Cumulative coverage with curative MDT has reached 80%, and WHO has set a target of reducing prevalence to one case in 10,000 of the population in each affected country by the year 2000.

The greatest challenges for leprosy control will come in India and Brazil. Improved, short-course drug regimens to increase patient compliance are being tested; vaccines are under development; and important studies on improving patient detection and compliance, particularly among women, are being extended.

The Onchocerciasis Control Programme (OCP) in West Africa has virtually eliminated transmission of onchocerciasis or river blindness in 11 countries (protecting 30 million of the 90 million people at risk), but problems remain in sustaining protection when vector control stops, and in establishing control outside the OCP area. However, a great tool, ivermectin, has been generously donated by Merck Pharmaceutical Co., and progress is now being made to improve its distribution.

Chagas disease is also now under heavy attack. The six 'southern cone' countries of Latin America are now attempting to interrupt transmission of the disease by the year 2000. Uruguay appears to have succeeded already. The main contribution of research in this disease has been to take the disease seriously and map its prevalence and socioeconomic impact; and then to encourage the view that the disease could be contained and controlled. A powerful research community was established in the affected countries, and fumigant cans and insecticidal paints were developed for families to control the vectors themselves, at half the cost of traditional spraying. But perhaps above all what the scientific community has done for Chagas disease, as well as for several other diseases including leprosy, has been to act as a goad, a prompt, and an encouragement to those responsible for control, not only to provide new means of control, but also to show faith that something can be done and to challenge those responsible until some action is taken.

Schistosomiasis, like onchocerciasis, has an effective treatment – the drug praziquantel which was developed and tested in the 1970s. Although the scale and frequency of distribution of the drug required is daunting to many countries, in 1993 a WHO expert committee concluded that in certain important areas the disease could now be controlled with praziquantel combined with selective snail control.

Lymphatic filariasis is a more recent success story. At a recent expert meeting in Penang, a global control strategy was defined around chemotherapy with single-dose diethyl carbamazine (DEC), or ivermectin, or both, or DEC-impregnated salt. This is to be combined with an extremely simple regimen of antibiotics and soap and water which has been shown to produce dramatic reductions in elephantiasis, and with vector control – either biologically with *Bacillus sphaericus* or with simple physical methods (such as polystyrene beads to cover the surface of water in latrines where the *Culex* mosquito vectors often breed).

These are some of the recent steps in science-based progress in control, but there has also been tremendous progress in the application of new scientific techniques to tropical diseases.

For example, there are now more than 1000 DNA sequences (the 'holy texts' of biology) from the *Plasmodium falciparum* parasite in our malaria DNA library; as a result, five candidate antigens are on line for trial as vaccines. And one of the most advanced techniques of molecular biology, the polymerase chain reaction (PCR), which can detect and 'amplify' tiny amounts of specified DNA, is now being used routinely in the field by OCP in West Africa. It is used to detect onchocercal parasites in blackflies in areas which have been cleared of the disease, to distinguish non-blinding and animal forms of the parasite from the blinding human form, helping OCP make substantial savings on larvicide costs.

Molecular biology is just coming of age in tropical diseases. In the past, we may have been overoptimistic at the speed with which the discipline would bring practical results, particularly with vaccines, because we underestimated the complexity of both the parasites and the human immune response. However, there is no doubt that these results will come over the next two decades.

In quite another direction, through applied field research, which has doubled at TDR in the last three to four years, there is also much hope, through the increasing engagement of the disease-affected communities themselves. In Africa, 80% of malaria cases are dealt with at home. At the same time, health systems are evolving to leave even more responsibility 'at the end of the track'. It is thus crucial to develop two-way means of communication with these communities, to learn more directly of the real problems people face and how they can be solved or ameliorated with the minimum of external resources. It is now clear that the latest 'qualitative methods' of social science can help to achieve this: for example, in several parts of Africa communication workshops have shown how with appropriate techniques to break down barriers of misunderstanding between women and health workers, communities can improve the quality of service for women at health posts essentially without cost.

Such social and economic studies are also providing us with important guidance to steer other kinds of research – into drugs, vaccines, diagnostics or vector control methods – in the most appropriate directions, increasing the likelihood of their future usefulness.

At the same time, any means for increasing the cost effectiveness of external intervention is also becoming crucial. Here again, though, there is hope to be found: for example, from new means of rapid epidemiological mapping of onchocerciasis, that will reduce costs of ivermectin distribution; and from a system of questionnaires posted to schools which have proved to be a highly effective and extremely cheap means of detecting those communities most affected with urinary schistosomiasis, in order to reduce costs of distribution of praziquantel.

Moreover, the very economic growth of several advanced developing countries, particularly in the East, has opened new opportunities for the development of tools for tropical diseases. Figure 1.1 shows that countries from Korea to Indonesia – all tropical disease endemic countries – have enjoyed growth rates from 4 to 9% ahead of the main industrialized countries.

Simultaneously, the rate at which totally new techniques of biological research and development – such as gene cloning, etc. – are being invented has been slowing down, as we move in biology from the era of research techniques to the era of application of those techniques to research. Figure 1.2 shows that this is allowing the more advanced developing countries to catch up with this technology.

We have seen the benefits of this combined economic growth and technological convergence in tropical disease research. Figure 1.3a shows where our main diagnostic kits were being developed and manufactured in 1987, while Fig. 1.3b shows the 1994 situation. As an example, a filariasis antigen detection technique was developed in the United States. Monoclonal antibodies were sent to India, and within 2 months the Indians were producing a prototype kit. Another Indian

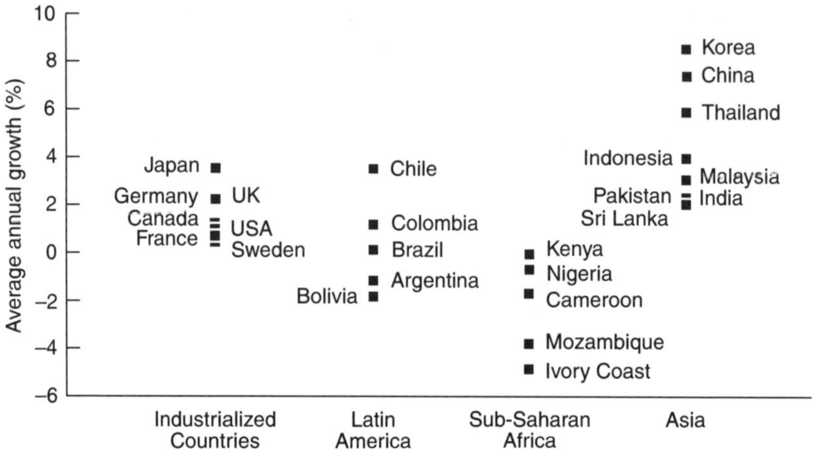

Fig. 1.1. Economic development, 1980–1992.

company has developed a cheaper method for producing eflornithine, the 'resurrection drug' for African trypanosomiasis. This encouraging trend is expected to continue; in particular, the new promise of naked DNA vaccines, and of combinatorial chemistry for testing millions of new drug candidates, may be fertile ground for rapid development in the newly industrializing countries.

We might envisage, then, sometime early in the 21st century, that if these achievements are maintained and extended, TDR's 'six diseases' may be reduced to three:

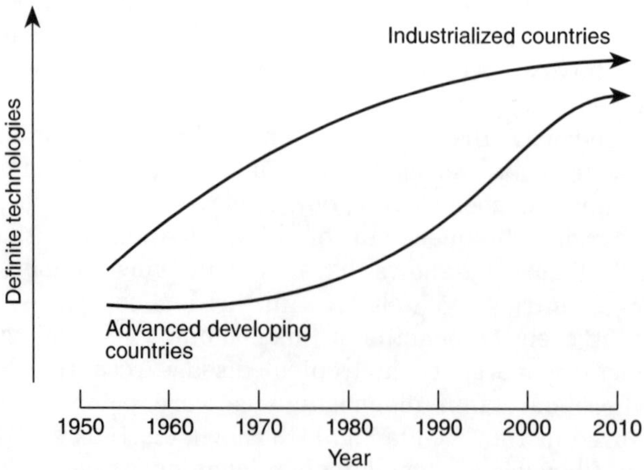

Fig. 1.2. Advances in R&D technology.

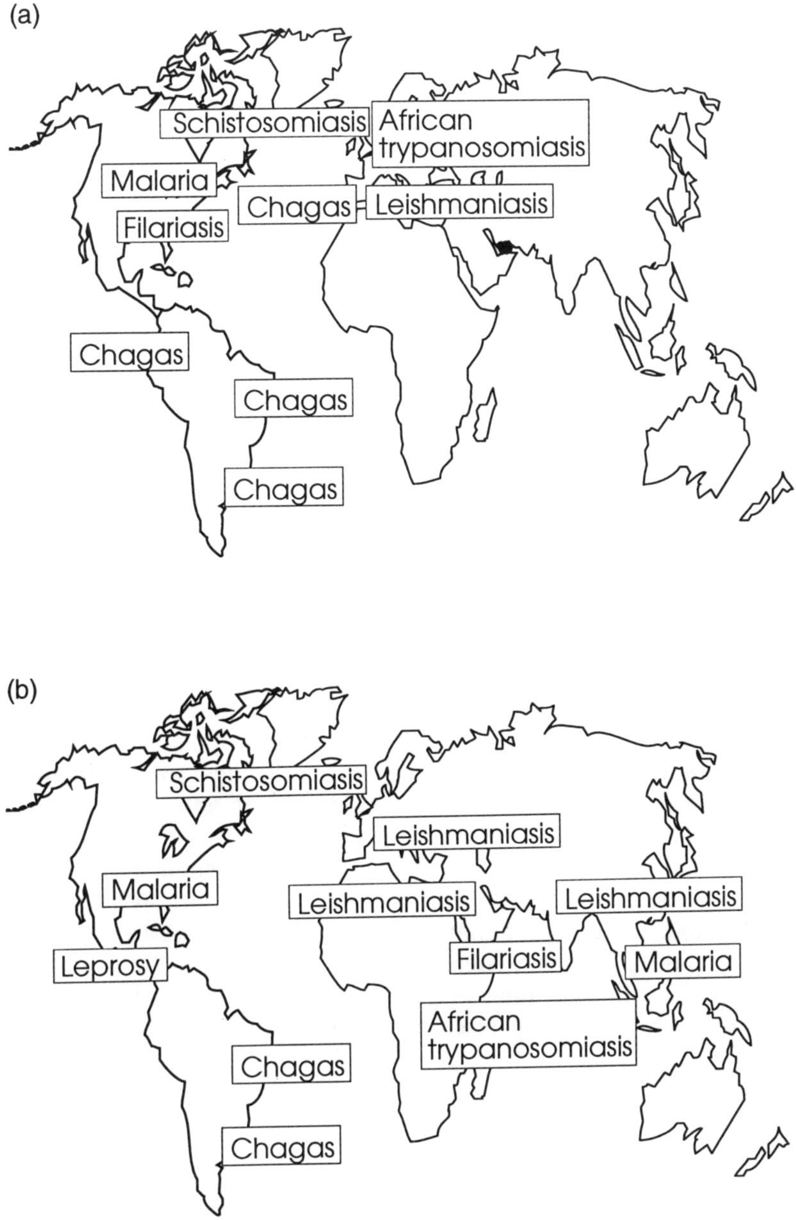

Fig. 1.3. Development and manufacture of diagnostic tests: (a) 1987; (b) 1994.

Box 1.2. Unanticipated scientific discoveries (during the last two decades).

1973 – Recombinant DNA
1975 – Monoclonal antibody
1987 – Polymerase chain reaction
1992 – Naked DNA vaccines

- malaria (which will need more than a vaccine to control it),
- leishmaniasis,
- African trypanosomiasis.

This is quite an achievement for the last couple of decades of parasitology as applied to tropical diseases.

However, the progress of science is unpredictable and the following uncertainty factors exist:

- scientific developments,
- economic developments (including international relations),
- sociocultural developments,
- emerging microbiological threats.

It may seem strange to non-scientists, but predictions in science are more risky than in other fields. Discoveries, by their very nature, are unexpected (see Box 1.2). We had no idea in the 1960s that the next decade would produce the revolution of recombinant DNA (1973) and monoclonal antibodies. And from the viewpoint of 1980, who could have predicted the power of the polymerase chain reaction (PCR) or the possibility of naked DNA vaccines?

Applied science – such as TDR's own field, which is goal-directed parasitology, science applied to tropical diseases – leads to further uncertainty: what use can be made of scientific discoveries? The late American physicist, Robert Oppenheimer, put it this way:

> it is a profound and necessary truth, that the deep things in science are not found because they are useful; they are found because it was possible to find them.

So the challenge to those who wish to use science for a purpose is a challenge of pragmatic opportunism.

Great promise for tropical diseases comes from the discipline of molecular genetics. The parasite, the vector and the host (man) will be taken in turn. In the parasite, molecular genetics will allow the production of stable genetic transformants of the organism – this can already be done with *Leishmania* and will be possible soon with *Plasmodium falciparum*. This will enable the identification, at a molecular level, of the factors which determine virulence and pathogenesis, and target specific pathways with drugs.

With genome mapping projects – efforts to read the complete 'holy books' – for many organisms, there are increasingly detailed maps on which to base genetic understanding. Because of widespread genetic homologies among all organisms on Earth, these maps will also allow us to draw on the immense amounts of research done on other creatures – such as the biologists 'model nematode', *Caenorhabdytis elegans*, and man himself through the human genome mapping project. Parasites will thus participate in the explosion of biological knowledge and understanding that is now under way.

Vectors also are being investigated by molecular geneticists. Some three to four years ago, the MacArthur Foundation and the TDR established their activities in the molecular entomology of the *Anopheles gambiae* mosquito, the cause of a million children's deaths in Africa each year through malaria. But as a result of this initiative, entomology, which was languishing, has been revived, and *A. gambiae* has become the most researched insect, second only to the scientists' old favourite, the fruitfly. So this killer mosquito has 'taken off' in the world's laboratories, so much so that scientists are using it to ask and answer fundamental questions in biology, bringing some of biology's brightest minds to bear on a subject that can really bring hope to Africa. Already there are prepared markers of transformation – for example a white eye gene – and low resolution maps of several chromosomes. Within two years we will have a detailed map down to 1–2 centimorgans of the whole genome. Factors determining the viability of the malaria parasite in the mosquito are being identified, and mechanisms of gene flow in wild populations elucidated.

We are still searching for a transformation technique, but there is a reasonable prospect that within 25 years we should be able to design a piece of transposable mosquito DNA – a self-propagating infection – that will stop the mosquito transmitting malaria. Once released, this artificial infection will spread naturally in the mosquito population until Africa is clear of malaria. Science fiction? Well, so was ridding the world of smallpox – until it was done.

Discussion of the potential of molecular genetics in the human host will be confined to the potential of naked DNA: that it could be possible to provide a continuous, lifetime dose of some required proteins (for example, to stimulate the immune system like a vaccine, or to interfere with parasite metabolism) with a single intramuscular injection. The injected, naked DNA would be targeted to integrate, say, into the cells of a certain tissue, where it would generate its gene product for the lifetime of the patient. For example, a naked gene designed for the liver

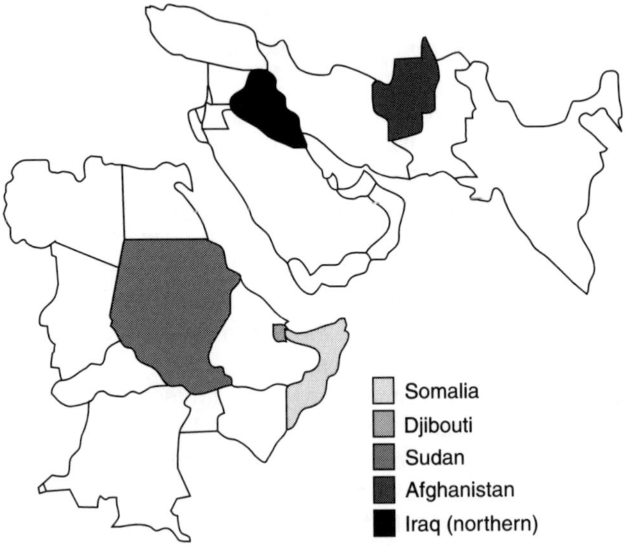

Fig. 1.4. Eastern Mediterranean region: major malaria problem areas.

could impinge on cholesterol or lipid metabolism, so you could eat as much fat as you liked! Or a naked gene might carry signals to trigger a parasite's apotheosis gene – the suicide genes that every cell seems to possess, and that are now of such interest in cancer research. Or a long-lived parasite, like the worms of filariasis, might be induced to die slowly, reducing the danger of extreme 'Mazotti' reactions in the patient when large worms are suddenly destroyed, releasing a flood of antigens into the body.

Gene therapy is also closely analogous. With this technique, genetic predispositions to pathogenic disease – for example, the tumour necrosis factor (TNF)-gene-linked sevenfold increase in susceptibility to cerebral malaria recently published in *Nature* – might be corrected.

Is naked DNA another dream? One signal that it might be a real prospect is the concern expressed about such products in the pharmaceutical industry – clearly one shot per life per patient is not going to sell many drugs!

So much for science. What of sociocultural trends? Here, there are several very troubling matters.

The first challenge in tropical diseases is the long-term consequence of conflict and instability, which seems to be so widespread and threatening. A recent presentation listed the worst regions for malaria in the WHO eastern Mediterranean region (Fig. 1.4): Sudan (continuing civil war), Somalia (anarchy), Iraq, Afghanistan and

Fig. 1.5. Sites of major epidemics of African trypanosomiasis in recent years.

Djibouti; and the sites of major epidemics of African trypanosomiasis (sleeping sickness) (Fig. 1.5): Uganda (after Idi Amin), Chad, Angola, Mozambique and Zaire (the latest epidemic, with some villages facing 70% and higher mortality). When disaster strikes the first effects reported are death from cholera and starvation; the collapse of tropical disease surveillance comes later, unheard and with long-lasting after-effects. This is an area where politics and medicine meet and parasitologists can only add their voices to the thousands of others pleading for peace.

There are four other outstanding trends with which we should be concerned:

- demographic transition,
- the environment,
- population growth,
- evolution of developing countries.

The demographic transition, in which the populations of developing countries are getting older, richer, smoking and eating more, will shift the attention of health systems to cardiovascular diseases and cancer, away from the poor – who form, and will always from, the vast majority of tropical disease sufferers.

Environmental change will affect tropical diseases profoundly. The increased energy in the atmosphere, the result of all that carbon dioxide, is causing increased instability and unpredictability in weather and rainfall patterns, and will increase the risk of epidemic malaria; global warming – the eight hottest years of this century have been in the last decade – will shift the boundaries of transmission zones, in particular increasing disease risk in the southern states of the former USSR. Leishmaniasis, filariasis and schistosomiasis will become increasingly urbanized as urban sprawl continues. Reforestation will bring the vectors and diseases to economic migrants. And population growth, greatest among the poor, will increase the total number of those with tropical diseases.

But the greatest sociopolitical threat is the evolutionary trend in certain parts of the Third, and even 'Fourth', World – the marginalization of people; their effective exclusion from the basic necessities of social and health care, whether it be in the Third World or on the streets of New York or London or Paris. In the United States, 30–40 million people are without access to healthcare, while in the Third World, many forces are driving towards increased inequities in what are already inefficient and unresponsive health services.

Also in the next century we will face accelerating biological change and new microbiological threats:

- evolution of drug resistance,
- response of climatic change,
- humanization of zoonotic infections.

Medical progress, in fact all material human progress, is a kind of conscious evolution of the human species in competition with the rest of the biosphere. But that biosphere hits back; it continues to compete and evolve with us, and will catch us whenever we drop our guard. Some organisms evolve faster than others: viruses (such as HIV); then the bacteria (tuberculosis); the protozoa (malaria); and perhaps next the nematodes, when *Onchocerca volvulus* evolves resistance to ivermectin, as its cousin *C. elegans* can be easily induced to do in the test-tube. Evolution is certain, and mankind is inflicting the most intensive evolutionary challenge that the biosphere has ever faced. But a single line of defence against many tropical diseases is not enough. Second, third and fourth lines of defence will be needed.

The impact of environmental change has already been described; but there is one particular danger that should be underlined – how increasing crowding and unplanned urbanization brings risks of the humanization of zoonotic infections. Some diseases, transmitted mainly via animals, are termed zoonotic. HIV probably began as an animal pathogen until it jumped the barrier and found an evolutionary opportunity to transmit itself among human beings. Human–human transmission presents the danger of quick-fire epidemics – as the example of the change from bubonic to pneumonic plague in India in 1994 bears witness.

Might such a transition occur in some tropical disease? The combination of zoonotic leishmaniasis and HIV in many Latin American cities is frightening; could the parasite evolve to some anthroponotic form, precipitating epidemic leishmaniasis in urban Latin America, as experienced in rural Sudan and India today?

There has been a decade or two of unprecedented scientific progress in parasitology, and there is great promise of more. But the parasitic tropical diseases are in danger of being forgotten by a rich world that has forgotten its poor, and they will be forgotten unless an aggressive, entrepreneurial approach is taken to grasp the scientific, political and economic opportunities that arise, and to set in place good defences against the evolution of our biological enemies.

2

Training in Parasitology

R. HOUIN

Faculté de Medécine, Laboratoire de Parasitologie-Mycologie, 8 rue du Général Sarrail, 94010 Créteil Cedex, France

At the last ICOPA in Paris in 1990, a symposium was held on training in parasitology, which is a real concern for most parasitologists. More recently, the last meeting of the American Society of Parasitology held a symposium entitled: 'Parasitology 2000'. A large part of that meeting was devoted to training, as the future of parasitology depends on how students are being trained now.

Five years ago, many attending the ICOPA symposium were pessimistic, and this view is still held by some in this field, who argue that programmes, courses and university positions are far from increasing. In 1990 the question being asked by J.R. Seed was 'is parasitology a relic of a heroic past, or will parasitologists be at the forefront of science?'

In fact, parasitologists are far from unanimous about the subject: in general, fundamental parasitologists are rather pessimistic, while those who work on applied parasitology are optimistic. The opening session of ICOPA VIII showed the concerns of medical parasitology, and the favourable position of official authorities. A similar optimism exists among veterinarians and, perhaps to a lesser degree, specialists in crop parasites.

During the last ICOPA the necessity was emphasized of improving the information about parasitology directed to universities, ministries, and other organizations involved in setting up education and research programmes. National societies were encouraged to carry on discussions on the subject and a World Federation resolution was agreed.

What is a parasitologist and what will a parasitologist be in 10 or 15 years time? Will such a general name still exist? If this question can be answered then the types of training to offer to students can be decided.

Fundamental parasitologists still appear the most unhappy about the evolution of their position, so the situation at this level will be examined first. It is often said that courses in parasitology and sometimes entire laboratories are suppressed in many universities. Does this mean that parasitology is no longer taught? Or does it mean that parasitology is taught in other ways, or at other levels? Perhaps by other people?

In 1990, Claude Combes said that one of the three major errors that must be absolutely avoided is fascination of the object, i.e. a tendency to cling to a familiar problem which has been studied for years. This tendency is particularly damaging at a time when parasitology is progressing so rapidly. It is evident that teaching parasitology without any knowledge of molecular biology or population genetics (among others) would make no sense. It is also evident that 'traditional parasitology' took all the techniques that were developed, and applied them to improve our knowledge of parasites.

In modern parasitology new developments come from all aspects of science, and not simply from techniques. New approaches are highly sophisticated, they need specific training to be used, and this training is quite difficult. People who benefit from this type of training have a considerable advantage over those who do not. Fashion multiplies this advantage, and this can lead to really frustrating situations. Combes defined a second error: fascination of the tool. To master a technique is never enough; it must be remembered that another technique will follow, and that progress can only be made by combining results obtained by numerous methods.

As parasites are useful models, scientists who have no knowledge of parasitology use parasites for their research. It was quickly realized that this situation was very awkward, and that knowledge of the study material cannot be avoided. If this is not the case, the conclusions of the work remain unapplicable, even if the research is carried out well.

With respect to training, this means that fundamental parasitology, especially all the aspects which contribute to defining parasites is, and will always remain, a high priority for students. However, methods of defining parasites have changed greatly over the last ten years. Microscopic examination was achieved using various tools, depending on the type of parasite and this information must now be passed on to students.

So, the need for academic courses for the teaching of this subject is raised. In Paris, at ICOPA VII, Hagai Ginsburg insisted that parasites can be used very early in academic courses to illustrate biological phenomena which are not specific to parasites, but for which parasites are particularly good models. This can be done in biochemistry and cell physiology courses, and also in immunology, nutrition, pharmacology and many others. This allows young students to hear about parasites at a time prior to finalizing their future careers, and can lead some to select parasitology for further study.

This is particularly interesting in developing countries, where parasites are such a heavy burden on society. Students in health care training, as well as those choosing science courses, should be aware of this field of knowledge early on. This is the only way to build the strong research teams which are necessary for the long, hard struggle against human parasites, and against all the other diseases which have economic consequences and contribute heavily in preventing development. This will also raise academic and governmental awareness of these problems.

Specific parasitology courses can only be proposed later, when students have acquired sufficient knowledge in general biology. It is now clear that parasites follow the same biological rules as free-living organisms, even though they have adapted to special conditions. Specific training must use the basics acquired earlier, and demonstrate how parasites can turn to their benefit conditions which are lethal for free-living organisms. This is probably the way to raise the interest of the best students, as it opens evolutionary concepts at their highest level. Understanding the mechanisms which allow *Leishmania* to live in a phagolysosome, a structure designed to destroy invaders is a great moment in science. Understanding how *Leucochloridium* develops in the tentacles of the snail *Succinea*, and gives them the appearance of a caterpillar, is another: it anticipates, in the intermediate host, the behaviour of the definitive host in which the worm has to develop, to continue its cycle, and finally reproduce. Such examples are numerous, e.g. the cell wall of *Trypanosoma* which is present in metacyclic forms at a time when it is not effective, but anticipates what will be necessary when the parasite is injected into vertebrates. Today parasites are no longer considered as 'degenerate' organisms; on the contrary, they are highly evolved. If any confirmation was necessary, it would be easy to say how quickly they beat all the defences set up by man (recently) or by nature (for millenia) against them.

Understanding this kind of concept also requires an in-depth knowledge of the systematics of parasites. It is certainly difficult to teach this to students who hardly know, when they arrive in universities, the bases of the classification of living organisms. But no precise knowledge can be acquired without defining precisely the organisms which are involved. It is known that, although parasites are closely related to each other and sometimes difficult to distinguish, they can still have quite different mechanisms of adaptation.

This level of training is certainly extremely important, and it is necessary for all students who will turn to research on parasites. This level, which could be defined as 'general parasitology', should also be achieved by students coming from health courses, as it would make no sense to teach them about human parasites only, for example. This general level of expertise should be aquired before embarking on post-graduate education.

The highest level of training can be associated with the preparation of a PhD thesis or equivalent. It is easier to attract excellent students to parasitology if they are to be part of an organization where modern research is being done.

At such a level, each laboratory can offer only a small number of places (with limited financial support as well), but students must continue to see parasitology as a whole, even if they work on an extremely narrow programme. This is the condition necessary for them to become parasitologists, to understand what other parasitologists do, and how each other's work can be related, even if they work in very different fields. This can also allow the possibility of changing their field of research if necessary.

In Paris, André Capron developed the idea that this level needs centres of excellence, grouping various teams, as it is impossible, at the present time, to have specialists in every field. In such groups, all the specialists must participate in training. Paradoxically, the least common area of expertise is the identification of parasites and hosts. This is because very few researchers are entering the field to fill the places of those senior scientists leaving it. Another danger stressed by André Capron is the tendency of these teams to be more and more experimental, and to stray from the realities of the field; parasites are, no doubt, among the most marvellous models for experimental research, but they are also fierce enemies, destroying herds and crops, and preventing a large area of the world from developing. C.P. Ramachandran, in his opening lecture of ICOPA VII, gave examples which must never leave the minds of any researcher in the field of parasitology.

One thing which must be developed is the distribution of information at this level. The natural tendency of students preparing their theses is to concentrate too much on their own subjects. Symposia specially designed for these students are a good solution in preventing such isolation. A theme is proposed each year, large enough to concern about a dozen students, and the students are asked to discuss their research and the research of the other participants. Senior researchers join in, introducing subthemes, and animate the discussion. Michael Kemp suggested that similar structures could be organized at the international level in which advanced students from developed countries, who have no direct perception of the human importance of parasites, would mix with students from developing countries, who would benefit from additional training at a high level. This group training would probably help to reduce the gap of incomprehension which sometimes appears between well-meaning people! It could also lead young scientists from developing countries to understand that high-level science, especially in the field of parasitology, is not removed from the realities of their countries. It is a waste when young scientists refuse to lend their skills to these problems, so severe in their native countries, and prefer to stay in parts of the world that are much less affected.

This discussion is applicable to the field of applied parasitology, which appears to be, in the opinions of colleagues, in a better situation from the point of view of institutional pressure. The needs in developing countries are extremely important, as already presented, but, even in industrialized regions, the profile of medical parasitology was recently raised by immunodepression, not only acquired immune deficiency syndrome (AIDS) (which is, however, an extraordinary opportunity for some parasites), but also iatrogenic immunosuppression. Transplantations are becoming more affordable treatments and they need long-term immunosuppression, which favours parasites (not always the same parasites as seen in AIDS patients).

It would, however, be unwise to antagonize developing and industrialized countries from the point of view of training in medical parasitology, but the priority should not be the type of parasite, or the number of trained specialists necessary, but the definition of levels of training. Here again, three levels must be distinguished, although they are quite different from the levels already described in fundamental parasitology.

The primary level concerns many students, as all those working in health organizations must know the bases of parasitology as they will meet parasitic diseases during their professional life. This means that health technicians, nurses, people involved in prevention programmes and laboratory diagnostic technicians as well as physicians, chemists and veterinarians, must receive an adapted training in parasitology. An error frequently made is to want to teach too much. These students want to learn medical or veterinary parasitology, but they are not concerned by the number of legs a tick has or by the structure of the hair on the 8th segment of a mosquito larva. Their predecessors were obliged to learn such notions, so parasitologists in some places were thought of as butterfly catchers! On the contrary, if these students receive a training adapted to their future profession, they will be deeply interested and can contribute to the good reputation of parasitology. This kind of mass training can easily be improved with slides or other items prepared at low cost by organizations like the World Health Organization (WHO).

The intermediate level is also specific to health education, at least in part, as it is the level of specialization in parasitology. As is the case for students doing fundamental courses, this intermediary level establishes the vocabulary of parasitology that is common to all scientists working on parasites. It also teaches in-depth on a relatively small number of parasites, e.g. human parasites on a medical course. There is a risk, at this level, of over-simplifying the subject, in which human parasites appear alone in a deserted world and students know some species (and their Latin binominal names) without realizing that other species exist.

Contrary to the case in fundamental parasitology, this level also focuses on professional positions, such as diagnostic medical biologists, or teachers who give training at the primary level. As for primary-level training it must always be kept in mind that intermediate-level training should be adapted to a particular profession and the fundamentals must be restricted to a level necessary to understand the proposed information. Practical training is also, in the case of biologists, an important part of any course. Parasitology is not yet an automatic part of biology, and students must be able to identify parasites that are likely to be present in a sample.

The third level is closely linked to the level of students coming from fundamental studies. There is no reason to follow one course rather than another to become a researcher in the field of applied parasitology, but from the beginning a course will concentrate more in one direction than another. At this level it is too late to acquire new bases, and cooperation is therefore essential. It is important to remember that fascination of the tool is a major danger, and that students must never forget the goal of their studies.

This being established, what different requirements do developing and industrialized countries have? It is clear that primary-level training is important in developing countries but does this mean that the requirements of developing countries are more extensive or is it simply that the current level of training is inadequate? The situation differs from one country to another, but it is clear there is a basic need for local training.

Does this mean that this is the only level at which students can be trained in developing countries? Certainly not, as is demonstrated in many places where academic institutions exist, and where the second, and sometimes third, levels are proposed. Concerning that last level, C.P. Ramachandran has stated that there has been a development in many countries of organizations able to develop new techniques for parasitic detection and new drugs to treat parasitic diseases. Such situations will probably become more and more frequent, as in industrial countries strict regulations discourage businessmen from investing in fields in which they are not directly concerned. This will provide excellent opportunities for local students to be trained at the highest levels.

However, even when such structures exist, their interest is to create, or to maintain links with, similar organizations in developed countries. As stated previously, even in these countries, it is difficult to group enough scientists to ensure good training. Both sides can improve the quality of training by exchanges, but the risk remains, of course, that young scientists from developing countries will not take their knowledge back to their native countries. Our possibilities of action in that field are really limited, as it generally depends on financial problems, which in turn depend mainly on political authorities. Another risk is inappropriate training. As already mentioned, it is a waste when young scientists return home with unapplicable knowledge, say, for example, heart transplantation in countries where basic surgery is barely possible.

In conclusion the parasitology which will exist at the turn of the century is already here. The researchers are already trained and it is certainly too late to change their minds!

The decisions made about training now will actually shape parasitology in the future and probably for the next 30 years. We must anticipate what will happen during that time, and give our students the tools they need to improve the field of parasitology, so that parasites keep their place of importance in the living world.

3

Immunity to Human Leishmaniasis

H. AKUFFO,[1] K. MAASHO[2] AND S. BRITTON[2]

[1]*Microbiology and Tumor Biology Center, Karolinska Institutet and Smittskyddsinstitutet, Box 280, S171 77 Stockholm, Sweden;* [2]*Department of Immunology, Microbiology, Pathology and Infectious Diseases, Division of Infectious Diseases, Karolinska Institute at Huddinge Hospital, 141 86 Huddinge, Sweden*

Introduction

The first part of this chapter will touch upon some of the problems associated with the study of immune mechanisms operating in human leishmaniasis, using a few published examples to illustrate the mechanisms generally believed to be operating. The second part of the chapter will discuss the authors' results using *Leishmania aethiopica* as a model for the study of natural immunity to *Leishmania* in man. Finally, the chapter will conclude with discussions of how subtle changes of the infecting organism can induce immune responses that could change the clinical features of the disease.

There is general consensus that *Leishmania* live and multiply within the macrophages of the infected vertebrae host. It is thus evident that elimination of *Leishmania* from macrophages will involve macrophage activation to kill. Furthermore, due to the intracellular site of *Leishmania* in the host, cellular mechanisms rather than antibody-mediated mechanisms are important for resistance to infection. Experimental mouse models of leishmaniasis have demonstrated that the outcome of infection with *Leishmania* spp. is genetically controlled and is governed by the type of T-cell immune response generated by the host. Murine CD4$^+$ cells have been elegantly divided into Th1 and Th2 phenotypes depending on the type of cytokines they produce (Mosmann *et al.*, 1986). The mouse model of *L. major* was able to illustrate the importance of the differential CD4$^+$ T-cell phenotype in an experimental model (Scott *et al.*, 1989; Coffmann *et al.*, 1991; Scott, 1991) where Th1 cell induction with the production of interferon-gamma (IFN-γ), interleukin (IL)-2 and lymphotoxin was associated with resistance

Table 3.1. Classification of leishmaniasis.

Tropism	Old World leishmaniasis (Africa, Asia, Europe)	New World leishmaniasis (America)
Visceral	*Leishmania donovani* *Leishmania infantum*	*Leishmania chagasi*
Cutaneous	*Leishmania major* *Leishmania tropica* *Leishmania aethiopica*	*Leishmania amazonensis* *Leishmania braziliensis* *Leishmania mexicana* *Leishmania panamensis* *Leishmania peruviana*, etc.

while Th2 responses were more prominent in susceptible strains which produced IL-4, IL-5 and IL-10 (Sadick *et al.*, 1990). The main role of these cytokines is believed to be to influence the activation of macrophages to kill intracellular amastigotes. This fundamental role of cytokine interplay has been corroborated in the human diseases, but due to the diversity of the *Leishmania* spp. causing disease in man, some detailed mechanisms may differ from one type of infection to another.

Complexities of *Leishmania* parasite tropism

The controlled experimental mouse models do not adequately address some of the complexities of the *Leishmania* parasite in terms of, for example, the tropism to different macrophages. Why do *Leishmania* spp. sometimes bypass a set of dermal phagocytic cells while others are apparently controlled by these cells? Classically leishmaniasis in man has been divided geographically and according to the tropism of the infecting organism (Table 3.1).

Visceralization must be under the control of fairly well-regulated mechanisms involving T-cells and probably other unknown mechanisms. Evidence of change in parasite tropism is illustrated by a number of observations.

1. Once the immune system is severely compromised as occurs in acquired immune deficiency syndrome (AIDS) the well-established site of proliferation of parasites which were believed to have a specific skin tropism are modified (Gramiccia *et al.*, 1992; Villanueva *et al.*, 1994). This would suggest that the choice of macrophages in which the various *Leishmania* spp. multiply is not absolute and may change following changes in the immune response of the host.

2. Suggestions that factors other than an innate tropism of the parasites are involved in the choice of macrophage in which multiplication occurs is illustrated in the change in tropism of *L. tropica* from a cutaneous nature to a visceralizing nature in apparently healthy US

Box 3.1. Characteristics of post kala azar dermal leishmaniasis.

- Occurs after inadequate treatment of visceral leishmaniasis
- Few parasites in lesions
- Hyperpigmented macules to infiltrated plaques and nodules
- Easily mistaken for lepromatous leprosy
- Change in organ tropism of parasite and/or change in host response?

soldiers returning from Operation Desert Storm, in the Middle East (Kreutzer *et al.*, 1993), which can not be explained by a breakdown in T-cell responses.

3. Even more fascinating is the apparent change in tropism that occurs during post kala azar dermal leishmaniasis (PKADL) (see Box 3.1), where an inadequately treated visceralizing *L. donovani* infection transforms to a dermotropic phenotype with no reports of conversion back to a visceralizing phenotype. The mechanisms operating in this situation are unknown.

The Role of Interferon-gamma

Despite the complexities associated with the host, the *Leishmania* species and tropism, one clear line of control of *Leishmania* multiplication in macrophages runs through all model systems: the role of IFN-γ in inducing the oxidative burst to kill *Leishmania*. Discussions relating to whether this is enough or if secondary signals are required have not been resolved. In the mouse system a prominent role of the nitric oxide pathway with enhancement by tumour necrosis factor-alpha (TNF-α) has been implicated (Liew *et al.*, 1990; Liew and Cox, 1991). However, the levels of nitric oxide induced *in vitro* by peripheral-blood-derived human macrophages in response to *Leishmania* parasite stimulation are much lower (if present at all) than induction in mouse peritoneal macrophages (Hatzigeorgiou *et al.*, 1993; and personal observation). Furthermore, a role for TNF-α could not be ascertained in the inhibition of *L. aethiopica* in cells of human monocyte–macrophage lineage (Mohammed *et al.*, 1992). If this is the case then it would imply an important difference in the mechanisms of *Leishmania* killing in mice and men.

IFN-γ is not only important for the direct activation of macrophages to kill. It has been shown that when $CD4^+$ cells mature in an environment rich in IFN-γ it leads to the development of Th1-type cytokines which in turn produce more IFN-γ and IL-2.

Box 3.2. Cytokines enhancing or diminishing the activating effect of IFN-γ on human macrophages to kill *Leishmania* amastigotes.

Up-regulation of human macrophage function to kill *Leishmania*
IL-2 (*L. aethiopica*), Akuffo *et al.* (1990)
IL-3 ± GM-CSF or M-CSF (*L. amazonensis*), Ho *et al.* (1992)
IL-12 (general effect on IFN-γ production, Manetti *et al.* (1994)

Down-regulation of human macrophage function to kill *Leishmania*
IL-10 (*L. donovani*), Ghalib *et al.* (1993)
TGF-β (*L. braziliensis*), Barral *et al.* (1993)

The prominent role of IFN-γ in human leishmaniasis has been illustrated in the various types of clinical manifestations of the disease. It would appear that in the localized, self-healing cutaneous lesions there is adequate IFN-γ production that soon leads to the elimination of the infecting organism (Pirmez *et al.*, 1993). In the mucocutaneous lesions where there is an exaggerated immune response without adequate elimination of the infecting organism there appears to be a mixture of Th1 and Th2 cell phenotypes (Pirmez *et al.*, 1993) and an apparent attack on both host cells and the infecting parasite. In lesions from diffuse cutaneous leishmaniasis where there is the antigen-specific cellular non-responsiveness there is little IFN-γ in response to the infecting organism (Akuffo, 1992) whereas the non-responsiveness in visceral leishmaniasis is more generalized with little or no IFN-γ generation (Carvalho *et al.*, 1985). Following cure of visceral leishmaniasis, however, there exist *Leishmania*-reactive clones of both Th1 and Th2 phenotypes (Kemp *et al.*, 1993).

While host factors are evidently important there appears to be an interplay between the down-regulatory mechanisms of the infecting parasite on the host immune response to induce IFN-γ to activate macrophages. Various cytokines have been shown to enhance the effects of IFN-γ on human macrophages to kill *Leishmania* parasites. These include IL-2 (Akuffo *et al.*, 1990) and IL-3 ± GM-CSF (granulocyte macrophage colony stimulating factor) (Ho *et al.*, 1990), and an effect of IL-12 in this system is also anticipated (Manetti *et al.*, 1993, 1994). Down-regulating cytokines such as IL-10 and transforming growth factor-beta (TGF-β) have been implicated in infections involving *L. donovani* (Ghalib *et al.*, 1993) and *L. braziliensis* (Barral *et al.*, 1993), respectively (see Box 3.2).

IL-10 is a cytokine which has down-regulating effects on many functions of macrophages leading to suppression of immune responses. A prominent role for IL-10 in human leishmaniasis is emerging. Generated in large enough amounts IL-10 seems to be able to efficiently down-

Table 3.2. Induction of IL-10, IFN-γ and IL-2.

Cytokine	Acute disease	Post-treatment	PKADL
IL-10	+++	+/−	++
IFN-γ	++	++	++
IL-2	++	++	

Source: Ghalib *et al.* (1993).

regulate the action of IFN-γ. This is well illustrated in studies of IL-10 induction in the cells of visceral leishmaniasis patients during acute disease and after treatment. In these studies IFN-γ and IL-2 levels did not change (see Table 3.2).

Despite the complexities associated with immunity to leishmaniasis in humans, a clear line remains for the importance of cytokines, especially IFN-γ. While this is a macrophage up-regulating cytokine other cytokines such as IL-10 are efficient in down-regulating the effect of IFN-γ. Tipping the balance towards one cytokine type or the other could determine whether a resistant or susceptible state is achieved.

The pathogenesis of *L. aethiopica* infection

Leishmania aethiopica is believed to induce both a controlled localized self-healing disease, local cutaneous leishmaniasis (LCL) and the disseminated diffuse cutaneous leishmaniasis (DCL). Epidemiological data show that even in an area of high transmission where all are potentially infected, some individuals stay healthy while some get the disease. *In vitro* results suggest that there exist natural immune mechanisms to leishmanial antigens, involving natural killer (NK) cells (Akuffo *et al.*, 1993). If the disease takes, then infection with *L. aethiopica* can lead to LCL or DCL and this we postulate is determined in part by the properties of the infecting organism.

If natural immunity to *Leishmania* exists, then Stockholm, in the far north of Europe, is a good place to study it since it is possible to find unexposed normals. *In vitro* stimulation of cells from normal non-exposed Swedish donors indicated that in some of these individuals there exist mononuclear cells able to proliferate and produce IFN-γ in response to leishmanial antigen (Akuffo, 1992; Akuffo and Britton, 1992). This response was not evident in all tested and was intermediate in intensity in other individuals. Testing the same individual a number of times resulted in reproducible results. These results posed many questions including:

- What antigens are the individuals responding to?
- Are we dealing with a generalized mitogenic response?
- Are we dealing with a superantigen?

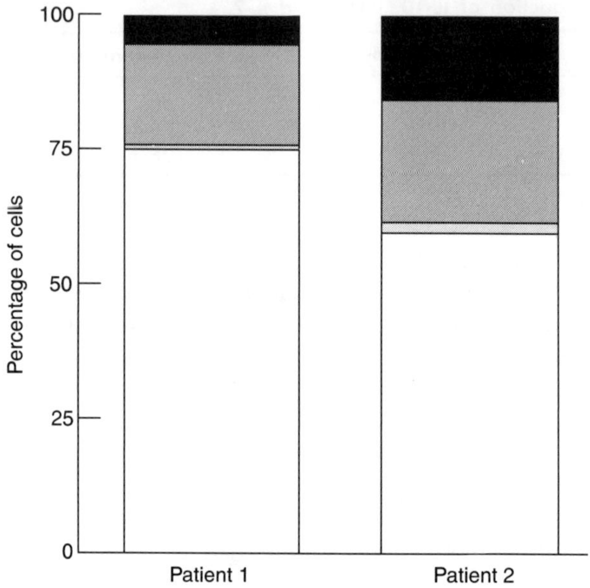

Fig. 3.1. FACScan analysis of *Leishmania*-stimulated cells from LCL patients. □ $CD4^+CD8^-$; ▒ $CD4^+CD8^+$; ▒ $CD4^-CD8^-$; ■ $CD4^-CD8^+$.

- Are these cross-reacting antigens?
- Which cell types are responding?

After ascertaining that the effect was not due to the presence of lipopolysaccharide, these questions were addressed. A cell analyser (FACScan) was used to determine the cells that were proliferating in the cultures of the normal Swedish donors and the results were compared with those of cells from LCL patients. *In vitro* stimulation of cells from *L. aethiopica*-infected LCL patients resulted in the proliferation of $CD4^+$ cells (see Fig. 3.1).

The proliferating cells to *L. aethiopica* antigen from the normal donors were shown to include some with CD4 and CD8 phenotypes but the large proportion of the cells were double negative cells ($CD4^-$ $CD8^-$). Since the other mononuclear cell types that induce IFN-γ are NK cells, the antibody against the surface marker of NK cells was also tested and the involvement of cells with the CD16/56 phenotype were found to proliferate in response to this (Akuffo *et al.*, 1993). The innate nature of part of this response has been further demonstrated using cord blood (unpublished results).

These findings have led to the postulation that the innate, NK response with its rapid production of IFN-γ may play a decisive role in the outcome of natural infection with *L. aethiopica* as illustrated in Fig. 3.2.

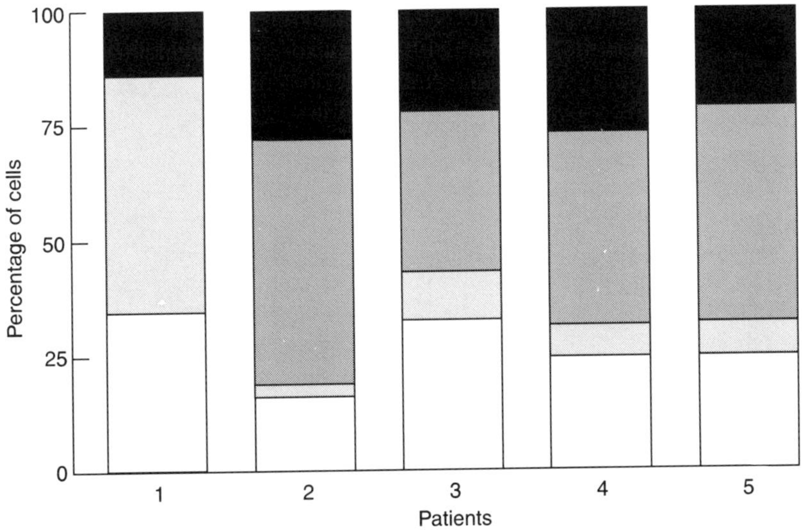

Fig. 3.2. FACScan analysis of *Leishmania*-stimulated cells in Swedish controls. □ $CD4^+CD8^-$; □ $CD4^+CD8^+$; ▩ $CD4^-CD8^-$; ■ $CD4^-CD8^+$.

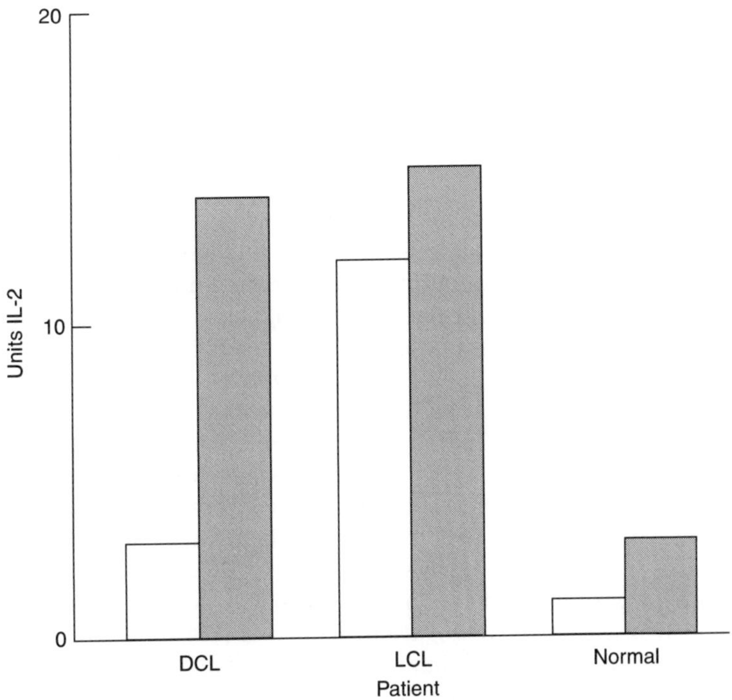

Fig. 3.3. Differential production of IL-2 following stimulation with LCL- or DCL-derived parasites. □ DCL IL-2; ▩ LCL IL-2.

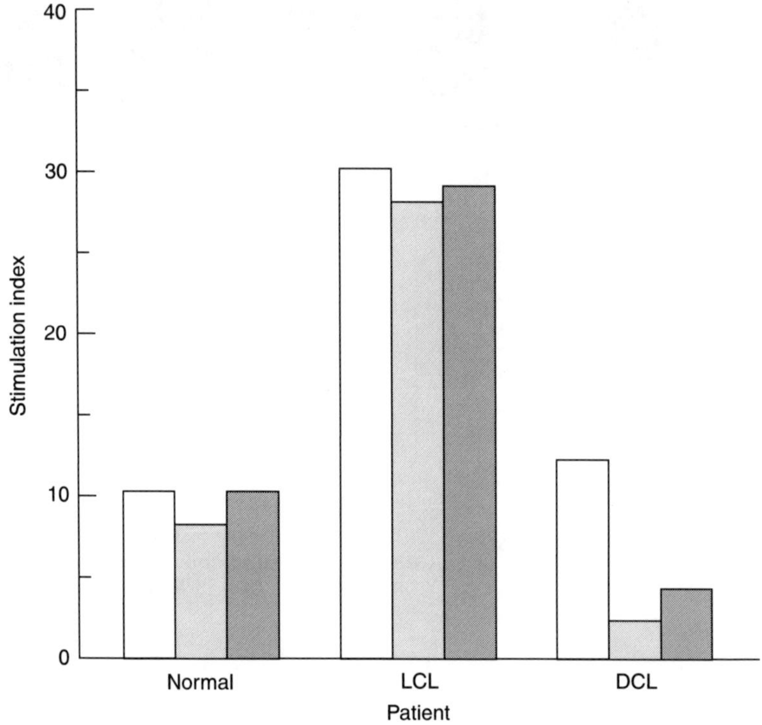

Fig. 3.4. Effect of DCL promastigotes on LCL proliferative responses. ☐ LCL Stimulation index (SI); ▨ DCL SI; ▩ DCL+LCL SI.

Other data from the authors' laboratory, using as a model the natural infection of *L. aethiopica* in man, has led to the postulation of a prominent role of the infecting organism in manipulating the immune system to produce down-regulating cytokines. The *L. aethiopica* that causes LCL and those that cause DCL have not been distinguished biochemically (Laskay *et al.*, 1991). However results suggest that the parasites derived from these two clinical forms may differ functionally (Akuffo *et al.*, 1987, 1988). In the first set of experiments the ability was studied of parasites derived from the two forms of leishmaniasis to induce IL-2 (Fig. 3.3) and subsequent proliferation to *L. aethiopica* antigen; with cells from DCL patients not being able to respond to their own parasites or those derived from other DCL patients, but able to respond to parasites derived from LCL patients. Mixing of DCL- with LCL-derived parasites abrogated this effect (Fig. 3.4).

Recent data has further illustrated that while LCL parasites preferentially induce messenger (m) RNA for IFN-γ, with little for IL-10, DCL parasites preferentially induce more mRNA for IL-10 in both

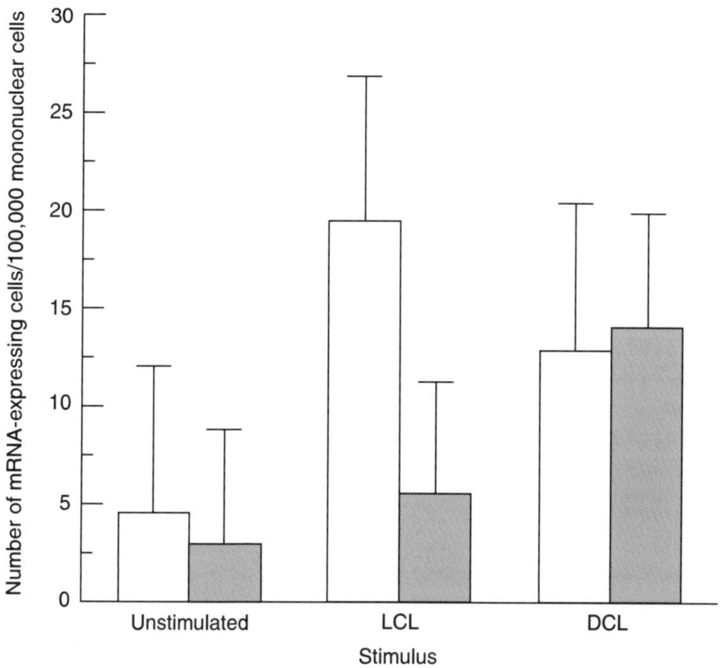

Fig. 3.5. Expression of mRNA for IFN-γ or IL-10 in cells stimulated with LCL- or DCL-derived parasites. □ IFN-γ; ▨ IL-10.

cells from patients and responding normals (Akuffo *et al.*, 1995) (see Fig. 3.5).

The ability of IL-10 to suppress many immune functions could explain the ability of DCL parasites to grow unheeded in the macrophages of the host. Thus even though the level of this difference in LCL and DCL is not known, apparently small (not easily detected) differences in the parasite can lead to different cytokine inductions that could change the clinical features of the infection. It may be prudent to take into consideration the potential role of natural immunity to *Leishmania* when designing vaccines against leishmaniasis so as not to compromise what protective benefit such immunity provides. Furthermore, the choice of parasite strains for producing heat-killed *Leishmania* vaccines may need to be selected wisely for the control of certain types of leishmaniasis.

References

Akuffo, H.O. (1992) Non-parasite-specific cytokine responses may influence disease outcome following infection. *Immunology Reviews* 127, 51–68.

Akuffo, H.O. and Britton, S.F.F. (1992) Contribution of non-*Leishmania*-specific immunity to resistance to *Leishmania* infection in humans. *Clinical and Experimental Immunology* 87, 58–64.

Akuffo, H., Schurr, E., Andersson, T. *et al.* (1987) Responsiveness in diffuse versus local cutaneous leishmaniasis is due to parasite differences. *Scandinavian Journal of Immunology* 26, 717–721.

Akuffo, H.O., Fehniger, T.E. and Britton, S. (1988) Differential recognition of *Leishmania* aethiopica antigens by lymphocytes from patients with local and diffuse cutaneous leishmaniasis. *Journal of Immunology* 141, 2461–2466.

Akuffo, H., Kaplan, G., Kiessling, R. *et al.* (1990) Administration of recombinant interleukin-2 reduces the local parasite load of patients with disseminated cutaneous leishmaniasis. *Journal of Infectious Diseases* 161, 775–780.

Akuffo, H., Maasho, K. and Howe, R. (1993) Natural and acquired resistance to *Leishmania*: cellular activation of *Leishmania aethiopica* of mononuclear cells from unexposed individuals is through the stimulation of NK cells. *Clinical and Experimental Immunology* 94, 516–521.

Akuffo, H., Maasho, K., Blostedt, M. *et al.* (1995) *Leishmania aethiopica* derived from diffuse cutaneous leishmaniasis patients preferentially induce mRNA for IL-10 while those from local cutaneous leishmaniasis patients induce IFN-γ. (In press).

Barral, A., Barral, N.M., Yong, E.C. *et al.* (1993) Transforming growth factor beta as a virulence mechanism for *Leishmania braziliensis*. *Proceedings of the National Academy of Sciences of the USA* 90, 3442–3446.

Carvalho, E.M., Badaro, R., Reed, S.G. *et al.* (1985) Absence of gamma interferon and interleukin 2 production during active visceral leishmaniasis. *Journal of Clinical Investigation* 76, 2066–2069.

Coffman, R.L., Varkila, K., Scott, P. *et al.* (1991) Role of cytokines in the differentiation of $CD4^+$ T-cell subsets *in vitro*. *Immunological Reviews* 123, 189–207.

Ghalib, H.W., Piuvezam, M.R., Skeiky, Y.A. *et al.* (1993) Interleukin 10 production correlates with pathology in human *Leishmania donovani* infections. *Journal of Clinical Investigation* 92, 324–329.

Gramiccia, M., Gradoni, L. and Troiani, M. (1992) HIV-*Leishmania* co-infections in Italy. Isoenzyme characterization of *Leishmania* causing visceral leishmaniasis in HIV patients. *Transactions of the Royal Society of Tropical Medicine and Hygiene* 86, 161–163.

Hatzigeorgiou, D.E., He, S., Sobel, J. *et al.* (1993) IL-6 down-modulates the cytokine-enhanced antileishmanial activity in human macrophages. *Journal of Immunology* 151, 3682–3692.

Ho, J.L., Reed, S.G., Wick, E.A. *et al.* (1990) Granulocyte–macrophage colony-stimulating factors activate intramacrophage killing of *Leishmania mexicana amazonensis*. *Journal of Infectious Diseases* 162, 224–230.

Kemp, M., Kurtzhals, J.A., Bendtzen, K. *et al.* (1993) *Leishmania donovani*-

reactive Th1- and Th2-like T-cell clones from individuals who have recovered from visceral leishmaniasis. *Infection and Immunity* 61, 1069–1073.

Kreutzer, R.D., Grogl, M., Neva, F.A. *et al.* (1993) Identification and genetic comparison of leishmanial parasites causing viscerotropic and cutaneous disease in soldiers returning from Operation Desert Storm. *American Journal of Tropical Medicine and Hygiene* 49, 357–363.

Laskay, T., Kiessling, R., De, W.T. *et al.* (1991) Generation of species-specific DNA probes for *Leishmania aethiopica. Molecular and Biochemical Parasitology* 44, 279–286.

Liew, F.Y. and Cox, F.E.G. (1991) Nonspecific defence mechanism: the role of nitric oxide. *Parasitology Today* 7, A17–A21.

Liew, F.Y., Li, Y. and Millott, S. (1990) Tumor necrosis factor-α synergizes with IFN-γ in mediating killing of *Leishmania major* through the induction of nitric oxide. *Journal of Immunology* 145, 4306–4310.

Manetti, R., Parronchi, P., Giudizi, M.G. *et al.* (1993) Natural killer cell stimulatory factor (interleukin 12 [IL-12]) induces T helper type 1 (Th1)-specific immune responses and inhibits the development of IL-4-producing Th cells. *Journal of Experimental Medicine* 177, 1199–1204.

Manetti, R., Gerosa, F., Giudizi, M.G. *et al.* (1994) Interleukin 12 induces stable priming for interferon gamma (IFN-gamma) production during differentiation of human T helper (Th) cells and transient IFN-gamma production in established Th2 cell clones. *Journal of Experimental Medicine* 179, 1273–1283.

Mohammed, A.M., Taye, T. and Akuffo, H.O. (1992) Mechanisms of resistance to *Leishmania aethiopica* 1. Interferon gamma in combination with a cytokine (not tumour necrosis factor alpha) is required, but cannot act alone in the inhibition of intracellular forms of *L. aethiopica* in THP1 cells. *European Journal of Immunology* 22, 2331–2337.

Mosmann, T.R., Cherwinski, H., Giedlin M.W. *et al.* (1986) Two types of murine helper T-cell clone – 1. Definition according to profiles of lymphokine activities and secreted proteins. *Journal of Immunology* 136, 2348–2357.

Pirmez, C., Yamamura, M., Uyemura, K., *et al.* (1993) Cytokine patterns in the pathogenesis of human leishmaniasis. *Journal of Clinical Investigation* 91, 1390–1395.

Sadick, M.D., Heinzel, F.P., Holaday, B.J. *et al.* (1990) Cure of murine leishmaniasis with anti-interleukin 4 monoclonal antibody. Evidence for a T cell-dependent, interferon gamma-independent mechanism. *Journal of Experimental Medicine* 17, 115–127.

Scott, P. (1991) IFN-gamma modulates the early development of Th1 and Th2 responses in a murine model of cutaneous leishmaniasis. *Journal of Immunology* 147, 3149–3155.

Scott, P., Pearce, E., Cheever, A.W. *et al.* (1989) Role of cytokines and $CD4^+$ T-cell subsets in the regulation of parasite immunity and disease. *Immunological Reviews* 112, 161–182.

Villanueva, J.L., Torre, C.J., Jurado, R. *et al.* (1994) *Leishmania* esophagitis in an AIDS patient: an unusual form of visceral leishmaniasis. *American Journal of Gastroenterology* 89, 273–275.

4

Immunoregulation in Visceral Leishmaniasis

E.M. Carvalho

Serviço de Imunologia, Universitaria Prof. Edgar Santos (HUPES), UFBA, Lab Immunologia 30 Rua and Rua Joao das Botas, a/n Canela 40.110.160, Salvador-Bahia, Brazil

Visceral leishmaniasis (VL), or kala azar, is the systemic form of *Leishmania* infection. The disease is caused by *Leishmania donovani*, *Leishmania chagasi* and *Leishmania infantum*. Fever, weight loss, hepatosplenomegaly, anaemia, leukopaenia and hyperglobulinaemia are the main clinical features of this disease. Immunologically, VL is is characterized by an absence of T-cell response to *Leishmania* antigen and polyclonal B-cell activation.

In endemic areas of VL, infected subjects may or may not develop classic signs and symptoms of VL. Capacity to produce interleukin (IL)-2 and interferon-gamma (IFN-γ) is associated with asymptomatic or subclinical self-healing VL. In contrast, individuals whose lymphocytes do not proliferate and, therefore, do not produce IFN-γ when stimulated by *Leishmania* antigen, will develop acute VL or a subclinical infection that progresses to classical VL. Immunological abnormalities can be found in monocyte and T-cell function, such as diminished production of tumour necrotic factor-alpha (TNF-α) and IL-1 after lipopolysaccharide (LPS) or *Listeria* stimulation, decreased production of IL-2 and IFN-γ by lymphocytes under *Leishmania* antigen stimulation, absence of delayed-type hypersensitivity to *Leishmania* antigen and a decreased capability of T-cells to activate macrophages and kill *Leishmania*. All these abnormalities may account for parasite multiplication and progression of the disease.

In the experimental model of leishmaniasis, the role of T-helper cell subsets in the pathogenesis of this disease have been evaluated in mice susceptible and resistant to *Leishmania major* infection. These studies show that activation of Th2 cells and consequently IL-4 production are associated with progression of the disease, while Th1 activation and IFN-γ production are associated with host capability to control *Leishmania* infection. In human VL the possibility that Th2 activation may be involved in progression of the disease has been suggested by the demonstration of elevated levels of IL-4 in VL sera, isolation of Th2-like T-cell clones from individuals who have been cured of visceral leishmaniasis and demonstration of messenger (m) RNA for IL-10 in bone marrow cells and lymph nodes from VL patients.

In this study, we evaluated both the pattern of cytokines secreted and the levels of cytokine mRNA in peripheral blood mononuclear cells (PBMC) of patients with VL, following *in vitro* stimulation with *Leishmania* antigen. Additionally, the role of various cytokines (IL-1, IL-2, IL-12, granulocyte macrophage colony stimulating factor (GM-CSF) and IFN-γ) or cytokine antagonists (α-IL-5 and α-IL-10 monoclonal antibodies (mAbs)) were evaluated as to their ability to restore *in vitro* lymphoproliferative responses and IFN-γ production, following stimulation with *Leishmania* antigen.

The present study includes 30 patients admitted to the Hospital Universitário Professor Edgard Santos and the Hospital Santo Antonio in Salvador, Bahia, Brazil. The patients had a history of fever and weight loss. Clinical examination revealed hepatosplenomegaly in all 30 patients. The diagnosis of VL was determined by detection of amastigote forms of *Leishmania* in spleen or bone marrow aspirates stained with Giemsa. All immunological studies were performed prior to therapy. The cytokine polymerase chain reaction (PCR) was determined both prior to and following therapy. The control group consisted of ten subjects cured of VL, seven individuals having asymptomatic *L. chagasi* infection (seroconverting individuals living in an endemic area that remain asymptomatic after three years of follow-up) and ten healthy subjects.

The mean [^{3}H]-thymidine uptake in cultures stimulated by *L. chagasi* (1340 ± 1110 cpm) was not statistically different ($P > 0.05$) from that observed in unstimulated cultures (917 ± 659 cpm), reflecting the low stimulation index (SI) found in these patients (SI ranged from 0.6 to 3.8). These numbers were lower ($P < 0.01$) than those observed in subjects cured of VL (8831 ± 7379 cpm) or in subjects with asymptomatic *L. chagasi* infection (8022 ± 6700 cpm). Absent or low IFN-γ production (levels ranging from less than 2 to 59 pg ml^{-1}) was observed in supernatants of lymphocyte cultures stimulated with *L. chagasi* antigen. In contrast, when mononuclear cells from VL patients were stimulated with phytohaemaglutinin (PHA) plus ionomicin, high levels of IFN-γ (980 ± 320 pg ml^{-1}) were detected. IFN-γ production in supernatants of lymphocytes from VL patients (9 ± 18 pg ml^{-1}) was lower ($P < 0.01$) than that observed in cured visceral leishmaniasis (148 ± 167 pg ml^{-1}) or in subjects with asymptomatic *L. chagasi* infection ($150 \pm 126\,pg\,ml^{-1}$). IL-4 levels were low or absent in culture supernatants of lymphocytes stimulated with *L. chagasi* antigen in all three groups studied. IL-4 levels were measured in sera of ten VL patients ($17 \pm 11\,pg\,ml^{-1}$), and this value did not differ ($P > 0.05$) from those found in the sera of ten healthy subjects ($16 \pm 12\,pg\,ml^{-1}$).

mRNA for IFN-γ, IL-4 and IL-10 was determined by PCR in extracts of PBMC from five VL patients prior to therapy and in two of such cases following antimonial therapy. mRNA was determined before therapy at time zero and in cells cultured for 72 h without stimulus or stimulated with *L. chagasi* antigen, and after therapy at time zero and in cells cultured for 72 h without stimulus or with *L. chagasi* antigen. In all patients, mRNA for IL-10 was observed prior to therapy, at time zero or after 72 h of incubation of cells with or without antigenic stimulation. After therapy, there was no IL-10 mRNA in one patient and a decrease in IL-10 mRNA for another patient. IL-4 mRNA was abundant at time zero and after 72 h in cultures with or without antigenic stimulation in four of the five patients, prior to therapy. After therapy in one case there was a decrease in IL-4 mRNA, while in another patient IL-4 mRNA was similar to that observed before therapy. There were marked variations in mRNA for IL-2 in the five patients tested. While mRNA for IL-2 was abundant in three patients mainly in stimulated cultures, in two patients the levels of IL-2 mRNA was very low regardless of the time point analysed. The mRNA for IFN-γ was detected primarily at time zero but also at 72 h in both unstimulated or stimulated cultures, before and after therapy.

To determine whether exogenous cytokines could restore *in vitro* T-cell responses in VL, PBMC from active VL patients were stimulated with soluble *L. chagasi* antigen in the presence of the following cytokines: IL-1, GM-CSF or IFN-γ at concentrations ranging from 1 to $100\,ng\,ml^{-1}$, IL-2 at concentrations ranging from 0.1 to $1\,ng\,ml^{-1}$ and IL-2 at concentrations ranging from 1 to $100\,\mu g\,ml^{-1}$. Cytokine combinations (such as IFN-γ plus IL-2) were also added to cultures. While IL-1, GM-CSF and IFN-γ enhanced lymphocyte proliferation by a maximum of 20% and IL-2 enhanced the lymphoproliferative response by 56%, the addition of IL-2 plus IFN-γ resulted in a 158% enhancement of lymphoproliferation in comparison with cultures stimulated with *L. chagasi*. These cytokines were not able to stimulate lymphoproliferative responses in the absence of *Leishmania* antigen. Although an increase in lymphocyte proliferation was observed in all patients ($P < 0.01$), after the addition of IL-2 plus IFN-γ this enhancement was variable from patient to patient. While in two patients the increase was 17% and 45%, respectively, in relation to cultures stimulated with *L. chagasi* alone, in the other three patients the enhancement of proliferation ranged from 87 to 283%.

To determine whether the monoclonal antibodies (mAbs) against α-IL-4 and α-IL-10 were capable of restoring *in vitro* lymphocyte proliferation and IFN-γ production, PBMC from patients with VL were stimulated with soluble *L. chagasi* antigen and different concentrations (25 to 250 $\mu g\ ml^{-1}$) of these mAbs. The α-IL-4 and α-IL-10 mAbs, alone and in combination, enhanced lymphocyte proliferation. The mean ± SEM (standard error from the mean) of [^{3}H]-thymidine uptake by lymphocytes from four VL patients stimulated with *L. chagasi* antigen was 792 ± 530 cpm. In cultures where α-IL-4, α-IL-10 or α-IL-4 plus α-IL-10 were added, the responses in six patients were 1602 ± 401, 2384 ± 567, 3990 ± 456, 4082 ± 1467, 6454 ± 268 and 8265 ± 1047 cpm. The enhancement of lymphocyte proliferation in cultures stimulated by *L. chagasi* antigen from these six patients was quite variable and primarily observed when mAbs were used at a concentration of 250 $\mu g\ ml^{-1}$.

The mAbs against IL-4 and IL-10 were also capable of restoring IFN-γ production in patients with VL. In six patients evaluated, IFN-γ levels in cultures stimulated with *L. chagasi* antigen alone were 0, 0, 0, 2, 18 and 59 $pg\ ml^{-1}$. Addition of α-IL-10 increased IFN-γ levels to 48, 97, 76, 140 and 223 $pg\ ml^{-1}$ in five of the six patients. In contrast with the results obtained from lymphocyte proliferation, where α-IL-4 associated with α-IL-10 had an additive or synergistic effect on restoration of lymphocyte proliferation, IFN-γ production in cultures containing α-IL-10 plus α-IL-4 mAbs was similar (45, 0, 102, 85, 146 and 292 $pg\ ml^{-1}$) to that observed in cultures with α-IL-10 mAb alone. The IFN-γ levels in cultures stimulated with *L. chagasi* plus α-IL-10 (97 ± 77 $pg\ ml^{-1}$) or *L. chagasi* plus α-IL-10 and α-IL-4 (112 ± 101 $pg\ ml^{-1}$) were higher ($P < 0.05$) than those observed in cultures stimulated with *L. chagasi* alone (13 ± 24 $pg\ ml^{-1}$) or *L. chagasi* plus α-IL-4 (21 ± 17 $pg\ ml^{-1}$) but did not differ statistically ($P < 0.01$) from those observed in cells from cured VL patients (148 ± 167 $pg\ ml^{-1}$). In unstimulated cultures, α-IL-4 and/or α-IL-10 mAbs induced no increased lymphocyte proliferation or IFN-γ production. Additionally, rat IgG-1 and IgG-2a isotype controls did not enhance lymphocyte proliferation or IFN-γ production in cultures stimulated with *L. chagasi* antigen.

The role of IL-10 in modulating the immune response in VL was further determined by evaluating the capability of IL-10 to suppress lymphocyte proliferation and IFN-γ production in three subjects cured of VL. In these experiments, at concentrations of 20 $ng\ ml^{-1}$, IL-10 suppressed lymphocyte proliferation and IFN-γ production by 80% and 90% respectively.

Since IL-10 and IL-12 are secreted by macrophages after interaction with intracellular parasites and IL-12 is a cytokine capable of a Th1-type of response, the capability of this cytokine in the restoration of T-cell response in VL was further studied. IL-12 used in concentrations ranging from 10 to 100 $\mu g\,ml^{-1}$ enhanced the SI from 1.4 (data of cultures stimulated with *L. chagasi*) to 19.2 (data observed in cultures stimulated by antigen plus IL-12). The IL-12 was also able to restore IFN-γ production. While IFN-γ levels in cultures stimulated with *L. chagasi* antigens was $< 8\,pg\,ml^{-1}$, in the presence of IL-12 the *L. chagasi* antigen induced high levels ($551 \pm 11\,pg\,ml^{-1}$) of IFN-γ. The effect of IL-12 was only observed in cultures stimulated with *L. chagasi* antigen since IL-12 alone was not able to induce lymphocyte proliferation or IFN-γ production. Indeed the capability of IL-12 in restoring T-cell function in VL was abrogated by addition of IL-10. Since IL-10 inhibited lymphocyte proliferation and IFN-γ production mediated by IL-12 and anti-IL-12 abrogated the effect of α-IL-10, it is likely that the disregulation in the secretion of such cytokines is the major factor involved in the absence of T-cell response in VL.

5

Epidemiology of Visceral Leishmaniasis and Its Control

R.C. MAHAJAN AND K. MOHAN

Department of Parasitology, Postgraduate Institute of Medical Education and Research, Chandigarh – 160 012, India

Introduction

Leishmaniasis appears to be far more abundant and of greater public health importance than was previously recognized. Although accurate quantitative evaluation is difficult, the World Health Organization (WHO) estimates that some 350 million people in the world are at risk of acquiring one or other form of leishmaniasis and that approximately 12 million people are currently infected (WHO, 1990). Visceral leishmaniasis (VL) has been reported from about 47 countries around the world which include 15 African countries, 16 Asian and eight each from the American and European continents. The annual VL cases have been estimated to be around 33,000 from Africa, 40,000 from Asia, 16,000 from the Americas and about 100 cases from Europe (Ashfold *et al.*, 1992).

Old World VL (*Leishmania donovani* complex)

The disease may be endemic, sporadic or epidemic with varying clinical forms in each situation. There is considerable under-reporting of VL cases, mainly due to lack of medical facilities, missed diagnoses and non-reporting despite clinical suspicion. Even today, in 52 countries the disease is not notifiable. It has been estimated that in Kenya and Italy every 1 out of 5 infected persons are recorded, while in Ethiopia 1 out of 7 and in India 1 out of every 10 infections is parasitologically confirmed and recorded (WHO, 1990).

In areas endemic to VL, the disease remains chronic and children are affected in particular. In North African and southwest Asian endemic foci, the commonest age group affected is 1–4 years. On the other hand, in southern Europe and China, people of all age groups are affected while in East Africa and the Indian subcontinent, the peak age is 5–9 years but the disease in teenagers is not uncommon. The male : female ratio is about 2 : 1.

Sporadic VL may occur in non-indigenous people of any age group once they enter an endemic area. The disease is usually acute in such cases and chances of developing rare complications like severe haemolytic anaemia, acute renal damage or severe mucosal haemorrhage are greater. In the epidemic form of VL, people of all age groups are susceptible except those who have acquired immunity during a previous epidemic. Males are infected more often than females in a ratio of 4 : 3. In recent years cases of VL are occurring in increasing numbers in people with acquired immune deficiency syndrome (AIDS) or in those with immunosuppression due to chemotherapy (WHO, 1990). Post kala azar dermal leishmaniasis (PKADL) due to *L. donovani* is more common in epidemic areas, particularly in the Indian subcontinent (Rees and Kager, 1987). It usually starts between 6 months and several years after the cure of VL. In some cases it may not be possible to find a history of VL. The chronic lesions of PKADL consist of multiple papular or nodular infiltrations of the skin, usually without ulceration. These nodules contain parasites and thus these patients may serve as reservoirs of infection.

New World VL

The disease is endemic or sporadic and the aetiological agent is *Leishmania chagasi*, although the disease shows clinical similarity to that caused by *Leishmania infantum* (Old World). These two parasites are indistinguishable by usual typing methods (WHO, 1990). Asymptomatic infections are very frequent with some developing classical kala azar and others resolve without specific treatment. Concomitant malnutrition or other infectious diseases appear to be decisive for the prognosis of New World VL infection. The New World VL is mainly seen in Central and South America.

VL in India

India along with Brazil, Sudan, Kenya, Bangladesh and Nepal contribute to more than 90% of all VL cases reported annually around the world (Ashfold *et al.*, 1992). The major VL focus in India is Bihar state where epidemics have been reported in 1891, 1917, 1933 and recently in 1978 and 1991. The 1991 epidemic recorded nearly 60,000 new cases; the actual figures could be much higher (0.25–0.3 million) considering the problem of under-reporting (Kaul *et al.*, 1993). Apart from Bihar, 10 districts of West Bengal and two districts of Uttar Pradesh are also endemic for VL. Assam, where more than 7500 deaths were recorded during 1987–1988, is free from kala azar at present (Choudhury and Saxena, 1987). Sporadic cases of VL are still reported from Tamil Nadu state and Delhi, most of which are in migrating populations (Sivaprakasam *et al.*, 1988; Mittal *et al.*, 1989). A study conducted in Bihar shows that in a lower age group (4–12 years) the VL incidence is similar in both sexes while incidence is quite low in females belonging to the reproductive age group (Sharma *et al.*, 1990). The household members of VL cases show seropositivity against *Leishmania donovani* in about 5–17.5% of cases (Saran *et al.*, 1992).

Animal Reservoirs and Zoonotic VL

Except for Afghanistan, Bangladesh, India and Nepal all other countries from where VL cases are reported also show one or more animal reservoirs harbouring the parasite (Ashfold and Bettini, 1987). Mere detection of parasites in any animal does not necessarily implicate it as a reservoir host and WHO has laid down certain criteria to implicate an animal species as a reservoir for VL (WHO, 1990). These include: the animal should be abundant, long lived and be a food source to sandflies; there should be intense host–vector contact with transmission among these hosts; more than 20% of specimens during the lifespan should be positive for parasites with abundant parasites in the skin/blood; the type of infection should be chronic, less pathogenic with a parasite indistinguishable from that found in man. In both New World and Old World VL, proven or suspected reservoir hosts show varying incidences of parasite detection depending on the area in question (Bray, 1974; Ashfold and Bettini, 1987; Desjeux, 1991).

In India, Nepal, Bangladesh and Afghanistan the disease is anthroponotic and has not been detected in any reservoir host to date. Patients with PKADL may serve as important reservoirs from one epidemic to another. In the transmission of Old World VL due to *L. infantum* in central, rural Asia, the jackal (*Canis aureus*), wolf (*Canis lupus*) and fox (*Vulpes vulpes*) play an important role as reservoirs (Ashfold and Bettini, 1987). In urban central Asia, extending up to southwest Asia, Turkey and the Mediterranean area, dogs and badgers (*Meles* sp.) serve as reservoirs of infection. Raccoon dogs (*N. procyanoides*) are reservoirs of infection in Miyum and China while rats are proven reservoir hosts in Spain and the Italian peninsula. In New World VL due to *L. chagasi*, foxes play a major role as a source of infection, *Lycalopex vetulus* in Ceará state and other parts of northeast Brazil and *Cerdocyon thous* in Amazonian Brazil (Shaw and Lainson, 1987). In addition, dogs are the main source for human infection in Venezula, Colombia,

Honduras, El Salvador and certain parts of Brazil. Opossums (*Didelphis* sp.) are also suspected VL reservoir hosts and in Colombia up to 32% of these animals are found to be infected (Desjeux, 1991). The infection rate of dogs in Brazil, Colombia and Venezuela ranges between 3 and 20%.

Proven and Principal Vectors of VL

The vectors of VL throughout the world belong to two principal genera, *Phlebotomus* and *Lutzomyia*, the former for Old World VL and the latter for New World VL (Lewis, 1974; Williams and Coelho, 1978). In anthroponotic *L. donovani* infections, *Phlebotomus argentipes* is the vector in India while *Phlebotomus martini* is the vector in Kenya. In zoonotic VL (*L. infantum*) transmission, many species of *Phlebotomus* are involved as vectors. For example, *Phlebotomus alexandri* in China, *Phlebotomus ariasi* in France and Spain, *Phlebotomus perniciosus* in Algeria, France, Italy and Spain, *Phlebotomus tobbi* in Cyprus, *Phlebotomus chinensis* in China and *Phlebotomus longiductus* in many parts of Russia. *L. infantum* has the remarkable property of adapting itself to vectors belonging to different species, subgenera and even genera, which is the reason why it can be transmitted through a wide range of vector species. Throughout the New World *Lutzomyia longipalpis* is the vector responsible for the transmission of zoonotic VL due to *L. chagasi* infection.

It is necessary for certain conditions to be satisfied before incriminating a species of sandfly as a vector for VL transmission. The vector species should be anthropophilic; it should also feed on reservoir hosts wherever VL is zoonotic; it should be infected in nature with the same parasite species which infects man; it should be able to transmit infection by biting; and lastly, it should have the same geographical distribution as that of the parasite. Usually the vectors supporting flourishing growth of the parasites and with a high anthropophilic index play a major role in intense VL transmission.

Transmission Sites

Zoophilic foci

In the Mediterranean basin where zoophilic VL by *L. infantum* is seen, the human–sandfly contact within the house leads to the formation of microfoci. This is more common in rural areas, villages in mountainous regions and periurban areas, as along the coast of North Africa. In these areas most people have dogs which are the proven reservoirs for *L. infantum*. In addition, Algerian houses have dog kennels where a high density of vectors like *P. perniciosus*, *Phlebotomus longicuspis* and *Phlebotomus perfiliewi* have been observed. A similar type of VL transmission also occurs in west and central Asia and China.

Anthroponotic foci

In India, Bangladesh and Nepal epidemics are usually seen in areas with heavy rainfall, a humidity of more than 70%, a temperature range of 15–38°C, abundant vegetation, subsoil water and alluvial soil, all of which facilitate high vector density (Bhattacharya, 1981). Rural peasants in India keep buffaloes and cows in the same living room or close to sleeping sites and these attract vectors. In Kenya humans live in close proximity to sandfly breeding and resting sites like termite hills, cowsheds and rodent burrows (Ashfold and Bettini, 1987). The grass-thatched huts provide easy access to sandflies and light clothing in hot climates increases the body exposure to possible sandfly bites.

The transmission sites in New World VL resemble those of the Mediterranean basin. Dogs and other animals stay indoors during the night and attract sandflies, especially *Lupus longipalpis*. In the Amazon region of South America the hen-house is built close to human dwellings, which is a favoured feeding and resting place for *Lu. longipalpis* (Lainson *et al.*, 1983). Moreover, chickens attract wild carnivorous animals, the sylvatic reservoir hosts for VL, thus increasing the transmission risk to domestic dogs and humans.

Other socioeconomic factors like primitive housing, poor hygiene, accumulation of rubbish and organic remnants, piles of bricks/stones around the house and cracks in the walls provide excellent breeding and resting sites for vectors thereby enhancing peridomestic transmission of VL. Low-income-group people live at the periphery of cities which could be the edge of natural foci (forest/desert), thereby becoming more prone to infection from the sylvatic cycle. Overcrowding and the use a single room by many people for sleeping attract peridomestic, anthropophilic vector species. Population movement is another important feature in VL transmission and it has been seen that many outbreaks are associated with agricultural development, construction of highways and railways and with labour movement for urban development. Such migration introduces non-immune people into zoonotic VL foci. Indian anthroponotic VL has shown a gradual spread in recent years mainly due to population movement (Sen Gupta, 1975).

VL Control Strategies

Like other vector-borne tropical diseases including malaria, control measures of VL mainly include early diagnosis, treatment and reduction in the source of infection; control of wild/domestic reservoir hosts wherever VL is zoonotic; vector control by chemical, biological, genetic and ecological methods along with personal protection. The methods vary according to the geographical area and type of infection in a given locality. Often integrated approaches with all the above methods are required as no single method is sufficient to contain the disease.

Diagnosis of VL, both clinical and laboratory based, plays an important role in obtaining an up-to-date picture of the epidemiological situation. Special care should be taken in tracing a history of travelling/residence in endemic areas whenever the disease is suspected on the basis of clinical signs and symptoms. Serology is useful in screening large populations at risk. Currently available direct agglutination test and dot-ELISA (enzyme-linked immunosorbent assay) are simple, reliable and applicable in field conditions on a large scale (WHO, 1990). Once suspected on the basis of clinical examination or serology, parasitological diagnosis from specimens from spleen, bone marrow or lymph node biopsy becomes a must for the initiation of treatment. Passive case detection could be done by trained staff at primary health centres and referred to higher levels for confirmation. In addition, active medical surveillance by regular and systematic screening of populations at risk in endemic areas appears to be the most efficient epidemiological method for early case detection.

Treatment of positive VL cases is done with pentavalent antimonials. The objectives of therapy are: to cure the patient and thereby reduce the source for transmission; to prevent relapses; to avoid development of unresponsiveness to drugs at a later stage of the disease and to keep hospitalization/treatment cost to a minimum. Initial therapy with meglumine antimonate or sodium stibogluconate is recommended for 20 days, though the duration of treatment varies in different geographical areas (Bryceson, 1987). In India a 40-day treatment schedule is required for the best results while in China a 6-day course has been shown to have a cure rate of more than 90%. Though PKADL cases should be treated in the same way as VL cases, the duration of treatment must be for at least 4 months (Sen Gupta, 1953). Recently, liposome preparations with antimonials have shown 200–700-fold more efficacy than free drugs against *L. donovani* (Croft, 1986). Apart from this, second line drugs like amphotericin-B, allopurinol in combination with pentamonials and pentamidine are also available, but care should be taken regarding their toxicity.

Control of reservoir hosts is essential in zoonotic VL foci to reduce human transmission. Elimination of stray dogs, if acceptable by the community; hygienic disposal of slaughterhouse remains which decrease dog–jackal contact; mass screening of domestic dogs by serological methods and prompt treatment of infected reservoirs are some of the methods which can be practised in order to control domestic/commensal reservoir hosts. No method is being allowed at present for the control of wild reservoirs like foxes and jackals. For opossums, trapping by baited pitfall traps is used in Colombia and for rats, filling/poisoning burrows have been tested in Italy, Spain and Russia with encouraging results.

Vector control is the most attractive target for control of many tropical diseases and VL is no exception. Many approaches are practised to either reduce vector density or to prevent vector–host contact. Chemical control using insecticides like DDT, BHC, malathione, fenitrothione or synthetic pyrethroids is most suited for domestic/peridomestic rural areas, construction sites and logging camps. DDT is the insecticide of choice because of low cost, high efficacy, long residual action and relative safety. For strictly endophilic vectors, indoor residual spraying is a cost-effective method. For partly exophilic or peridomestic vectors, spraying should be done on the outer surface of shelters, crevices in stone walls and other outdoor resting sites. In addition, selective outdoor spraying is also practised for tree trunks and termite hills in Kenya. Space spraying with thermal fogs, aerosols or mist is not evaluated sufficiently for efficacy in sandfly control (WHO, 1990).

Biological control of sandflies is not practised at present though a fungal agent *Entomophthoraceous* was found to infect 20% of sandfly larvae and some predatory mites were found to kill larvae *in vitro*. Further research is required to investigate an effective biological control method. Recently, the bacterial toxins (biocides) from *Bacillus sphaericus* and *Bacillus thuringiensis* have been found to be useful in killing early-stage sandfly larvae in laboratory conditions. Environmental modifications like clearing of forests around the endemic villages have proved to be effective in reduction/elimination of man–vector contact in Colombia, French Guyana and Panama. Location of labour camps should be away from man–vector contact sites, and rubble and rubbish around the houses should be removed, preferably with community participation.

Individual protection is another important approach in reducing infective sandfly bites. Application of chemical repellents on the skin/clothing, like diethyltoluamide, diethylphthalate or hexamide can protect against sandfly bites for a period of 8, 6.5 and 7.3 h, respectively. Neem oil (2% in coconut/mustard oil base) was found to protect more than 86% individuals for 8 h in an experiment conducted in India. Insecticide-treated (diethylphthalate, diethyltoluamide) curtains and bed nets can provide 98% protection for a period of 17–20 days. Even for torn or improperly tucked-in bed nets, impregnation with permethrin or deltamethrin is found to be very effective in repelling the vectors. Vaccination for individual protection is still in the experimental stage; more progress has been made with cutaneous leishmaniasis than with VL.

In many endemic areas no single control method is effective or sufficient to reduce transmission. In such areas integrated approaches are required where simultaneous application of two or more of the above described methods, on a priority basis, is essential (WHO, 1990). Health education, awareness among population and community participation in control programmes could help in effective prevention of epidemics and reduce morbidity and mortality due to VL.

References

Ashfold, R.W. and Bettini, S. (1987) Ecology and epidemiology: old world leishmaniasis. In: Peters, W. and Killick-Kendrick, R. (eds) *The Leishmaniasis in Biology and Medicine*, Vol. I. Academic Press, London, pp. 370–375.

Ashfold, R.W., Desjeux, P. and deRaadt, P. (1992) Estimation of population at risk of infection and number of cases of leishmaniasis. *Parasitology Today* 8, 104–105.

Bhattacharya, P.K. (1981) Distribution of Kala-azar in India. *Transactions of the Royal Society of Tropical Medicine and Hygiene* 75, 333–334.

Bray, R.S. (1974) Zoonosis in leishmaniasis. In: Soulsby, E.J.L. (ed.) *Parasitic Zoonosis*. Academic Press, New York, pp. 65–67.

Bryceson, A. (1987) Therapy in man. In: Peters, W. and Killick-Kendrick, R. (eds) *The Leishmaniasis in Biology and Medicine*, Vol. II. Academic Press, London, pp. 848–895.

Choudhury, N. and Saxena, N.B.L. (1987) Visceral leishmaniasis in India – a brief review. *Journal of Communicable Diseases* 19, 332–340.

Croft, S.L. (1986) Liposomes in the treatment of parasitic diseases. *Pharmacy International* 7, 229–233.

Desjeux, P. (1991) *Information on the Epidemiology and Control of the Leishmaniasis by Country or Territory*. Document WHO/LEISH/91.30, World Health Organization, Geneva, pp. 5–47.

Kaul, S.M., Das, R.K., Shiv Raj, Saxena, N.B.L. and Narasimham, M.V.V.L. (1993) Entomological monitoring of Kala-azar control in Bihar State. *Indian Journal of Communicable Diseases* 25, 101–106.

Lainson, R., Shaw, J.J., Silveria, F.T. and Fraiha, H. (1983) Leishmaniasis in Brazil: XIX. Visceral leishmaniasis in the Amazon region. *Transactions of the Royal Society of Tropical Medicine and Hygiene* 77, 323–330.

Lewis, D.J. (1974) The biology of phlebotomidae in relation to leishmaniasis. *Annual Review of Entomology* 19, 363–384.

Rees, P.H. and Kager, P.A. (1987) Visceral leishmaniasis and post kala azar dermal leishmaniasis. In: Peters, W. and Killick-Kendrick, R. (eds) *The*

Leishmaniasis in Biology and Medicine, Vol. II. Academic Press, London, pp. 584–615.

Mittal, V., Bhatia, R. and Sehgal, S. (1989) Clinicoepidemiological profile of kala azar patients in Delhi. *Journal of Communicable Diseases* 21, 255–261.

Saran, R., Gupta, A.K. and Sharma, M.C. (1992) Evidence of *L. donovani* infection in household members residing with VL patients. *Journal of Communicable Diseases* 24, 242–244.

Sen Gupta, P.C. (1953) Chemotherapy of leishmanial diseases: a resume of recent researches. *Indian Medical Gazette* 88, 20–35.

Sen Gupta, P.C. (1975) Return of kala azar. *Journal of the Indian Medical Association* 65, 89–90.

Sharma, M.C., Gupta, A.K., Saran, R. and Sinha, S.P. (1990) The effect of age and sex on incidence of kala azar. *Journal of Communicable Diseases* 22, 277–278.

Shaw, J.J. and Lainson, R. (1987) Ecology and epidemiology: new world leishmaniasis. In: Peters, W. and Killick-Kendrick, R. (eds) *The Leishmaniasis in Biology and Medicine*, Vol. I. Academic Press, London, pp. 293–297.

Sivaprakasam, P., Padmanabhan, B. and Sadanand, A.V. (1988) Recent trends in the incidence and epidemiology of kala azar in Madras city. *Journal of Communicable Diseases* 20, 185–195.

WHO (1990) *Control of the Leishmaniasis*. Technical report series-793. World Health Organization, Geneva, pp. 9–117.

Williams, P. and Coelho, M. de V. (1978) Taxonomy and transmission of *Leishmania*. *Advances in Parasitology* 16, 1–42.

6

Vaccination against Visceral Leishmaniasis Using a Pure Parasite Protein

N. RACHAMIM,[2] R. SARFSTEIN[1] AND C.L. JAFFE[1*]

[1]Department of Parasitology, Hebrew University-Hadassah Medical School, P.O. Box 12272, Jerusalem 91120, Israel and [2]Department of Membrane Research and Biophysics, Weizmann Institute of Science, Rehovot, Israel

Extensive investigations have been carried out into the immunological mechanisms controlling cutaneous disease caused by *Leishmania major* in mice. Resistance or susceptibility to disease is controlled by the genetic predilection of the host to develop a specific cell-mediated response to parasite antigen upon infection, culminating in either a T-cell-specific $CD4^+$ Th1 or Th2 response (Locksley and Louis, 1992; Liew and O'Donnell, 1993). The former response leads to healing and the latter to chronic disease (Locksley and Louis, 1992; Liew and O'Donnell, 1993). Specific $CD8^+$ T-cell responses are involved in controlling reinfection (Milon and Louis, 1993). An extensive interplay of host cytokines, including interleukin (IL)-4, IL-6, IL-10, IL-12, interferon-gamma (IFN-γ) and others (Reiner and Locksley, 1993), are important in directing the development of specific T-cell responses and the outcome of the disease. Antibodies to *Leishmania* appear to play at most a very minor role in host resistance to this parasite (Mitchell, 1984).

Visceral leishmaniasis (VL) primarily caused by the *Leishmania donovani* complex and occasionally by *Leishmania tropica* results in death if not promptly diagnosed and treated (Peters and Killick-Kendrick, 1987; Kreutzer *et al.*, 1993). Recent reports of epidemics and drug-resistant parasite strains in India and Africa (Seaman *et al.*, 1992; Thakur and Kumar, 1992) have highlighted the importance of an effective vaccine for this fatal disease. However, little is known about the immunological mechanisms controlling the development of VL following

* Corresponding author.

Box 6.1. Properties of the protective protein dp72.

- Present in *L. donovani* and other species
- Molecular mass varies between strains and species of *Leishmania*:
  - *L. donovani,* 72–80 kDa
  - *L. major,* 60 kDa
  - *L. tropica,* 60 kDa
  - *L. aethiopica,* 60 kDa
- Present in both promastigotes and amastigotes
- Primarily localized in endoplasmic reticulum with small amount on the surface membrane
- Labelled metabolically with orthophosphate, fatty acids, methionine and glucose
- Complexed with DNA-antigen is sensitive to treatment with deoxyribonuclease I, copper/phenanthroline and phosphodiesterase I
- No N- or O-linked glycosylation. Not affected by tunicamycin
- Partitions differently in Triton X-114 when isolated in the presence or absence of $ZnCl_2$
- Antibodies to protein in patients with VL

parasite infection. Longitudinal studies in endemic areas have shown that significant levels of asymptomatic and subclinical infections exist, many which self-cure and fail to develop into clinical disease (Badaro *et al.*, 1986; Holaday *et al.*, 1993a). Recent studies in humans suggest that the development and resolution of cutaneous leishmaniasis (CL) and VL are controlled by similar immunological mechanisms (Ghalib *et al.*, 1993; Holaday *et al.*, 1993b; Karp *et al.*, 1993; Pirmez *et al.*, 1993; Melby *et al.*, 1994). However, infection of mice with *L. donovani* differs from acute human VL, as the mice do not die, and the parasite burden eventually diminishes in some strains. This behaviour is similar to asymptomatic or subclinical human disease.

Animals vaccinated with pure or crude parasite proteins are good models for studying the immunological mechanisms involved in establishing a protective response. One such protein, dp72, purified from *Leishmania donovani* (*L. d.*) *infantum* has been shown to give good protection in the BALB/c mouse model when challenged with either *L. d. chagasi* or *L. d. donovani*. The properties of this protein are summarized in Box 6.1. Briefly, this protein is found in amastigotes and promastigotes of the *L. donovani* complex. Patients with VL, but not other diseases, produce antibodies to dp72 allowing the pure protein to be used for the specific serodiagnosis of VL. Marked molecular mass microheterogeneity has been found for dp72 within the donovani complex and strains of the same subspecies. As shown by immunolocalization using electron microscopy it is primarily an intracellular protein and a related protein of smaller molecular mass has been identified in *L. tropica*, *L. major* and *Leishmania aethiopica*. dp72 is not a glycoprotein, but contains several post-translational modifications and can be metabolically labelled with myristic or palmitic acid, and orthophosphate. Partitioning experiments in TX-114 suggest that dp72 is amphipathic and may contain a glycolipid anchor. Digestion with deoxyribonuclease I, phenanthroline/$CuSO_4$ and snake venom phosphodiesterase I destroy the antigenic epitope recognized by monoclonal antibodies to dp72 and suggest that it may be part of a DNA–protein complex (Jaffe and Zalis, 1988a,b; Jaffe *et al.*, 1990 and unpublished data).

Mice immunized with dp72 (6 μg total given in two or three injections) are protected against an intravenous (i.v.) challenge with either *L. d. donovani* (Khartoum strain) amastigotes (10^6) or promastigotes (10^7). *Corynebacterium parvum* (100 μg first injection only) was used as adjuvant. After 18–21 days the mice were sacrificed and the liver parasite burden (LPB) determined. Immunization with dp72 reduced the LPB by 60–81% compared to the control mice which only received adjuvant (Jaffe *et al.*, 1990; Rachamim and Jaffe, 1993; and Fig. 6.1). Importantly, it appears that live promastigotes are also capable of stimulating protection in mice immunized with dp72 (Fig. 6.1). Mice immunized twice with the pure protein (total 4 μg) and boosted with irradiated promastigotes (100 krad) showed protection comparable to mice receiving three boosts of dp72 (total 6 μg) (81 and 76% respectively).

Even though the immunized mice have antibody titres of $>10^{-4}$ against the pure protein, antibodies do not mediate protection. This was shown by two different approaches. First, the LPB 1 day after challenge of immunized and control mice was evaluated by touch biopsy and limiting dilution. The number of liver amastigotes was identical for both groups of animals evaluated by either technique (Rachamim and Jaffe, 1993). Limiting dilution measures viable intracellular parasites. This would eliminate any potential error in counting stained dead amastigotes in the touch biopsy preparations. Therefore, the initial inoculum of live parasites infecting the liver Kupffer cells was the same for both groups of mice. Second, transfer of hyperimmune serum against dp72 1 day prior to parasite challenge had no significant effect on the LPB. Indeed, a small but consistently higher LPB was observed in mice receiving the hyperimmune serum compared to mice receiving normal mouse serum (Fig. 6.2). These experiments suggest that mechanisms which reduce the initial parasite inoculum, such as antibody-mediated complement lysis, prevention of binding to macrophages or activation of phagocytotic cells, are not involved in the protection of animals immunized with dp72.

Mice immunized with dp72 also show specific T-cell responses to

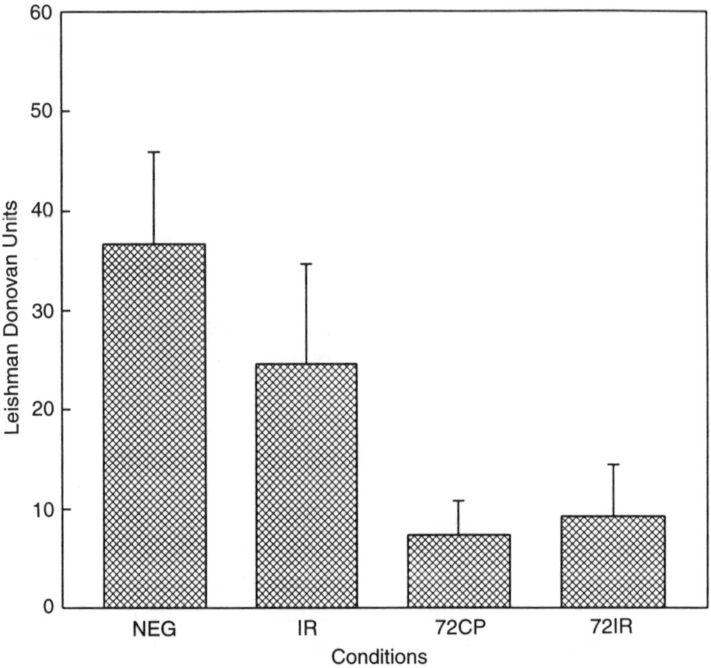

Fig. 6.1. Protection of mice immunized with dp72 antigen alone or in combination with irradiated promastigotes. BALB/c mice were immunized with either dp72 (6 μg total, administered in three boosts) or dp72 (4 μg total in two boosts) followed by 10^7 irradiated *L. donovani* promastigotes. *C. parvum*, 100 μg first boost only, was used as adjuvant with some groups. Twenty days after the last boost the animals were challenged with *L. donovani* promastigotes (10^7 parasites per mouse, i.v.) and sacrificed 18 days later. Leishman Donovan units ± SE were calculated from liver biopsies. NEG, mice injected with *C. parvum* ($n = 6$); IR, mice injected with irradiated promastigotes at the same time as the 72IR group ($n = 6$); 72CP, mice receiving dp72 + *C. parvum* ($n = 6$); 72IR, mice receiving dp72 + *C. parvum* and irradiated promastigotes ($n = 6$).

the pure protein and not to other pure antigens, including gp63, gp70 and LPG (Rachamim and Jaffe, 1993). Good lymphocyte proliferation was also observed when crude *L. donovani* antigen was used instead of dp72. In order to identify the T-cell subset proliferating to dp72 spleen cells were treated with antibodies to either $CD4^+$ or $CD8^+$ T-cells. Lysis of $CD4^+$ and not $CD8^+$ T-cells abolished proliferation, showing that $CD4^+$ T-cells proliferate to the pure antigen (Rachamim and Jaffe, 1993). Analysis of lymphokine secretion by the total T-cell blasts showed the production of IL-2 and IFN-γ. No IL-4 was measured indicating that the proliferating $CD4^+$ T-cells belong to the Th1 subset.

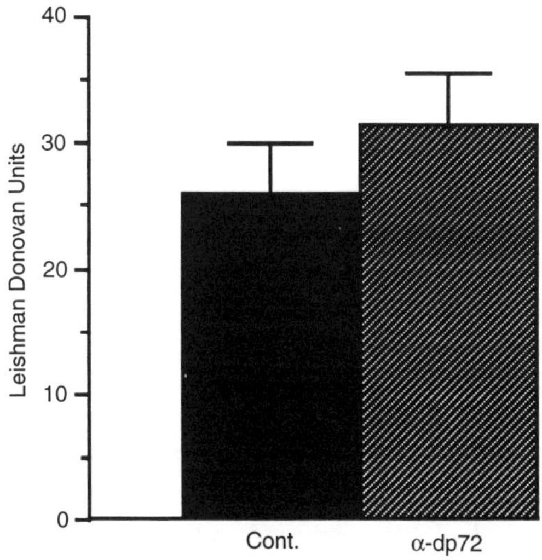

Fig. 6.2. Transfer of dp72-specific serum to naive mice does not protect. Pooled sera from mice immunized with dp72 (α-dp72, $n = 6$) or from normal non-immune mice (Cont., $n = 6$) were injected (1 ml total per mouse, twice, i.v.) into naive BALB/c mice. Twenty-four hours later the mice were challenged, i.v., with 10^7 *L. donovani* promastigotes. After 18 days the mice were sacrificed and the Leishman donovan units ± SE were calculated from the liver biopsies.

Production of tumor necrosis factor-alpha (TNF-α) was also examined and not found in the supernatants of the proliferating cells (Table 6.1 and unpublished data).

The importance of T-cells in protecting dp72-immunized mice against *L. donovani* was evaluated by two approaches. First, the adoptive transfer of dp72-specific T-cells to naive mice was examined. Spleen cells from mice immunized with the pure protein were stimulated *in vitro* for 5 days, purified on Percoll gradients and expanded

Table 6.1. Secretion of lymphokines from dp72-immunized lymphocytes.

	FACS		IL-2 (cpm)*		INF-γ (U ml^{-1})	
Treatment	%CD4$^+$	%CD8$^+$	–	+K	–	+K
Non-immune	14.2	3.8	10	11	0	0.3
Immune	18.8	2.9	12	1318	0.7	15.0

* Spleen cells for fluorescent-activated cell sorting (FACS) were taken 2 weeks after the last boost with the pure protein dp72. Spleen cells were incubated either in the absence of crude parasite antigen (–) or with *L. donovani* crude antigen (+K) (10^6 lysed cells ml^{-1}). Supernatants were collected after 24 or 48 h and used to measure IL-2, IL-4, INF-γ and TNF-α.

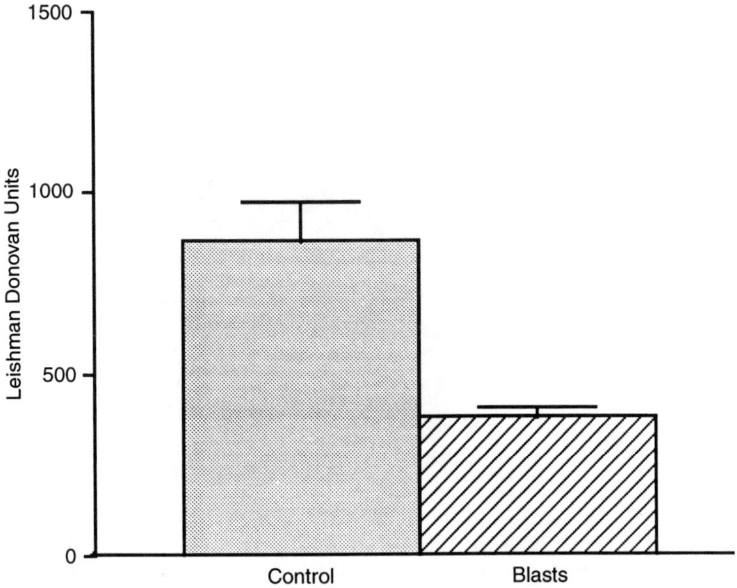

Fig. 6.3. Adoptive transfer of dp72 antigen-specific blasts to naive mice. Spleen cells from mice immunized with dp72 (7.5 μg plus 100 μg *C. parvum* per mouse) were incubated with pure antigen in complete T-cell culture medium containing 5% fetal calf serum. After 5 days the blasts were purified on Percoll gradients and cultured for 3 additional days in the same medium plus 5% EL-4 cell supernatants. Naive BALB/c mice were injected i.v. with the Percoll-purified blasts (10^7 cells per mouse). All mice were challenged 24 h later, i.v., with 10^7 *L. donovani* promastigotes. After 18 days the animals were sacrificed and the Leishman donovan units ± SE were calculated from the liver biopsies. Control, naive mice (n = 4); Blasts, mice injected with the antigen-specific T-cell blasts (n = 4).

with IL-2 for 3 more days. Naive BALB/c mice were injected with 10^7 blasts per animal and 24 h later challenged i.v. with 10^7 *L. donovani* promastigotes. After 18 days the mice were sacrificed and the LPB determined (Fig. 6.3). Mice receiving the anti-dp72 T-cell blasts showed a 57% reduction in the LPB compared to the control animals. Furthermore, this protection was significantly reduced if the T-cell blasts were first treated with an antibody to the $CD4^+$ T-cells which eliminated 97% of the cells belonging to this subset (unpublished data). Treatment with either a non-specific rat IgG or antibodies to $CD8^+$ T-cells had no effect on the protection (unpublished data).

The finding that $CD4^+$ T-cells can transfer protection was corroborated using an antigen-specific T-cell line, T(30)6, produced from mice immunized with dp72 by stimulation with crude *L. donovani* antigen. Fluorescent-activated cell sorting (FACS) analysis showed that

T(30)6 was CD4$^+$. The line proliferated very strongly in the presence of both crude antigen of *L. donovani* and *L. infantum*, and secreted IL-2 and IFN-γ. Secretion of IFN-γ was similar to that found when T(30)6 was stimulated with Concavalin A. No IL-4 was found in the culture supernatants. Prior to adoptive transfer the blasts were expanded with IL-2 and purified on Percoll gradients. Twenty-four hours before being challenged as above with promastigotes, one group of animals received 10^6 blasts. Eighteen days later the mice were biopsied and the LPB determined by touch biopsies. Mice receiving the CD4$^+$ cell line showed a 60% reduction in LPB.

Second, the effect of antibodies on different T-cell subsets on the LPB was examined in mice immunized with dp72. Immunized and control mice were injected with antibodies to either IFN-γ, CD4$^+$, CD8$^+$ or CD4$^+$ plus CD8$^+$ T-cells. Mice injected with either phosphate-buffered saline (PBS) or normal rat IgG served as controls. Treatment was begun 5 days prior to challenge with promastigotes and continued for an additional 2 weeks. As previously shown mice immunized with dp72 and receiving injections of either PBS or rat IgG had significantly reduced LPB compared to the control mice injected only with PBS (Rachamim and Jaffe, 1993) (Fig. 6.4). However, treatment of the immunized animals with either antibody to T-cells or IFN-γ reversed the protection and resulted in a LPB either equal to or greater than the control non-immunized animals. As expected the passive transfer of antibodies to CD4$^+$ T-cells reversed the protection in the immunized mice. Unexpectedly, antibodies to CD8$^+$ T-cells also abrogated protection, even though these cells do not proliferate *in vitro* to dp72 and the elimination of CD8$^+$ T-cells from antigen-specific T-cell blasts had no effect on the protection. These results suggest that both subsets of T-cells are involved in reducing the LPB of immunized mice.

A similar depletion of the control non-immunized mice with either antibodies to IFN-γ or the T-cell subsets resulted in extremely high LPB (Fig. 6.4). Indeed control mice given these antibodies had LPB approximately twice that of immunized mice receiving identical treatments or control non-immunized animals treated with PBS.

Homologues of the protein dp72 purified from *L. donovani* were found in other species of *Leishmania* (Jaffe *et al.*, 1990). Monoclonal antibodies specific for dp72 react with proteins of about 60 kDa in *L. tropica*, *L. major* and *L. aethiopica*. Serum and T-cells from mice immunized with the pure protein recognized antigens and proliferate, respectively, when incubated with crude antigens from *Leptomonas* and *L. major* (Jaffe *et al.*, 1990; Rachamim and Jaffe, 1993). This suggested that immunization with the *L. donovani* protein might also protect against other species of *Leishmania*. Mice were immunized, as above, with dp72 and challenged in the rear left foot with 10^7 promastigotes

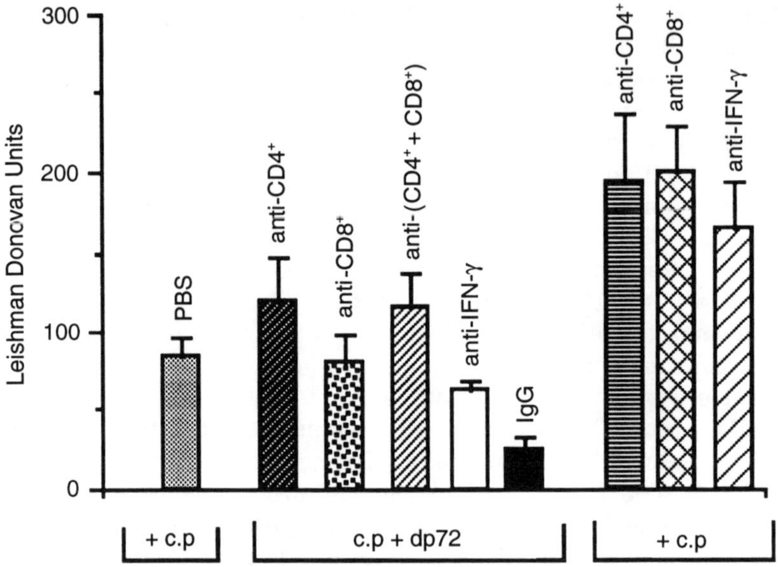

Fig. 6.4. Passive transfer of antibodies to T-cell subsets or IFN-γ and their effect on VL in naive and immune mice. Five days before the challenge with 10^7 *L. donovani* promastigotes, i.v., and 2 weeks after the last boost, a regime of antibody injections was carried out. Immunized mice (6 μg total dp72 + 100 μg *C. parvum*, i.p.) received either PBS, rat immunoglobulin (IgG), anti-CD4$^+$ (GK1.5), anti-CD8$^+$ (H35), anti CD4$^+$ plus anti-CD8$^+$, or anti-IFN-γ (HB170) antibodies. Treatment with PBS, anti-CD4$^+$, anti-CD8$^+$ or anti-IFN-γ was done in parallel on non-immunized mice (100 μg *C. parvum* alone). At 18 days all the mice were sacrificed and the Leishman donovan units $\pm$ SE were calculated from liver biopsies. n = 6 mice for each group. c.p, *Corynebacterium parvum*.

of *L. major* (WR309). As seen in Fig. 6.5, no lesion developed in the immunized mice. These mice were not examined by limiting dilution assays, therefore the presence of amastigotes in apparently normal skin cannot be excluded. However, in experiments where *L. major* (Fredlin) was used to challenge the mice no lesion was apparent, even after 6 months. When these animals were examined by limiting dilution no parasites were found at the site of injection (Rachamim and Jaffe, 1993).

In summary, mice immunized with pure protein isolated from promastigotes of either *L. infantum* or *L. chagasi* are protected against challenge by either stage of the *L. donovani* parasite. In addition, immunization with the donovani protein gives cross-species protection against challenge by *L. major*. This is especially striking considering how susceptible BALB/c mice are to infection with *L. major*. The studies described above showed that antibodies to dp72 do not mediate the protection, and as expected the protection is cell mediated. Antigen-

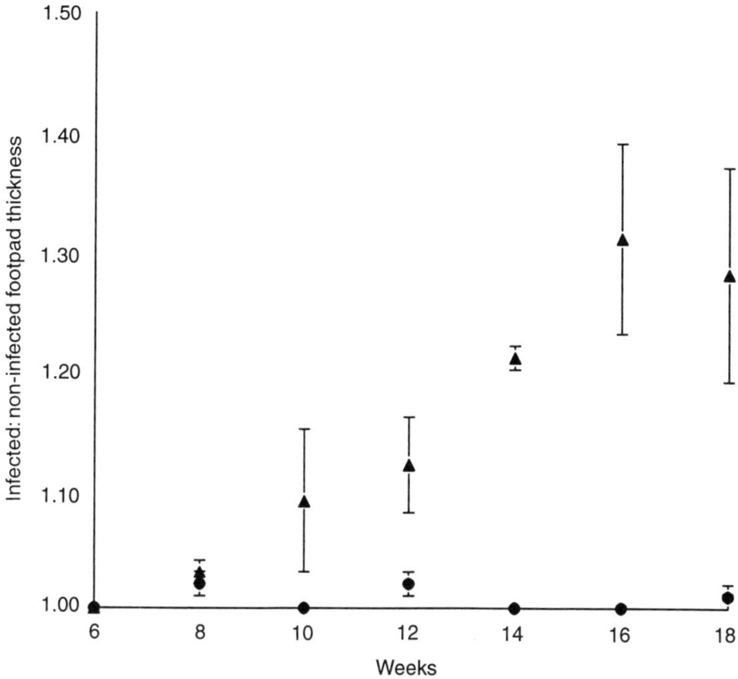

Fig. 6.5. Immunization with dp72 from *L. donovani* protects against a challenge by *L. major*. BALB/c mice were immunized with dp72 (6 μg total plus 100 μg *C. parvum*) (●). Control mice received adjuvant only (▲). Mice were challenged, s.c., in the left rear foot with 10^7 stationary phase *L. major* (WR309, MHOM/IL/79/Perlstein) promastigotes 2 weeks after the last boost. The thickness of both footpads was measured (mm) and the ratio of infected to non-infected footpad ± SE is shown (n = 6).

specific $CD4^+$ T-cells, Th1 subset, from mice immunized with dp72 proliferate to the pure and crude antigen. Given alone just prior to challenge these specific T-cell blasts and lines can protect naive hosts against *L. donovani*. Surprisingly, depletion *in vivo* of the $CD8^+$ T-cells in addition to the $CD4^+$ T-cells also reversed the protection observed in immunized mice, suggesting that an intact immune system is required for protection against *L. donovani*. These results, as well as finding that naive mice depleted of either T-cell subset had higher parasite burdens, suggest that additional protective mechanisms are active in immunized mice depleted of T-cells and IFN-γ. Studies are in progress to identify these factors.

Acknowledgements

This work was supported in part by the UNDP/WHO/TDR Programme, the MERC Programme N01–A0–45186 and the AVICENNE Programme AVI–CT92–0003.

References

Badaro, R., Jones, T.C., Carvalho, E.M., Sampaio, D., Reed, S.G., Barral, A., Teixeira, R. and Johnson, W.D. Jr (1986) New perspectives on a subclinical form of visceral leishmaniasis. *Journal of Infectious Diseases* 154, 1003–1011.

Ghalib, H.W., Piuvezam, M.R., Skeiky, Y.A.W., Siddig, M., Hashim, F.A., El-Hassan, A.M., Russo, D.M. and Reed, S.G. (1993) Interleukin-10 production correlates with pathology in human *Leishmania donovani* infections. *Journal of Clinical Investigation* 92, 324–329.

Holaday, B.J., Pompeu, M.M.L., Evans, T., Braga, D.N.M., Texeira, M.J., Sousa, A.Q., Sadick, M.D., Vaconcelos, A.W., Abrams, J.S., Pearson, R.D. and Locksley, R.M. (1993a) Correlates of *Leishmania*-specific immunity in the clinical spectrum of infection with *Leishmania chagasi*. *Journal of Infectious Diseases* 167, 411–417.

Holaday, B.J., Pompeu, M.M.L., Jeronimo, S., Texeira, M.J., Sousa, A.Q., Vaconcelos, A.W., Pearson, R.D., Abrams, J.S. and Locksley, R.M. (1993b) Potential role for interleukin-10 in the immunosuppression associated with kala azar. *Journal of Clinical Investigation* 92, 2626–2632.

Jaffe, C.L. and Zalis, M. (1988a) Purification of two *Leishmania donovani* membrane proteins recognized by sera from patients with visceral leishmaniasis. *Molecular and Biochemical Parasitology* 27, 53–62.

Jaffe, C.L. and Zalis, M. (1988b) Use of purified parasite proteins from *Leishmania donovani* for the rapid serodiagnosis of visceral leishmaniasis. *Journal of Infectious Diseases* 157, 1212–1220.

Jaffe, C.L., Rachamim, N. and Sarfstein, R. (1990) Characterization of two proteins from *Leishmania donovani* and their use for vaccination against visceral leishmaniasis. *Journal of Immunology* 144, 699–706.

Karp, C.L., El-Safi, S.H., Wynn, T.A., Satti, M.M.H., Kordofani, A.M., Hashim, F.A., Hag-Ali, M., Neva, F.A., Nutman, T.B. and Sacks, D.L. (1993) *In vivo* cytokine profiles in patients with kala-azar. *Journal of Clinical Investigation* 91, 1644–1648.

Kreutzer, R.D., Grogl, M., Neva, F.A., Fryauff, D.J., Magill, A.J. and Aleman-Munoz, M.M. (1993) Identification and genetic comparison of leishmanial parasites causing viscerotropic and cutaneous disease in soldiers returning from Operation Desert Storm. *American Journal of Tropical Medicine and Hygiene* 49, 357–363.

Liew, F.Y. and O'Donnell, C.A. (1993) Immunology of leishmaniasis. *Advances in Parasitology* 32, 161–259.

Locksley, R.M. and Louis, J.A. (1992) Immunology of leishmaniasis. *Current Opinion in Immunology* 4, 413–418.

Louis, J. and Milon, G. (1993) CD8$^+$ T-cells and immunity to intracellular pathogens. *Parasitology Today* 9, 196–197.

Melby, P.C., Andrade-Narvaez, F.J., Darnell, B.J., Valencia-Pacheco, G., Tryon, V.V. and Palomo-Cetina, A. (1994) Increased expression of proinflammatory cytokines in chronic lesions of human cutaneous leishmaniasis. *Infection and Immunity* 62, 837–842.

Milon, G. and Louis, J.A. (1993) CD8$^+$ T-cells and immunity to intracellular pathogens. *Parasitology Today* 9, 194–197.

Mitchell, G.F. (1984) Host-protective immunity and its suppression in a parasitic disease: murine cutaneous leishmaniasis. *Immunology Today* 5, 224–226.

Peters, W. and Killick-Kendrick, R. (eds) (1987) *The Leishmaniasis in Biology and Medicine*. Academic Press, London.

Pirmez, C., Yamamura, M., Uyemura, K., Paes-Oliveira, M., Conceicao-Silva, F. and Modlin, R.L. (1993) Cytokine patterns in the pathogenesis of human leishmaniasis. *Journal of Clinical Investigation* 91, 1390–1395.

Rachamim, N. and Jaffe, C.L. (1993) Pure protein from *Leishmania donovani* protects mice against both cutaneous and visceral leishmaniasis. *Journal of Immunology* 150, 2322–2331.

Reiner, S.L. and Locksley, R.M. (1993) Cytokines in the differentiation of Th1/Th2 CD4$^+$ subsets in leishmaniasis. *Journal of Cellular Biochemistry* 53, 323–328.

Seaman, J., Ashford, R.W., Schorscher, J. and Dereure, J. (1992) Visceral leishmaniasis in southern Sudan: status of healthy villagers in epidemic conditions. *Annals of Tropical Medicine and Parasitology* 86, 481–486.

Thakur, C.P. and Kumar, K. (1992) Post kala-azar dermal leishmaniasis: a neglected aspect of kala-azar control programmes. *Annals of Tropical Medicine and Parasitology* 86, 355–359.

7

Visceral Leishmaniasis in Iran and the Role of Serological Tests in Diagnosis and Epidemiological Studies

Gh.H. Edrissian

School of Public Health and Institute of Public Health Researches, Tehran University of Medical Sciences, P.O. Box 6446–14155, Tehran, Iran

Introduction

The first case of human visceral leishmaniasis (VL) in Iran was reported by Pouya (1949a) in a boy from the Caspian area in 1949. Afterwards more cases of kala azar were found in different parts of the country. Up to the end of 1975 about 120 laboratory-confirmed cases had been reported from all over the country except the southeast (Nadim *et al.*, 1978).

From 1976, when specific serological techniques, particularly the indirect fluorescent antibody test (IFAT), were applied in the diagnosis of VL in Iran (Behforouz *et al.*, 1976; Edrissian *et al.*, 1977, 1981; Edrissian and Darabian, 1979; Kohanteb *et al.*, 1984) the number of diagnosed cases increased rapidly. The more frequent discovery of cases of kala azar and their report by investigators at several academic centres, raised the awareness in clinicians and other medical groups, particularly paediatricians, about epidemiology and practical laboratory diagnosis of the disease in Iran.

At the present time kala azar is known as an endemic disease in some areas of at least two provinces: Firooz-Abad and Jahrum in Fars Province in the south and Meshkin-Shahr and Moghan areas in Ardabil Province in northwest Iran (Edrissian *et al.*, 1988, 1993) where about 1800 cases of kala azar had been diagnosed up to the end of 1993 in each province.

In all other provinces the number of diagnosed cases varies from a single case in a few provinces to 88 in Boushehr and Khusistan. Altogether more than 4300 cases had been diagnosed in at least 113 cities and districts in the 25 provinces of Iran up to the end of 1993

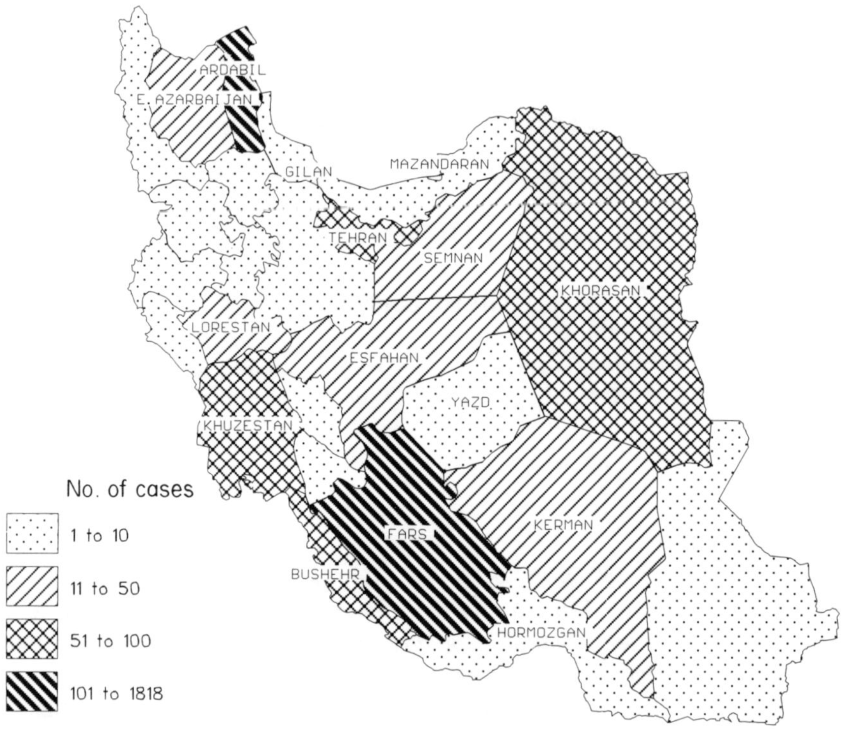

Fig. 7.1. Geographical distribution of the diagnosed cases of kala azar in Iran during 1949–1993.

(Fig. 7.1). Over the 5 years from 1989 to 1993 about 3000 cases of kala azar were been diagnosed; more than double the 1361 cases that were found during the previous 40 years (1949–1988). The steady increase of diagnosed cases is shown in Fig. 7.2.

The high number of cases in Bushehr, Khusistan (Maraghi *et al.*, 1992), Khorasan (Tabarestani *et al.*, 1991), East Azerbaijan and Kerman (Niknaphs *et al.*, 1993) indicates that the disease may also be endemic in some areas of these provinces.

The results of the sero-epidemiological studies of VL in human as well as in canine reservoirs in the endemic foci of Ardabil and Fars Provinces have shown the high endemicity of the disease in these areas (Edrissian *et al.*, 1988, 1993).

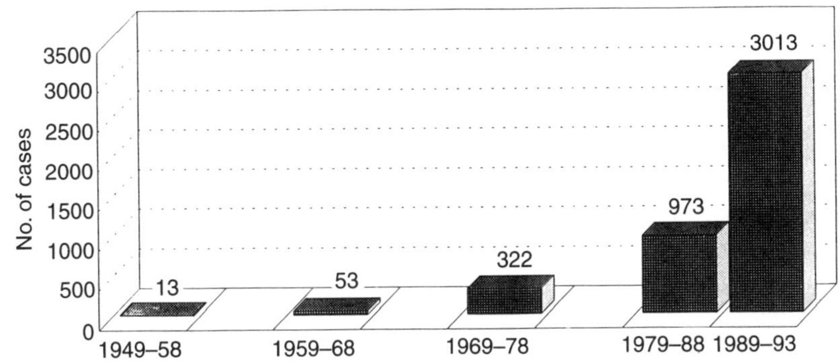

Fig. 7.2. Rapid increase in the diagnosed cases of kala azar in Iran during 1949–1993.

Causative Agent

The epidemiological evidence of kala azar in Iran has shown that the disease is Mediterranean in type and the agent is most probably *Leishmania infantum* (Edrissian *et al.*, 1988). Three *Leishmania* isolates collected from kala azar patients in the Meshkin-Shahr endemic focus were sent to the Montpellier and London Centres for Identification of *Leishmania*, through the World Health Organization (TDR Programme). These isolates were identified and reported as *L. infantum* MON-1.

Vectors

Among the sandflies of subgenus *Phlebotomus*, the most suspected vector of kala azar in Iran, on the basis of epidemiological evidence, was *Phlebotomus major*.

In recent years *Phlebotomus kandelakii* and *Phlebotomus perfiliewi* infected with promastigotes have been found in the Meshkin-Shahr and Moghan areas in Ardabil Province (Nadim *et al.*, 1992). Infected *P. major* was also found in Firooz-Abad in Fars Province (Sahabi *et al.*, 1992).

Reservoir Hosts

When Neligan in 1913 observed amastigote forms of *Leishmania* in the internal organs of a dog in Tehran which also had cutaneous leishmaniasis (CL) on its head, the possibility of the existence of zoonotic VL in the dog was propounded by Wenyon (1914).

Canine kala azar was reported by Pouya in 1949 in the Caspian area, where he also reported the first human case of the disease in Iran (Pouya 1949b). Later a few other infected dogs were found in the Tehran area (Mobedi, 1968; Rafyi, 1968). Wild carnivores, such as jackals and foxes, have been found infected with *Leishmania* in their visceral organs (Hamidi *et al.*, 1982). These animals are probably the original sources of infection for human VL, particularly in the sporadic type of kala azar, as well as for canine VL, which is itself the main source of infection for human VL, mostly in the endemic foci such as Meshkin-Shahr and Moghan. In these areas in a few cases foxes have also been found infected with *Leishmania* (Abai, 1980).

Serological (direct agglutination test (DAT) and IFAT) and parasitological examination (PE) of suspected reservoirs of VL, including dogs, jackals and foxes in the Firooz-Abad and Jahrum areas, Fars Province and dogs in the Meshkin-Shahr and Ghermi areas in Ardabil Province, showed that these animals are highly infected (Edrissian *et al.*, 1993; Bokai, 1994).

Human Infection

Age and sex distribution

Kala azar in Iran is the infantile type and is found mostly in children. The age of kala azar patients in the endemic foci of Ardabil Province ranges from 3 months to 40 years in males and to 35 years in females (Edrissian, 1988). Of the 493 patients in the study, 53.3% were in the age group of 1–2 years and 93.4% of the patients were children of ≤12 years old (Fig. 7.3).

In the endemic foci of Fars Province the age range of kala azar patients was 1.5 months to 40 years (Alborzi, 1993); and of the 275 cases reported, 54.5% of the patients were in the age group of 1–2 years and 99.3% of the patients were ≤12 years (Edrissian *et al.*, 1993).

The ratio of male to female kala azar patients from the two abovementioned endemic foci were 1.34 and 1.37, respectively. However, in sero-epidemiological studies, using IFAT and DAT, in children of the two above endemic foci, sero-positive rates (SPR) in males were considerably less than SPR in females and the ratio of male SPR to female SPR were 0.38 and 0.76 by DAT and 0.17 and 0.80 by IFAT, in the two foci, respectively.

This suggests that a more subclinical form of the infection occurs in females than males.

In 240 sporadic cases of kala azar found in non-endemic areas of

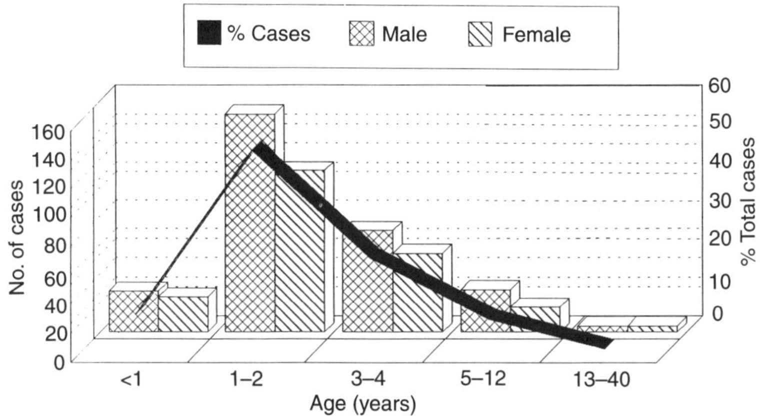

Fig. 7.3. Age and sex distribution of 493 kala azar patients in the endemic foci of Meshkin-Shahr and Ghermi, Ardabil Province, northwest Iran, 1980–1989. Of the patients studied 57.2% are male while 42.8% are female.

Iran the age ranges were 5 months to 50 years in males and to 52 years in females. The peak percentage of cases was 30.8% in the age group of 3–4 years and 85.9% of the patients were ≤12 years old. The ratio of males to females in these patients was 1.79 (Fig. 7.4).

These data show, with respect to age and sex, that there are considerable differences between kala azar patients of the endemic

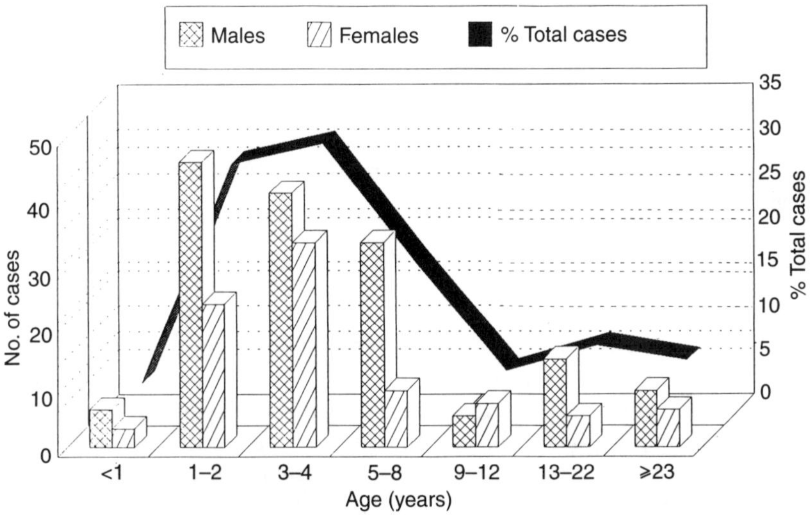

Fig. 7.4. Age and sex distribution in 240 sporadic cases of kala azar in Iran. Of the patients studied 64% were male while 36% were female.

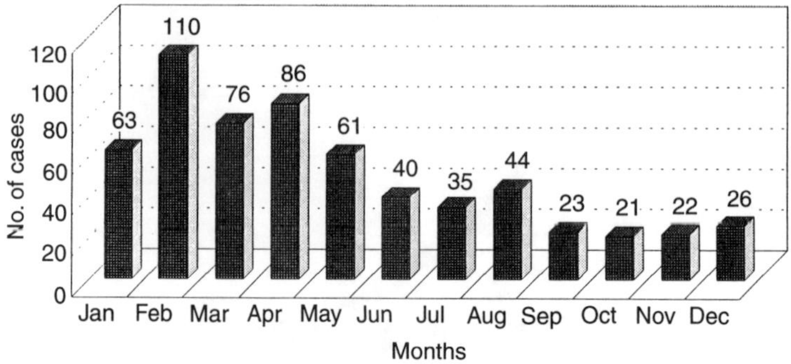

Fig. 7.5. Monthly distribution of 607 diagnosed kala azar cases in passive case detection in the Meshkin-Shahr district hospital during 1991–1993.

foci and the sporadic cases found in apparently non-endemic areas. In sporadic cases, the disease occurs in older age groups and the ratio of males to females is also higher compared with endemic cases.

In the endemic focus of VL in Ghermi, Ardabil Province, a few cases of CL have been observed in children, who usually had small lesions on their faces. These patients had previous histories of kala azar and they had been treated with glucantine (A. Nadim, 1993, Tehran, personal communication). As oriental sores are not prevalent in this area, it seems that the causative agent of such CL is *L. infantum*. Such cases need further study.

The annual monthly distribution of 607 kala azar patients in their first visit to the Meshkin-Shahr district hospital between 1991 and 1993 is shown in Fig. 7.5. The majority of the cases are diagnosed during the colder months of January to May. As the second phase of the activity of sandflies in this area is in August this distribution shows that the average incubation period for human VL is about 6 months in these areas (Soleimanzadeh *et al.*, 1993).

Mortality rate

There is no regular record for the mortality rate of kala azar in Iran. In a few old reports the mortality rates of the disease among patients who were usually hospitalized varied from 10 to 30% (Tehernia and Jalayer, 1968; Nadim *et al.*, 1978; Hashemi-Nassab and Zadeh-Shirazi, 1982). Such high mortality rates in patients who were mostly under specific treatment were mainly due to the delay in referral to well-equipped and experienced medical centres. In recent years, in spite of the increase in diagnosed cases of kala azar in Iran, the mortality of the disease particularly in the endemic foci is very low. This is probably due

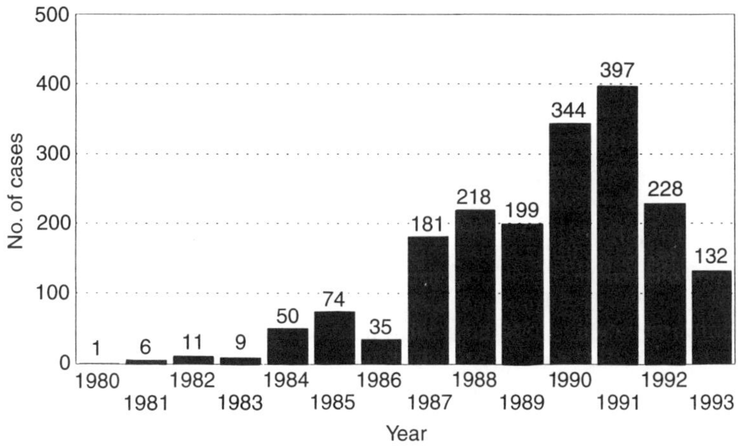

Fig. 7.6. Number of kala azar cases by year, mostly seen in Meshkin-Shahr district hospital, during 1980–1993.

to the availability of specific serological techniques and to better knowledge of the epidemiology, diagnosis and treatment of the disease in some endemic areas. Furthermore, residents of the endemic foci are more familiar with the symptoms of the disease in children, so they are referred to medical centres earlier on, thus reducing the mortality of kala azar in Iran.

Improved facilities for the diagnosis and treatment of kala azar in Meshkin-Shahr city caused a rapid annual increase in the number of diagnosed cases during 1980–1991 in this area. However, starting interventions, including case finding, treatment, control of sandflies, elimination of stray and infected dogs, and health education carried out in the 'pilot project of the control of kala azar' in May 1991, caused a reduction in number of the diagnosed cases in the study area after 1991 (Fig. 7.6).

Treatment of kala azar patients is usually carried out with meglomine antimonate (glucantine) based on daily intramuscular (i.m.) injections of 20 mg of Sb^{5+} for 20 days (Soleiman-Zadeh *et al.*, 1993).

Serodiagnosis of Kala Azar

The simple non-specific serological Napier aldehyde or formol gel test (FGT) have been occasionally used in the diagnosis of VL in Iran.

Comparisons of FGT with IFAT and (or) parasitological examination (PE) of the bone marrow smears were carried out in 96 suspected kala azar patients. In 84 cases (87.5%) the results of FGT corresponded with IFAT and (or) PE. But in 12 patients (12.5%) where amastigote forms were observed (11 cases) or IFAT was positive in titres ≥1 : 296 (six cases) FGT was negative. FGT usually becomes positive in the advanced or chronic stages of the disease. Therefore FGT with clinical symptoms and epidemiological evidence would be helpful in the diagnosis of VL in areas where no facilities for specific serological tests or PE are available.

Specific serological tests such as IFAT and enzyme-linked immunosorbent assay (ELISA) were applied and evaluated in the serodiagnosis of VL in suspected kala azar patients in Iran during 1976–1980 (Behforouz *et al.*, 1976; Edrissian *et al.*, 1977, 1981; Edrissian and Darabian, 1979) with successful results. These studies showed that there are some cross-reactions in both IFAT and ELISA in a few infectious diseases, such as malaria, typhoid and liver abscess, with *Leishmania* antigen, mostly in low titres, but in kala azar patients the *Leishmania* antibody titres are usually enough high to be affected by such cross-reactions.

As kala azar in Iran is prevalent in rural and remote areas where facilities for IFAT and ELISA are not available, we tried to use a practical and reliable test such as DAT which was suggested by Harith *et al.* (1986) for serodiagnosis and sero-epidemiological studies of VL.

DAT has been used on a large scale in serodiagnosis and sero-epidemiological studies of VL in human and animal reservoir hosts in Iran (Edrissian *et al.*, 1990, 1993; Soleiman-Zadeh *et al.*, 1993). It has been also evaluated and compared with IFAT, ELISA using soluble (S-ELISA) and intact or figurated (F-ELISA) antigens prepared from promastigote forms of *L. donovani*, in kala azar suspected patients referred from Tehran medical centres and the Meshkin-Shahr district hospital (Fig. 7.7). This study showed that DAT is the most specific but the least sensitive in serodiagnosis of VL. In contrast IFAT was the least specific and the most sensitive.

There was no cross-reaction in DAT with *L. donovani* antigen in about 100 blood samples taken from toxoplasmosis, tuberculosis, brucellosis, malaria and typhoid patients. However, in the three other tests cross-reactions were observed in three to five of the above infectious diseases. Of these tests, after DAT, F-ELISA seems to be the most practical and reliable (Khorshidian *et al.*, 1994).

A comparison of IFAT and DAT in the serodiagnosis of VL in 2016 suspected kala azar patients and 317 dogs mostly from the Ardabil endemic foci is shown in Fig. 7.8.

Among 265 suspected kala azar patients, mostly referred from different medical centres of Tehran to the Protozoology Unit, School of Public Health for IFAT and DAT, 38 cases of bone marrow puncture had been carried out and in 18 of these cases (47.4%) amastigotes

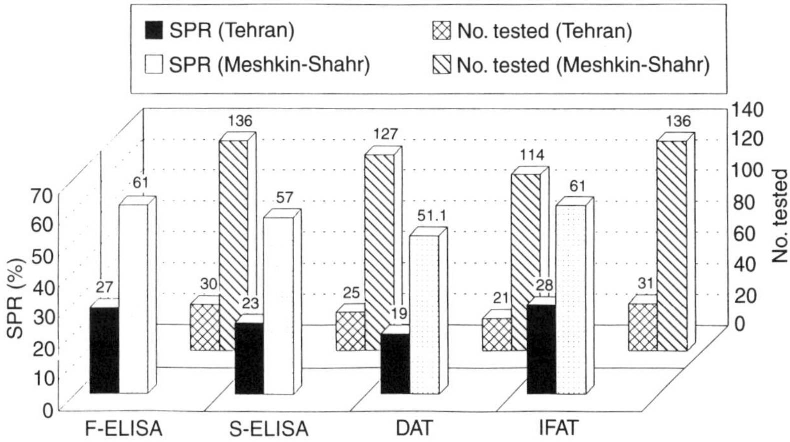

Fig. 7.7. *Leishmania* sero-positive rates in serum samples collected from patients with suspected kala azar in Tehran and Meshkin-Shahr areas in Iran and tested by four different serological tests. SPR, sero-positive rates in values or titres; F-ELISA, OD = >0.10; S-ELISA, OD = >0.09; DAT titre = >1 : 3200; IFAT titre = >1 : 160.

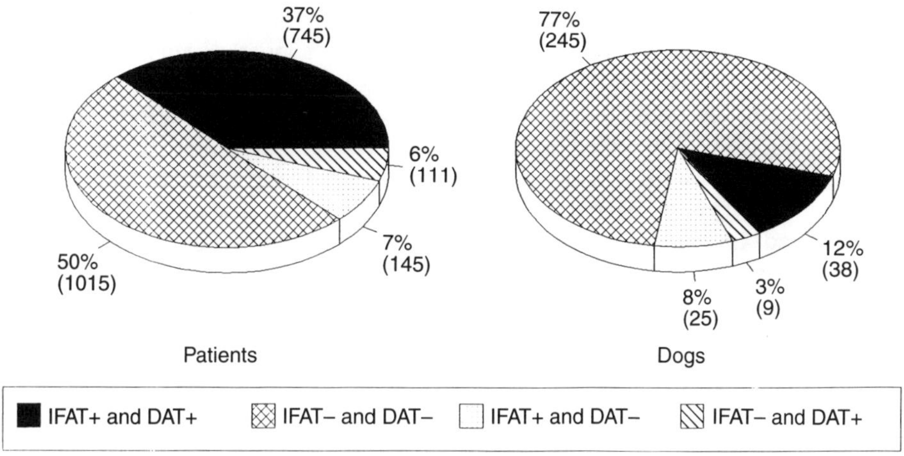

Fig. 7.8. Comparison of IFAT and DAT in the serodiagnosis of VL in 2016 patients with suspected kala azar and 317 dogs in Meshkin-Shahr, Ardabil Province, 1993.

were found in microscopical examination of the stained smears. All of these 18 parasite-positive cases were also serologically positive in IFAT and DAT. But in four cases where the IFAT and DAT were both positive in rather high titres, no parasites were found in the bone marrow smears (Edrissian *et al.*, 1990).

From the DAT seropositive (in titres of ≥1 : 3200), suspected kala azar patients in Meshkin-Shahr district hospital, 479 cases of bone marrow puncture were carried out for direct parasitological examination during 1985–1990. Only in 69% of cases were amastigotes found. As the DAT sero-positive patients are mostly children with clinical symptoms of kala azar, in this area, diagnosis and treatment are usually based on clinical signs and symptoms plus positive DAT (Soleimand-Zadeh *et al.*, 1993). In the above studies amastigotes were found in only a few cases during PE of suspected kala azar patients or in dogs, whereas in IFAT or DAT the antibody titres were low or no antibody was detected. Therefore, specific serological tests such as DAT, F-ELISA and IFAT are more practical and helpful in laboratory diagnosis of VL, particularly in the kala azar endemic areas, than the parasitological examination.

Sero-epidemiological Studies of Visceral Leishmaniasis

In humans

A preliminary IFAT serological survey of VL was carried out in 1977 in the Izeh area of Khusistan Province in the southwest part of Iran where sporadic cases of kala azar had been reported, mostly among Bakhtiary tribes (Nadim *et al.*, 1978; Maraghi *et al.*, 1992). In 1230 blood samples collected from different age groups the SPR in titres of 1 : 80 to 1 : 320 was, in total, 3.4%. As the minimum (0.4%) and maximum (4.7%) SPR were in the youngest (0–4 years) and oldest (≥20 years) age groups, respectively, it seems that such low antibody titres observed in this study were mostly cross-reactions of some other infectious disease such as malaria which was more or less prevalent in the area at that time (Edrissian *et al.*, 1981; Edrissian, 1986).

In the IFAT serological surveys which were done during 1986–1989, in the endemic foci of Meshkin-Shahr and Ghermi, Ardabil Province, the SPR varied from 0.80 to 3.19% (in titres of 1 : 80 to 1 : 640) in different villages and they were proportional to the number of diagnosed cases in these villages. In 395 treated kala azar patients the SPR was 23.3% in titres ≥1 : 320 (Edrissian *et al.*, 1988).

Active and passive case detection (ACD and PCD) of kala azar, using serological DAT, was carried out in the pilot control project of

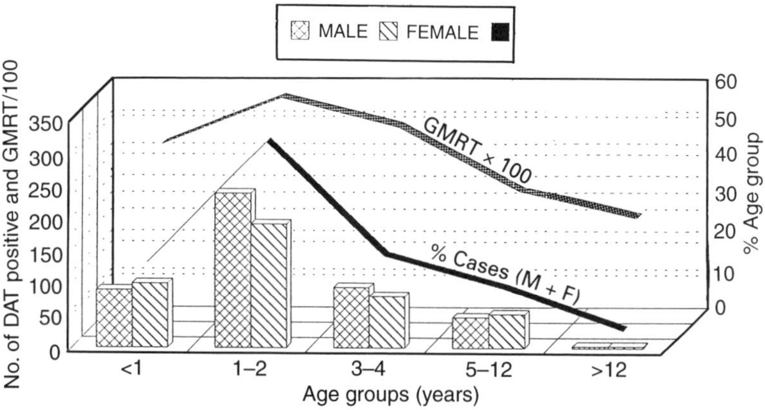

Fig. 7.9. Age and sex distribution in 850 DAT leishmania sero-positive cases among patients with suspected kala azar in Ardabil Province, northwest Iran, 1991–1993. Of the patients studied 54.5% were male while 45.5% were female. GMRT, geometric mean of reciprocal titres (⩾1 : 3200).

kala azar in the endemic focus of Meshkin-Shahr during 1991–1993. Of 18,974 collected samples in ACD, 855 cases (4.5%) were DAT-positive in titres ≥1 : 3200. From these sero-positive cases, 140 (0.74%) were new cases without previous kala azar history. In the Ghermi area, from 3043 samples 96 cases (3.15%) were DAT-positive and 34 cases (1.12%) were newly detected.

In PCD from 1791 suspected kala azar patients referred to the Meshkin-Shahr district hospital 687 cases (38.3%) were sero-positive and mostly diagnosed as kala azar patients. Of these patients, 607 (33.9%) had no previous history of the disease.

The age and sex distribution of DAT sero-positive cases and geometric mean of reciprocal titres (GMRT) among the children tested in the Meshkin-Shahr endemic focus is shown in Fig. 7.9. Figure 7.9 is very similar to Fig. 7.3 which shows the age and sex distribution in confirmed kala azar patients diagnosed in this area.

IFAT and DAT sero-epidemiological studies of VL have been also carried out in Firooz-Abad and Jahrum, Fars Province, in the southern part of Iran in 1990–1991 (Edrissian *et al.*, 1993). In total, from 1766 blood samples, the SPR was 4.2% in IFAT and 1.8% in DAT. In those individuals who had a previous history of kala azar and who had been treated, the SPR was 20.3% in IFAT and 15.2% in DAT. This study showed that the decrease of the SPR in DAT over 8 months was slower than in IFAT. GMRT in DAT sero-positive cases were more

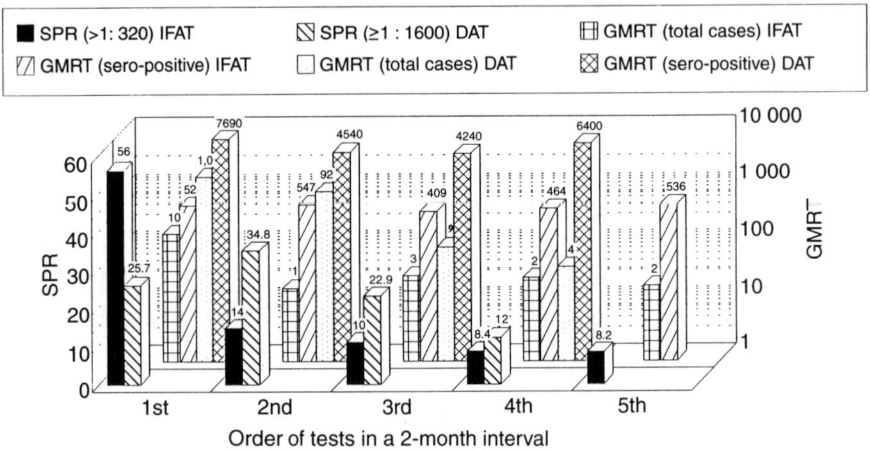

Fig. 7.10. The trends of *Leishmania* antibody titres in IFAT and DAT sero-positive individuals in Firooz-Abad and Jahrum, Fars Province, Iran, 1990–1991.

stable than GMRT in IFAT sero-positive cases (Fig. 7.10). Of 606 treated kala azar patients in the Meshkin-Shahr endemic focus the SPR decreases over 72 months were from 73.7% to 2.0% in IFAT and from 78.9% to 12.0% in DAT (Fig. 7.11). These findings also indicate that the leishmanial antibodies detectable by DAT are more stable than those detectable by IFAT. In both tests the antibodies in some treated patients persisted for quite a long time. In some of the treated kala azar patients of the Meshkin-Shahr area who were followed up every 2 months, the DAT *Leishmania* antibody titres did not decrease even after 2 years. These children did not have any clinical symptoms and in a few cases, where bone marrow puncture was performed to search for amastigotes, no parasites were found.

In animal reservoirs

Serological (DAT and IFAT) examination and PE of 66 carnivores suspected to be reservoirs of VL (including 24 dogs, 30 jackals, ten foxes and two mongooses), shot in the Firooz-Abad and Jahrum endemic foci of kala azar in 1990–1991, showed that dogs, jackals and foxes are highly infected (Fig. 7.12) (Edrissian *et al.*, 1993).

In the endemic foci of Meshkin-Shahr and Ghermi from 317 random blood samples collected from dogs, 38 (12.0%) cases were sero-positive in both IFAT (titres ≥1 : 640) and DAT (titres ≥1 : 320). The SPR in IFAT and DAT were 19.9 and 14.8%, respectively. A comparison of the two tests is shown in Fig. 7.8.

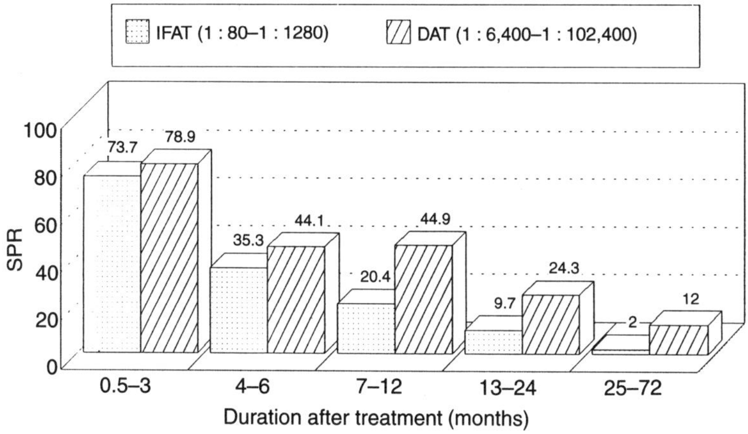

Fig. 7.11. The trend of sero-positive cases in treated kala azar patients in the Meshkin-Shahr area, northwestern Iran.

In the dogs which had clinical symptoms of VL, biopsy of the enlarged lymph nodes was carried out for detection of the parasite. Amastigotes were detected in five dogs, four of which were also serologically positive in both tests. But in one parasite-positive dog the IFAT and DAT tests were both negative. Only 13.6% of the DAT sero-positive dogs had clinical symptoms of VL (Bokai, 1994). Infection in dogs is the main problem in the control of kala azar in the Meshkin-Shahr endemic focus.

The results of the above studies show the efficiency of DAT and

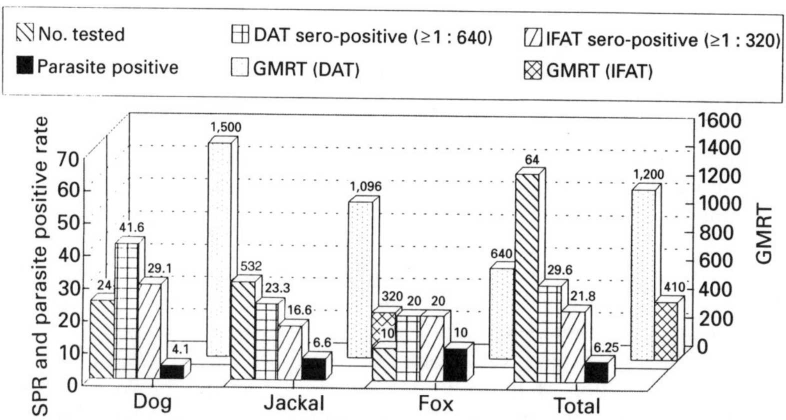

Fig. 7.12. Serological and parasitological examinations of 64 suspected animal reservoirs of kala azar in Firooz-Abad and Jahrom areas, Fars Province, Iran, 1990–1991.

IFAT in sero-epidemiological surveys of VL in human and animal reservoirs. DAT is more practical and specific than IFAT.

Acknowledgements

Some of the unpublished findings in this chapter are the results of two research projects that received financial support from the UNDP/World Bank/WHO special programme for Research and Training in Tropical Diseases (TDR) as well as from the School of Public Health and Institute of Public Health Researches, Tehran University of Medical Sciences, and also CDC, Ministry of Health and Medical Education.

References

Abai, M.R. (1990) Determination of the founa, seasonal activity of sandflies and the animal reservoirs of visceral leishmaniasis in Meshkin-Shahr area. MSPH thesis, School of Public Health, Tehran University of Medical Sciences, No. 1223.

Alborzi, A.V. (1993) Kala azar in south part of Iran. *IV Iranian Congress of Infectious and Tropical Diseases*, Tehran, October 12–14, 1993 [in Persian].

Behforouz, N., Rezai, H.R. and Gettner, S. (1976) Application of immunofluorescence to detection of antibody in *Leishmania* infection. *Annual of Tropical Medicine and Parasitology* 70, 293–301.

Bokai, S. (1994) Sero-epidemiological studies of canine visceral leishmaniasis in Meshkin-Shahr areas, Ardabil province, north-west Iran. Unpublished PhD thesis, School of Public Health, Tehran University of Medical Sciences, No. 1662.

Edrissian Gh.H. (1986) Sero-epidemiology in leishmaniasis. *2nd Asian Congress of Parasitology*, Lucknow, 13–16 February 1988.

Edrissian, Gh.H. (1990) Kala azar in Iran. *Medical Journal of Islamic Republic of Iran* 4, 235–238.

Edrissian, Gh.H. and Darabian, P. (1979) A comparison of enzyme linked immunosorbent assay and indirect fluorescent antibody test in the serodiagnosis of cutaneous and visceral leishmaniasis in Iran. *Transactions of the Royal Society of Tropical Medicine and Hygiene* 73, 289–292.

Edrissian, Gh.H., Akhavan, S., Samar, G. and Zavar, A. (1977) Immunofluorescence as a method of choice in diagnosis of visceral leishmaniasis. *Journal of the Iranian Medical Council* 3, 185–190 [in Persian].

Edrissian, Gh.H., Darabian, P., Zovein, Z., Seyedi-Rashti, M.A. and Nadim, A. (1981) Application of the indirect fluorescent antibody test in the serodiagnosis of cutaneous and visceral leishmaniasis in Iran. *Annual of Tropical Medicine and Parasitology* 75, 19–24.

Edrissian, Gh.H., Hafizi, A., Afshar, A., Soleiman-Zadeh, Gh., Movahed-

Danesh, A.M. and Garossi, A. (1988) An endemic focus of visceral leishmaniasis in Meshkin-Shahr, East Azarbaijan province, north-west part of Iran and IFA serological survey of the disease in this area. *Bulletin de la Societe Pathologie Exotique* 81, 238–248.

Edrissian, Gh.H., Hajaran, H., Ataian, A., Kanani, A., Afshar, A., Sarkissian, M., Soleiman-Zadeh, Gh., Movahed-Danesh, A.M., Hafizi, A. and Ghorbani, M. (1990) Application of direct agglutination test (DAT) in sero-diagnosis and sero-epidemiological studies of kala azar in Iran. *The First Congress of Parasitic Diseases in Iran Guilan (IRAN)*, 11–13 December 1990 [in Persian].

Edrissian, Gh.H., Ahanchin, A.R., Gharachahi, A.M., Ghorbani, M., Nadim, A., Ardehali, S., Hafizi, A., Kanani, A., Sarkissian, M., Hajaran, H., Tahvildar-Bidruni, Gh. (1993) Sero-epidemiological studies of visceral leishmaniasis and search for animal reservoirs in Fars province, southern Iran. *Iranian Journal of Medical Sciences* 18, 99–105.

El-Amine, R.M., Wright, E.P., Kager, P.A., Laarman, J.J. and Pondman, K.W. (1985) ELISA using intact promastigotes for immunodiagnosis of Kala azar. *Transactions of the Royal Society of Tropical Medicine and Hygiene* 79, 344–350.

Hamidi, A.N., Edrissian, Gh.H., Tahvildar-Bidruni, Gh., Javadian, E. (1982) Visceral leishmaniasis of jackals and dogs in northern Iran. *Transactions of the Royal Society of Tropical Medicine and Hygiene* 76, 756–757.

Harith, A.E., Kolk, A.H.J., Kager, A., Leeuwenburg, J., Muigai, R., Kiugu, S. and Laarman, J.J. (1986) A simple and economical direct agglutination test for sero-diagnosis and sero-epidemiological studies of visceral leishmaniasis. *Transactions of Royal Society of Tropical Medicine and Hygiene* 80, 583–587.

Hashemi-Nassab, A. and Zadeh-Shirazi, H. (1982) Viseral leishmaniasis (kala azar) in Fars province, Iran: study of 130 cases. *Transactions of the Royal Society of Tropical Medicine and Hygiene* 76, 756–757.

Khorshidian, S., Hajaran, H., Sarkissian, M.T. and Edrissian, G.H. (1994) Evaluation of ELISA using intact promastigotes as antigen, for the diagnosis of visceral leishmaniasis and its comparison with IFAT. *Iranian Journal of Medical Sciences* 19, 15–18.

Kohanteb, J., Rezai, H.R. and Ardehali, S. (1984) Application of electro-immunodiffusion and crossed electroimmunodiffusion for the diagnosis of visceral leishmaniasis (kala azar). *Journal of Tropical Medicine and Hygiene*, 87, 201–205.

Maraghi, S., Edrissian, Gh.H. and Ebrahimzadeh, A. (1993) Kala azar in Khusistan. *Journal of Medical Faculty, Guilan University of Medical Sciences* 1, 1–7 [in Persian].

Mobedi, I., Nadim, A. and Sadeghi, V. (1968) Canine visceral leishmaniasis in Tehran. *8th International Congress of Tropical Medicine and Malaria*, Tehran, 7–15 September 1968.

Nadim, A., Navid-Hamidi, A., Javadian, E., Tahvildar-Bidruni, Gh. and Amini, H. (1978) Present status of kala azar in Iran. *American Journal of Tropical Medicine and Hygiene* 27, 25–28.

Nadim, A., Javadian, J., Tahvildar-Bidruni, Gh., Mottaghi, M. and Abai, A. (1992) Epidemiological aspects of kala azar in Meshkin-Shahr, Iran: investigation on vector. *Iranian Journal of Public Health* 21, 61–68.

Niknaphs, P., Daee Parizi, M.H. and Ahmadi, A. (1993) Report of 40 cases of kala azar from Kerman province. *Journal of Kerman University of Medical Sciences* 1, 30–37 [in Persian].

Pouya, Y. (1949a) Kala azar in Iran. *Journal of Faculty of Medicine of Teheran* 7, 115–121 [in Persian].

Pouya, Y. (1949b) Study of visceral leishmaniasis in Guilan and Mazandaran provinces of Iran. *Journal of Faculty of Medicine of Teheran* 7, 359–391 [in Persian].

Rafyi, A., Niak, A. and Naghshineh R. (1968) Canine visceral leishmaniasis in Iran. *Veterinary Record* 83, 269–710.

Sahabi, Z., Seyedi-Rashti, M.A., Nadim, A., Javadian, E. and Kazemaini, M.R. (1992) Preliminary report on the natural leptomonad infection of *Phlebotamus major* in an endemic focus of visceral leishmaniasis in Fars province, south of Iran. *Iranian Journal of Public Health* 21, 87–93.

Soleiman-Zadeh, G., Edrissian, G.H., Movahed-Danesh, A.M. and Nadim, A. (1993) Epidemiological aspects of kala azar in Meshkin-Shahr, Iran: human infection. *Bulletin of the World Health Organization* 71, 759–762.

Tabarestani, M., Elahi, R. and Fata, A.M. (1991) Retrospective and prospective studies of kala azar in 40 patients in Mashhad University Hospitals. *Medical Journal of Mashhad* 37, 5–20 [in Persian].

Tehernia, A.C. and Jalayer, T. (1968) Visceral leishmaniasis in children in southern Iran. *Annals of Tropical Medicine and Parasitology* 62, 171–173.

Wenyon, C.M. (1914) Some aspects of surgery in the tropics. *Transactions of the Royal Society of Tropical Medicine and Hygiene* 7, 214–216.

8

Differential Cause of T-cell Non-responsiveness to *Plasmodium falciparum* Antigens in Healthy Individuals and in Acutely Ill Malaria Patients

L. Hviid

Centre for Medical Parasitology, Department of Infectious Diseases M 7641, National University Hospital (Rigshospitalet) and Institute for Medical Microbiology and Immunology, University of Copenhagen, Copenhagen, Denmark

Introduction

Studies from experimental malaria models have demonstrated clearly the involvement of T-cells both in the protective immune response to malaria and in immunopathology. In humans, the mechanisms involved in the acquisition and maintenance of antimalarial immunity are much less clear, but there is general consensus that T-cells are important, and numerous studies of human T-cell responses to malaria antigens have been published.

It is a consistent finding from studies of T-cell responses in endemic populations that peripheral T-cells from a considerable proportion fail to recognize malaria-specific antigen *in vitro*. The background of this widespread non-responsiveness, which has been observed in healthy, aparasitaemic individuals as well as in acutely ill patients has been debated for a long time. In this chapter it is suggested that non-responsiveness in healthy and in acutely ill individuals may have different causes, and the evidence in favour of this hypothesis is summarized.

T-cell Non-responsiveness to Antigenic Stimulation *in vitro*

T-cells from a considerable proportion of healthy individuals from malaria-endemic areas fail to recognize malaria-specific antigen *in vitro* (Good *et al.*, 1988; Riley *et al.*, 1988b; Hoffman *et al.*, 1989; Troye-Blomberg *et al.*, 1989; Hviid *et al.*, 1990; Mshana *et al.*, 1990). Responses to unrelated antigens are unaffected (Riley *et al.*, 1989; Mshana *et al.*, 1990). While the non-responsiveness to short malaria-derived peptide sequences may relate to diversity in the ability of major histocompatibility complex (MHC) molecules to present antigen, there is little evidence from human studies that this is a major explanatory factor (Riley *et al.*, 1992). Furthermore, the fact that non-responsiveness is also observable in studies employing large, native antigens, which must be assumed to contain a number of T-cell epitopes, is also difficult to explain by genetic restriction.

In addition to the non-responsiveness in healthy individuals, peripheral T-cells from *Plasmodium falciparum* malaria patients are characterized by a marked inability to respond when stimulated by malaria-specific antigen, and to a lesser degree by non-malarial antigens (Bygbjerg *et al.*, 1986; Ho *et al.*, 1986, 1988; Theander *et al.*, 1986; Riley *et al.*, 1988; Hviid *et al.*, 1991b). Finally, T-cells from individuals having low-grade, asymptomatic parasitaemia tend to be non-responsive (Theander *et al.*, 1986b; Petersen *et al.*, 1989).

The Role of CD8$^+$ Cells in Peripheral T-cell Non-responsiveness

A number of studies have investigated the possible causes for the observed T-cell anergy. These studies have revealed a number of interesting differences between healthy individuals and acutely ill patients, and some of these findings are summarized in Table 8.1.

T-cell responsiveness can be restored in non-responsive, healthy individuals by the addition of exogenous T-cell growth factors such as interleukin (IL)-2 (Mshana *et al.*, 1990). Removal of supposedly immunosuppressive CD8$^+$ T-cells has a similar effect (Riley *et al.*, 1989; Mshana *et al.*, 1990; Theander *et al.*, 1993). Finally, more soluble CD8 (sCD8) can be detected in supernatants of malaria antigen-stimulated T-cells from non-responsive donors than in T-cell supernatants from responding individuals (Mshana *et al.*, 1990). Taken together, these observations suggest that the functional anergy of T-cells from healthy individuals is the consequence of activation of regulatory CD8$^+$ T-cells *in vitro*.

In contrast to the above, studies of T-cells from malaria patients have failed to revert the non-responder phenotype by either exogenous IL-2 or $CD8^+$ cell depletion (Troye-Blomberg *et al.*, 1985; Ho *et al.*, 1986, 1988). It thus seems unlikely that the peripheral T-cell dysfunction in such patients is mediated by $CD8^+$ T-cells. In fact, acutely ill *P. falciparum* patients appear to have reduced rather than increased suppressor cell function (Gilbreath *et al.*, 1983).

Acute *P. falciparum* Malaria is Characterized by T-cell Activation

There is considerable indirect evidence that acute malaria is characterized by T-cell activation, despite the marked *in vitro* non-responsiveness of peripheral T-cells. Thus, plasma/serum levels of T-cell cytokine interferon-gamma (IFN-γ) (Rhodes-Feuillette *et al.*, 1985), as well as other indicators of T-cell activation such as soluble IL-2 receptor (sIL-2R) (Ho *et al.*, 1988; Nguyen-Dinh and Greenberg, 1988; Hviid *et al.*, 1991b; Jakobsen *et al.*, 1994) and soluble β_2-microglobulin (Chiewsilp *et al.*, 1988; Elhassan *et al.*, 1994), are all increased in patient plasma/serum. Recent reports of increased plasma sCD4 (Elhassan *et al.*, 1994) and sCD8 (Kremsner and Bienzle, 1989; Elhassan *et al.*, 1994) indicate that activation is present both in the CD4 and in the CD8 subset.

Acute *P. falciparum* Malaria is Characterized by Depletion of T-cells from the Peripheral Circulation

Several studies have demonstrated malaria-associated lymphopaenia, including reduction both in $CD3^+$ T-cells in general (Ho *et al.*, 1986; Chougnet *et al.*, 1992), and in the $CD4^+$ subset responsible for the majority of T-cell proliferation *in vitro* (Ho *et al.*, 1986). In contrast, normal CD3 and CD4 values have been reported in healthy individuals with non-responsive T-cells (Petersen *et al.*, 1989; Mshana *et al.*, 1990).

In addition to the reduction in $CD3^+$ and $CD4^+$ T-cells in acute malaria, we have shown a disease-induced selective depletion from the peripheral circulation of T-cells characterized by high surface expression of the T-cell adhesion molecule leucocyte function-associated antigen 1 (LFA-1) (LFA-1^{hi} cells) (Hviid *et al.*, 1991; Elhassan *et al.*, 1994). A similar phenomenon has since been described in acute Kawasaki disease (Furukawa *et al.*, 1993).

The reduction in peripheral T-cells, the depletion of LFA-1^{hi} T-cells, and the T-cell non-responsiveness lead us to suggest that acute *P. falciparum* malaria episodes are associated with a temporary

Table 8.1. Immunology differences in healthy and acutely ill T-cell non-responders.

	Healthy non-responders[1]	Acutely ill non-responders[1]
Reduced proliferation to unrelated antigens	No (Riley *et al.*, 1989; Mshana *et al.*, 1990)	Yes (Ho *et al.*, 1986, 1988; Hviid *et al.*, 1991b)
Proliferation restored by exogenous IL-2	Yes (Mshana *et al.*, 1990)	No (Troye-Blomberg *et al.*, 1985; Ho *et al.*, 1988)
Proliferation restored by depletion of $CD8^+$ cells	Yes (Riley *et al.*, 1989; Mshana *et al.*, 1990; Theander *et al.*, 1993)	No (Ho *et al.*, 1986)
Proliferation restored by depletion of adherent cells	n.a.	No (Ho *et al.*, 1988)
Increased *in vitro* production of sCD8	Yes (Mshana *et al.*, 1990)	n.a.
Increased plasma sIL-2R	n.a.	Yes (Ho *et al.*1988; Nguyen-Dinh and Greenberg, 1988; Hviid *et al.*, 1991b; Jakobsen *et al.*, 1994)
Increased plasma sCD4	n.a.	Yes (Elhassan *et al.*, 1994)
Increased plasma sCD8	n.a.	Yes (Kremsner and Bienzle, 1989; Elhassan *et al.*, 1994)
Increased plasma soluble β_2-microglobulin	n.a.	Yes (Chiewsilp *et al.*, 1988; Elhassan *et al.*, 1994)
Increased plasma IFN-γ	n.a.	Yes (Rhodes-Feuillette *et al.*, 1985)
Reduced $CD3^+$ subset[2]	No (Riley *et al.*, 1989; Mshana *et al.*, 1990)	Yes (Ho *et al.*, 1986; Chougnet *et al.*, 1992)
Reduced $CD4^+$ subset[2]	No (Riley *et al.*, 1989; Mshana *et al.*, 1990)	Yes (Ho *et al.*, 1986)
Depletion of LFA-1^{hi} cells	No (unpublished observation)	Yes (Hviid *et al.*, 1991; Elhassan *et al.*, 1994)
Increased plasma sICAM-1	n.a.	Yes (Hviid *et al.*, 1993, 1994; Elhassan *et al.*, 1994; Wenisch *et al.*, 1994)

Table 8.1. *(continued)*

	Healthy non-responders[1]	Acutely ill non-responders[1]
Increased plasma sE-selectin	n.a.	Yes (Hviid *et al.*, 1993, 1994; Elhassan *et al.*, 1994; Wenisch *et al.*, 1994)
Increased plasma sVCAM-1	n.a.	Yes (Elhassan *et al.*, 1994; Hviid *et al.*, 1994)
Increased plasma sTNFr-I	n.a.	Yes (Kern *et al.*, 1992; Wenisch *et al.*, 1994)

n.a., No information available.
[1] Relative to healthy responders.
[2] In some studies results were presented as absolute counts, in others as percentage of peripheral blood mononuclear cells.

redistribution of antigen-reactive T-cells away from the peripheral circulation, presumably to sites of endothelial activation, and likely to involve LFA-1 (Hviid *et al.*, 1991a,b). Indeed, there is ample evidence of endothelial inflammation in acute *P. falciparum* malaria, with increased plasma/serum levels of a number of soluble endothelial inflammation markers such as sICAM-1, sE-selectin, sVCAM-1, and sTNFr-I (Kern *et al.*, 1992; Hviid *et al.*, 1993, 1994; Elhassan *et al.*, 1994; Wenisch *et al.*, 1994). In further support of the reallocation hypothesis, a study in mice infected with *Plasmodium chabaudi* showed T-cell reactivity in the spleen at a time where no such reactivity could be shown in the peripheral circulation (Langhorne and Simon-Haarhaus, 1991), and we have found splenic accumulation of LFA-1^{hi} T-cells using the same model (Felsing and Hviid, unpublished).

The Temporal Stability of the Non-responder Phenotype

The inability of peripheral T-cells from acutely ill *P. falciparum* patients to respond to antigen-stimulation *in vitro* is alleviated after drug cure, and is generally normal after 1 month in the absence of reinfection or recrudescence (Hviid *et al.*, 1991b).

In a study from an area of unstable, seasonal malaria transmission we found that T-cell responses were significantly depressed during the rainy season in healthy, aparasitaemic individuals, compared to the responses from the same individuals during the malaria-free dry season (Theander *et al.*, 1990). This variation in peripheral T-cell function was later directly demonstrated to be due to seasonal differences in the suppressive capacity of $CD8^+$ T-cells (Theander *et al.*, 1993). Data from another study supports this finding (Riley *et al.*, 1993).

Concluding Remarks

Peripheral T-cell non-responsiveness to malaria antigens is widespread in endemic populations, both in healthy individuals and in acute malaria patients. The data summarized here suggest that the cause of this non-responsiveness is different according to parasitological status, and emphasizes the importance of careful clinical and parasitological characterization of donors in studies of cellular immune reactivity.

Acknowledgements

The research at the Centre for Medical Parasitology is supported by the Danish International Development Agency (Danida) and the Danish Medical Research Council. L.H. is supported by a senior research fellowship from the University of Copenhagen.

References

Bygbjerg, I.C., Jepsen, S. and Theander, T.G. (1986) Lymphocyte response to purified *Plasmodium falciparum* antigens during and after malaria. *Acta Tropica (Basel)* 43, 55–62.

Chiewsilp, D., Chiewsilp, P., Boonpucknavig, S. and Rugsakul, T. (1988) Elevation of β_2-microglobulin in malaria. *Transactions of the Royal Society of Tropical Medicine and Hygiene* 82, 688–689.

Chougnet, C., Tallet, S., Ringwald, P. and Deloron, P. (1992) Kinetics of lymphocyte subsets from peripheral blood during a *Plasmodium falciparum* malaria attack. *Clinical and Experimental Immunology* 90, 405–408.

Elhassan, I.M., Hviid, L., Satti, G., Akerström, B., Jakobsen, P.H., Jensen, J.B. and Theander, T.G. (1994) Evidence of endothelial inflammation, T cell activation, and T cell reallocation in uncomplicated *Plasmodium falciparum* malaria. *American Journal of Tropical Medicine and Hygiene* 51, 372–379.

Furukawa, S., Matsubara, T., Tsuji, K., Okumura, K. and Yabuta, K. (1993) Transient depletion of T-cells with bright CD11a/CD18 expression from peripheral circulation during acute Kawasaki disease. *Scandinavian Journal of Immunology* 37, 377–380.

Gilbreath, M.J., MacDermott, R.P., Pavanand, K., Phisphumvithi, P., Kongchareon, S. and Wimomwattrawatee, T. (1983) Deficiency of Con A-induced suppressor cell activity in peripheral blood mononuclear cells from Thai adults naturally infected with *Plasmodium falciparum* and *Plasmodium vivax*. *Parasite Immunology* 5, 431–440.

Good, M.F., Pombo, D., Quakyi, I.A., Riley, E.M., Houghten, R.A., Menon, A., Alling, D.W., Berzofsky, J.A. and Miller, L.H. (1988) Human T-cell recognition of the circumsporozoite protein of *Plasmodium falciparum*: immunodominant T-cell domains map to the polymorphic regions of the molecule. *Proceedings of the National Academy of Sciences of the United States of America* 85, 1199–1203.

Ho, M., Webster, H.K., Green, B., Looaresuwan, S., Kongchareon, S. and White, N.J. (1988) Defective production of and response to IL-2 in acute human falciparum malaria. *Journal of Immunology* 141, 2755–2759.

Ho, M., Webster, H.K., Looareesuwan, S., Supanaranond, W., Phillips, R.E., Chanthavanich, P. and Warrell, D.A. (1986) Antigen-specific immunosuppression in human malaria due to *Plasmodium falciparum. Journal of Infectious Diseases* 153, 763–771.

Hoffman, S.L., Oster, C.N., Mason, C., Beier, J.C., Sherwood, J.A., Ballou, W.R., Mugambi, M. and Chulay, J.D. (1989) Human lymphocyte proliferative response to a sporozoite T cell epitope correlates with resistance to falciparum malaria. *Journal of Immunology* 142, 1299–1303.

Hviid, L., Theander, T.G., Abdulhadi, N.H., Abu-Zeid, Y.A., Bayoumi, R.A. and Jensen, J.B. (1991) Transient depletion of T cells with high LFA-1 expression from peripheral circulation during acute *Plasmodium falciparum* malaria. *European Journal of Immunology* 21, 1249–1253.

Hviid, L., Theander, T.G., Abu-Zeid, Y.A., Abdulhadi, N.H., Jakobsen, P.H., Saeed, B.O., Jepsen, S., Bayoumi, R.A.L. and Jensen, J.B. (1991) Loss of cellular immune reactivity during acute *Plasmodium falciparum* malaria. *FEMS Microbiology and Immunology* 76, 219–228.

Hviid, L., Theander, T.G., Elhassan, I.M. and Jensen, J.B. (1993) Increased plasma levels of soluble ICAM-1 and ELAM-1 (E-selectin) during acute *Plasmodium falciparum* malaria. *Immunology Letters* 36, 51–58.

Hviid, L., Theander, T.G., Jakobsen, P.H., Abu-Zeid, Y.A., Abdulhadi, N.H., Saeed, B.O., Jepsen, S., Bayoumi, R.A.L., Bendtzen, K. and Jensen, J.B. (1990) Cell-mediated immune responses to soluble *Plasmodium falciparum* antigens in residents from an area of unstable malaria transmission in the Sudan. *APMIS* 98, 594–604.

Jakobsen, P.H., Morris-Jones, S., Rønn, A., Hviid, L., Theander, T.G., Elhassan, I.M., Bygbjerg, I.C. and Greenwood, B.M. (1994) Increased plasma concentrations of sICAM-1, sVCAM-1 and sELAM-1 in patients with *Plasmodium falciparum* or *P. vivax* malaria and association with disease severity. *Immunology* 83, 665–669.

Jakobsen, P.H., Morris-Jones, S., Theander, T.G., Hviid, L., Hansen, M.B., Bendtzen, K., Ridley, R.G. and Greenwood, B.M. (1994) Increased plasma levels of soluble IL-2R are associated with severe *Plasmodium falciparum* malaria. *Clinical and Experimental Immunology* 96, 98–103.

Kern, P., Hemmer, C.J., Gallati, H., Neifer, S., Kremsner, P., Dietrich, M. and Porzsolt, F. (1992) Soluble tumour necrosis factor receptors correlate with parasitemia and disease severity in human malaria. *Journal of Infectious Diseases* 166, 930–934.

Kremsner, P.G. and Bienzle, U. (1989) Soluble CD8 antigen in *Plasmodium falciparum* malaria. *Journal of Infectious Diseases* 160, 357–358.

Langhorne, J. and Simon-Haarhaus, B. (1991) Differential T cell responses to *Plasmodium chabaudi chabaudi* in peripheral blood and spleens of C57BL/6 mice during infection. *Journal of Immunology* 146, 2771–2775.

Mshana, R.N., McLean, S. and Boulandi, J. (1990) *In vitro* cell-mediated immune responses to *Plasmodium falciparum* schizont antigens in adults from a malaria endemic area: CD8$^+$ T lymphocytes inhibit the response of low responder individuals. *International Immunology* 2, 1121–1132.

Nguyen-Dinh, P. and Greenberg, A.E. (1988) Increased levels of released interleukin-2 receptors in *Plasmodium falciparum* malaria. *Journal of Infectious Diseases* 158, 1403–1404.

Petersen, E., Høgh, B., Perlmann, H., Kabilan, L., Troye-Blomberg, M., Marbiah, N.T., Hanson, A.P., Björkman, A. and Perlmann, P. (1989) An epidemiological study of humoral and cell-mediated immune response to the *Plasmodium falciparum* antigen Pf155/RESA in adult Liberians. *American Journal of Tropical Medicine and Hygiene* 41, 386–394.

Rhodes-Feuillette, A., Bellosguardo, M., Druilhe, P., Ballet, J.J., Chousterman, S., Canivet, M. and Péiriès, J. (1985) The interferon compartment of the immune response in human malaria: II. Presence of serum-interferon gamma following the acute attack. *Journal of Interferon Research* 5, 169–178.

Riley, E.M., Andersson, G., Otoo, L.N., Jepsen, S. and Greenwood, B.M. (1988) Cellular immune responses to *Plasmodium falciparum* antigens in Gambian children during and after an acute attack of falciparum malaria. *Clinical and Experimental Immunology* 73, 17–22.

Riley, E.M., Jepsen, S., Andersson, G., Otoo, L.N. and Greenwood, B.M. (1988) Cell-mediated immune responses to *Plasmodium falciparum* antigens in adult Gambians. *Clinical and Experimental Immunology* 71, 377–382.

Riley, E.M., Jobe, O. and Whittle, H.C. (1989) CD8$^+$ T cells inhibit *Plasmodium falciparum*-induced lymphoproliferation and gamma interferon production in cell preparations from some malaria-immune individuals. *Infection and Immunity* 57, 1281–1284.

Riley, E.M., Morris-Jones, S., Taylor-Robinson, A.W. and Holder, A.A. (1993) Lymphoproliferative responses to a merozoite surface antigen of *Plasmodium falciparum*: preliminary evidence for seasonal activation of CD8$^+$/HLA-DQ-restricted suppressor cells. *Clinical and Experimental Immunology* 94, 64–67.

Riley, E.M., Olerup, O., Bennett, S., Rowe, P., Allen, S.J., Blackman, M.J., Troye-Blomberg, M., Holder, A.A. and Greenwood, B.M. (1992) MHC and malaria: the relationship between HLA class II alleles and immune responses to *Plasmodium falciparum*. *International Immunology* 4, 1055–1063.

Theander, T.G., Bygbjerg, I.C., Andersen, B.J., Jepsen, S., Kharazmi, A. and Ødum, N. (1986) Suppression of parasite-specific response in *Plasmodium falciparum* malaria. A longitudinal study of blood mononuclear cell proliferation and subset composition. *Scandinavian Journal of Immunology* 24, 73–81.

Theander, T.G., Bygbjerg, I.C., Jacobsen, L., Jepsen, S., Larsen, P.B. and Kharazmi, A. (1986) Low parasite specific T cell response in clinically immune individuals with low grade *Plasmodium falciparum* parasitaemia.

Transactions of the Royal Society of Tropical Medicine and Hygiene 80, 1000–1001.

Theander, T.G., Hviid, L., Abu-Zeid, Y.A., Abdulhadi, N.H., Saeed, B.O., Jakobsen, P.H., Reimert, C.M., Jepsen, S., Bayoumi, R.A.L. and Jensen, J.B. (1990) Reduced cellular immune reactivity in healthy individuals during the malaria transmission season. *Immunology Letters* 25, 237–242.

Theander. T.G., Hviid, L., Abu-Zeid, Y.A., Abdulhadi, N.H., Saeed, B.O., Jakobsen, P.H., Jepsen, S., Bayoumi, R.A.L. and Jensen, J.B. (1993) Activation of CD8-positive suppressor cells during the malaria season. *Immunology of Infectious Diseases* 3, 97–102.

Troye-Blomberg, M., Andersson, G., Stoczkowska, M., Shabo, R., Romero, P., Patarroyo, E., Wigzell, H. and Perlmann, P. (1985) Production of IL-2 and IFN-γ by T cells from malaria patients in response to *Plasmodium falciparum* or erythrocyte antigens *in vitro*. *Journal of Immunology* 135, 3498–3504.

Troye-Blomberg, M., Riley, E., Perlmann, H., Andersson, G., Larsson, Å., Snow, R.W., Allen, S.J., Houghten, R.A., Olerup, O., Greenwood, B.M. and Perlmann, P. (1989) T and B cell responses of *Plasmodium falciparum* malaria-immune individuals to synthetic peptides corresponding to sequences in different regions of the *P. falciparum* antigen Pf155/RESA. *Journal of Immunology* 143, 3043–3048.

Wenisch, C., Varijanonta, S., Looareesuwan, S., Graninger, W., Pichler, R. and Wernsdorfer, W. (1994) Soluble intercellular adhesion molecule-1 (ICAM-1), endothelial leukocyte adhesion molecule-1 (ELAM-1), and tumour necrosis factor receptor (55 kDa TNF-R) in patients with acute *Plasmodium falciparum* malaria. *Clinical Immunology and Immunopathology* 71, 344–348.

9

Iron, Oxidant Stress and Malaria

J. GOLENSER,[1] M. CHEVION[2] AND Z.I. CABANTCHIK[3]

Departments of [1]Parasitology and [2]Cellular Biochemistry, Hadassah Medical School, The Hebrew University, Jerusalem, Israel and [3]Department of Biological Chemistry, The Hebrew University, Jerusalem, Israel

Introduction

Malaria, a disease caused by various species of *Plasmodium*, is a major health problem in many developing nations. More than 400 million people suffer from the disease, resulting in over 1.5 million annual deaths, primarily from *Plasmodium falciparum* (Struchler, 1989).

Plasmodium, which is transmitted to humans by mosquitoes, invades first the liver cells and subsequently the erythrocytes. The major medical symptoms are associated with the erythrocytic stages, in which the intracellular parasite profoundly alters both the structures and the functions of the host cell (Cranston *et al.*, 1984; Cabantchik, 1992; Elford and Pinches, 1992). Some of the alleged alterations may subserve parasite development, survival and propagation. Others are the result of parasite physiological mechanisms, which only indirectly affect host functions, or vice versa. Thus the parasite–host coexistence demands a delicate balance between acting and counteracting factors. One such example is oxidant stress, which encompasses the formation of reactive oxygen species (ROS) which are mostly in the form of free radicals and result primarily from parasite metabolic activities. In the absence of protective anti-oxidative mechanisms against increased fluxes of free radicals, the host cell could eventually fail to support parasite growth leading to cell lysis and premature release of parasites (Schirmer *et al.*, 1987; Golenser *et al.*, 1991).

In this chapter the relationship that exists between plasmodia, oxidant stress and iron metabolism is discussed. Redox active iron might be released during haemoglobin digestion by parasites (Atamna and Ginsburg, 1993; Har-El *et al.*, 1993) and mediate ROS-induced damage of the parasite's development. Conditions prevailing in blood genetic traits, physiological factors (e.g. vitamin C) and interaction of parasitized cells with effecter cells (e.g. polymorphonuclear (PMN) cells) might induce excessive production of reactive hydroxyl (OH·) radicals and lead to disruption of the host–parasite oxido-reductive balance and to plasmodial death. The following topics will be briefly described:

1. Oxidant stress and ROS
2. Role of ROS in inhibiting plasmodial development in erythrocytes with abnormal genetic traits
3. Role of individual ROS in parasite development
4. Role of ROS in cell-mediated immunity against malaria
5. Redox-reactive iron and induction of parasite oxidant stress
6. Role of protective enzymes
7. Iron chelators as antimalarial drugs

Oxidant Stress and ROS

Oxidant stress describes a set of intracellular or extracellular conditions leading to chemical or metabolic generation of oxygen-derived free radicals, which cause a variable degree of toxic effects in cells (Halliwell and Gutteridge, 1984). A free radical is an atom or molecule which contains at least one unpaired electron in its outer shell, such as the superoxide radical O_2. Another well-known oxygen-containing free radical is the lipid peroxy radical which induces lipid peroxidation. Hydrogen peroxide, which is metabolically produced by all living cells, including plasmodia, can be converted to a highly reactive (and biologically damaging) OH· radical in the presence of trace metals (the Haber–Weiss reaction or Fenton reaction):

$$2O_2^{\cdot-} + 2H^+ \xrightarrow{\text{SOD}} H_2O_2 + O_2$$

Haber–Weiss reaction:

$$O_2^{\cdot-} + H_2O_2 \xrightarrow{\text{Cu/Fe}} OH\cdot + OH^- + O_2$$

$$2AH_2 + 2O_2 \xrightarrow{\text{Cu/Fe}} 2A + H_2O_2$$

$$AH_2 + 2Fe(III)/Cu(II) \longrightarrow A + 2Fe(II)/Cu(I) + 2H^+$$

Fenton reaction:

$$2Fe(II)/Cu(I) + 2H_2O_2 \longrightarrow 2Fe(III)/Cu(II) + 2OH\cdot + 2OH^-$$

$$3AH_2 + 2O_2 \xrightarrow{\text{Cu/Fe}} 3A + 2H_2O + 2OH\cdot$$

(A = reducing biomolecule)

This process probably occurs in the infected red blood cell in the presence of redox active iron (Har-El *et al.*, 1993). OH· radicals are capable of causing damage to a variety of cellular components, including proteins and nucleic acids. They can also extract H atoms from methylene groups resulting in the formation of carbon radicals, which upon reaction with O_2 form peroxy radicals. The peroxy radical can react with H atoms from neighbouring lipid molecules and yield lipid hydroperoxides, which in turn initiate chain reactions leading to chemical changes in membrane structure and function (Saltman, 1989).

Malaria and Blood Genetic Disorders

It is widely accepted that some genetic disorders of erythrocytes can be expressed as balanced polymorphism. On the basis of that view rests the assumption that the alleged polymorphism stems from an evolutionary force which provides the heterozygote a selective advantage (Allison, 1954). A unique example of a how a genetic trait can provide at least partial protection from a fatal disease such as malaria is found in glucose-6-phosphate dehydrogenase (G6PD)-deficient carriers. Red blood cells of G6PD-deficient carriers have a demonstrable propensity to undergo oxidant stress by a variety of physiological and pharmacological factors, due to their restricted anti-oxidant buffer capacity (Friedman, 1979; Luzzatto, 1979). As previously shown by us (Golenser *et al.*, 1988, 1992) and others (Nagel and Roth, 1989), that type of erythrocyte may fail to support intracellular growth of parasites.

The experiments to be described were carried out *in vitro* with defined growth stages of *P. falciparum* cultures: young forms (rings, 10 h post-invasion) and advanced forms (trophozoites, about 35 h post-invasion). They were basically of two types: (i) erythrocytes infected after chemical treatment and (ii) erythrocytes treated after infection. The treatment of G6PD-deficient erythrocytes with inducers of oxidant stress such as diamide or isouramyl is known to alter erythrocyte membrane properties and markedly decrease the rate of invasion. Treatment of parasitized erythrocytes with the same compounds or alternatively incubation with PMN cells (which also induce oxidant stress) reduced the intracellular parasite gowth, most profoundly in G6PD-deficient cells. Similar results were described in other blood genetic traits, including thalassaemia and ovalocytosis (reviewed by Hunt and Stocker, 1990; Golenser *et al.*, 1991). G6PD-deficient erythrocytes and certain thalassaemic and haemoglobin variants are also more sensitive to phagocytosis, owing to the alterations in their surface membranes (P. Arese, personal communication: Yuthavong and Wilairat, 1993). Moreover, these changes occur in the deficient

Table 9.1. The effect of ascorbate on hypoxanthine (Hx) incorporation into erythrocytes parasitized with *P. falciparum*.

Time of harvest	Ascorbate (mM)	Hx incorporation (cpm)	Change (%)
19 h	0	24,904	–
	1	26,544	+6
	3	27,736	+11
	9	28,361	+14
43 h	0	20,958	–
	1	19,151	−9
	3	16,704	−21
	9	2,016	−91

erythrocytes at an earlier parasitic stage than in the normal erythrocytes (P. Arese, personal communication). This reduces the toxification of macrophages which occurs after phagocytosis of parasitized erythrocytes (Arese *et al.*, 1991). These findings indicate that genetic variants of apparently normal behaviour (at least in heterozygotes form), when challenged with certain promoters of oxidant stress or dietary factors, can confer partial resistance against malaria because of the apparent sensitivity of these erythrocytes to oxidant stress.

Role of OH Radicals

Formation of OH radicals and ensuing cell damage were followed after treatment of parasitized erythrocytes with ascorbate or exposure to PMN cells (see next section). Although ascorbate is an important antioxidant, in the presence of redox active iron or copper it becomes a potent pro-oxidant. Thus ascorbate causes oxidation of haemoglobin to methaemoglobin, membrane lipid peroxidation of G6PD-deficient parasitized erythrocytes, and even inhibition of polyamine incorporation (Marva *et al.*, 1992). As depicted in Table 9.1, ascorbate (3 and 9 mM), while slightly (11–14%, respectively) facilitating development of young parasites, was demonstrably detrimental to the mature forms (21–91% inhibition, respectively).

The production of OH· radicals was followed by the fast reaction with salicylic acid which produces 2,3- and 2,5-dihydroxybenzoic acid derivatives (DHBAs). These can be quantified with a high degree of sensitivity, in the range of ng ml^{-1} by high-performance liquid chromatography (HPLC), coupled with an electrochemical detector. When non-parasitized red blood cells were incubated for 1 h with salicylic acid, a basal level of DHBAs was produced. We compared the levels of DHBAs produced under different treatments, as indicated by the ratio between DHBA levels obtained with cells treated under a given

Table 9.2. The production of DHBA in erythrocytes parasitized with plasmodial stages, treated with ascorbate.

Ascorbate (mM)	Non-parasitized DHBA		Parasitized DHBA	
	ng ml^{-1}	r	ng ml^{-1}	r
			Ring forms:	
0	65	1.0	68	1.0
1	68	1.0	72	1.0
3	73	1.1	75	1.1
			Advanced forms:	
0	65	1.0	187	2.9
1	74	1.1	277	4.3
3	75	1.1	315	4.8

Erythrocytes with 50% parasitaemia were employed.
r is the ratio between DHBA level under given conditions and that in the non-infected cells in the absence of ascorbate.

set of conditions and those found in untreated non-infected (in the absence of ascorbate) cells. As can be seen in Table 9.2, parasitized cells with young forms of parasites (50% parasitaemia) produced approximately the same level of DHBA as non-infected cells. In contrast, parasitized cells with advanced stages produced 2.9 times more DHBA (i.e. OH radicals), a level which was raised to 4.8 in the presence of ascorbate. When DHBA levels are normalized for 100% parasitaemia, the figures for OH radical production in parasitized cells (mature parasites) become even more profound. However, more significantly, the results indicate an apparent correlation between OH radical production and inhibition of parasite growth (Table 9.1).

The attribution of increased DHBA production to OH radical production was based on the assumption that the intracellular concentration of salicylate, the scavenger, was not rate limiting. Although reports in the literature indicated a major increase in anion permeability of infected cells (Kutner *et al.*, 1982), we opted for directly assessing salicylate concentrations in infected and normal erythrocytes. For that purpose we examined the time dependence of scavenger intraerythrocytic concentration under various conditions. The observed kinetics of uptake of salicylate into infected red blood cells was essentially parallel to that of non-infected cells, reaching plateau levels within 5 min, although the total amount of salicylate attained within infected cells was approximately 20% higher than in normal cells. This apparent 1.2-fold increase might have resulted from either binding or adsorption to intracellular membranes, chemical modification or increased volume of distribution. However, irrespective of the nature of the slight increase, it could not possibly explain the 10-fold increase in DHBA production in ascorbate-primed infected cells. Increased uptake of inducers of oxidant stress may affect the parasite. Likewise, ascorbate uptake into infected cells was 2.5-fold higher than in non-infected cells (Marva *et al.*, 1992).

Role of ROS in Cell-mediated Immunity

Another means to induce oxidant stress in parasitized erythrocytes and inhibit their development is by inducing a respiratory burst of PMN cells. PMN cells are non-dividing cells with an *in vitro* life span of 6–7 h. The PMN cells have an important role in defence against various pathogens which is achieved by processes involving: (i) phagocytosis, (ii) external degranulation and release of destructive enzymes, (iii) release of ROS (and other free radicals). PMN cells were administered to parasites developing in normal or G6PD-deficient erythrocytes for 4 h, and both nucleic acid synthesis by the parasites and production of OH radicals were measured. Parasites within G6PD-deficient erythrocytes were more sensitive to the PMN cells than normal parasitized erythrocytes, a phenomenon which could be correlated with production of ROS, especially OH radicals (Golenser *et al.*, 1992). The results indicate that PMN-mediated killing of malaria parasites via OH radical production, requires direct contact between the effector cell and the target. It is also conceivable that PMN cells may have had a role in the geographical spread of G6PD deficiency. This is also in accordance with the hypothesis that some inherited traits of red blood cells confer protection against malaria by their increased sensitivity to oxidant stress (see previous sections).

Redox Active Iron and Induction of Oxidant Stress

We assessed the possibility that mature parasite experience increased oxidant stress because of increased release of redox active iron was considered. Supernatants from media of cultures of non-infected or infected cells were assessed in terms of their capacity to induce DNA fragmentation in the presence of ascorbic acid. Redox active iron was found essential for inducing OH-mediated breakage of test DNA (Har-El *et al.*, 1993). When M-13 DNA (7200 bases) was incubated with a haemolysate from non-infected cells, fragmentation of DNA was dismal. A small but significant increase in DNA degradation could be seen after incubation with supernatants from infected cells with young parasites. However, a marked increase in DNA breakage was evident when exposed to supernates of cells infected with mature forms (Fig. 9.1), a

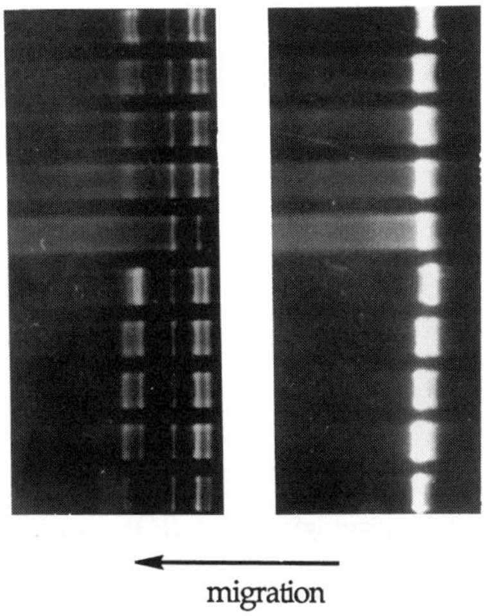

RBC lysate	incubation time
infected with mature parasite	0'
infected with mature parasite	10'
infected with mature parasite	20'
infected with mature parasite	30'
infected with mature parasite	60'
infected with ring form parasite	0'
infected with ring form parasite	10'
infected with ring form parasite	20'
infected with ring form parasite	30'
infected with ring form parasite	60'

Fig. 9.1. Degradation of single- and double-stranded DNA in the presence of parasitized erythrocytes. About 2 μg of single-stranded DNA or about 7 μg of double-stranded DNA of the baceriophage M13 were incubated with lysates (25-fold diluted) of erythrocytes infected with either the mature (lanes 1–5) or the ring form (lanes 6–10) of the parasites (at 15% parasitaemia) in phosphate-buffered saline (pH = 7.2), for different time intervals, as indicated, at 37°C. At the end of incubation, samples were transferred to an ice-water bath, then run on an agarose gel.

process which was augmented in the presence of ascorbate. The fact that exposure of supernatants to 65°C for 10 min (causes complete inactivation of pancreatic DNAase I) did not change the fragmentation patterns, showed that most of the effect on DNA was not due to DNAase activity.

The idea that iron mediated the above DNA fragmentation was assessed in terms of DNA-induced fragmentation and oxidation of ascorbate in the presence of haemin or iron chelators. The results led us to conclude that the level of the redox active (chelatable) form of iron is indeed substantial in mature parasites (Har-El *et al.*, 1993).

In summary, we have shown in the above sections a correlation between increased fluxes of OH radicals and inhibition of parasite growth, an ascorbate pro-oxidative action on infected red blood cells (unlike in other cells, where the action is pro-reductive) and the involvement of redox active forms of iron (liberated during parasite growth) in the increased oxidant stress to the host–parasite system.

Table 9.3. The effect of catalase on the inhibition of *P. falciparum* development by ascorbate. Number in parentheses indicate percentage inhibition of hypoxanthine incorporation.

Ascorbate (mM)	No catalase (cpm)*	Catalase (30 μg ml^{-1})	
		Active	Inactivated
–	81,010 (–)	81,400 (0)	80,270 (–)
2.5	36,454 (55)	77,330 (5)	58,597 (27)
5.0	17,012 (79)	48,026 (41)	16,054 (80)

* Catalase was inactivated by incubation in 100°C for 3 min.

Protective Enzymes

Various mechanisms may prevent the damaging effect of ROS, mainly free radical scavengers and protective enzymes (Halliwell, 1990). Superoxide dismutase (SOD), catalase and glutathione (GSH) peroxidase are considered to be the most important enzymes. Catalase can also protect *P. falciparum in vitro* against ascorbate-induced damage (Table 9.3). The hexose monophosphate shunt (HMS) produces NADPH for reductive antioxidant protection. Atamna *et al.* (1994) found that the HMS activity in parasitized erythrocytes is 78 times higher than that of normal erythrocytes. Reduction of endogenous oxidative stress by elimination of redox active iron or scavengers was correlated with the decreased activity of the HMS. The results imply that HMS has a significant role in counteracting increased oxidant stress in parasitized erythrocytes.

We investigated specifically the role of superoxide dismutation. The prevention of biological damage can be achieved by SOD which causes dismutation of the superoxide to H_2O_2, which in turn, is removed by catalase and GSH peroxidase. However, redox active iron released during the development of plasmodia in the erythrocyte may convert the H_2O_2 to the deleterious OH radical. We therefore examined the effect of elevated intracellular SOD (3–5 times the normal level) on the development of *Plasmodium berghei* in transgenic mice with human-expressed SOD gene. Parasites grown *in vivo* were treated *in vitro* with inducers of oxidant stress such as paraquat or ascorbate. All treatments equally reduced the incorporation of tritiated hypoxanthine by parasites from either normal or transgenic mice. However, the course of infection in infected mice was such that the transgenic mice succumbed faster (Fig. 9.2). The accelerated mortality could be attributed to tissue damage, caused either by ROS or nitrogen reactive intermediates, combined with increased levels of tumor necrosis factor (TNF) (Clark *et al.*, 1989). Indeed, treatment of human erythrocytes with tempol, an SOD-mimicking agent, inhibited the *in vitro* develop-

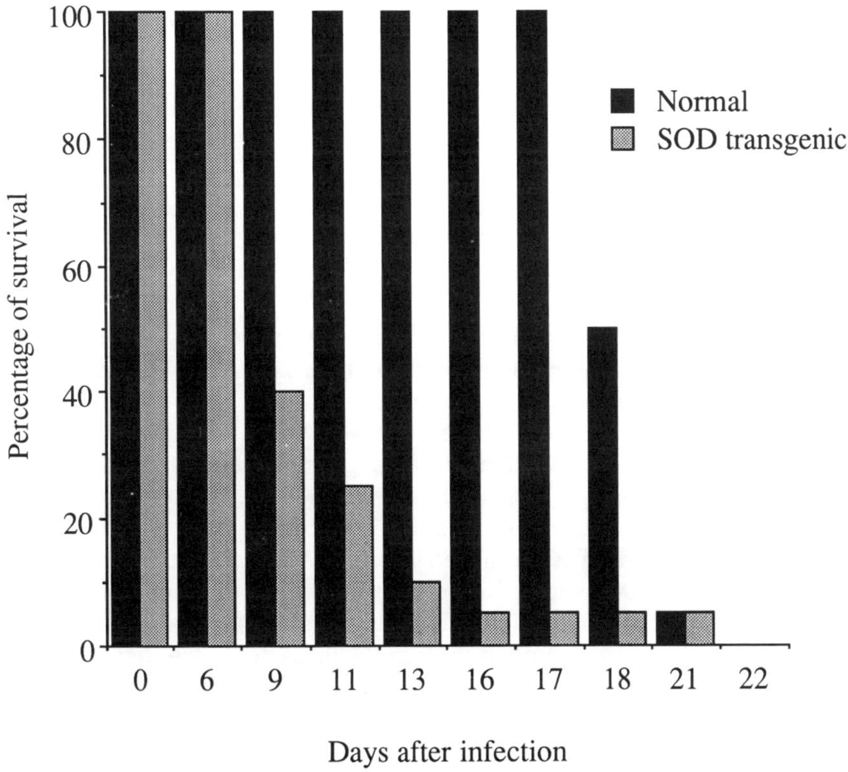

Fig. 9.2. Survival rates of normal and SOD transgenic mice infected with *P. berghei*.

ment of *P. falciparum* (Table 9.4), in correlation with the results obtained with the SOD overexpressing animal model of malaria (J. Golenser *et al.*, unpublished results).

Iron Metabolism and Antimalarial Drugs

Vennerstrom and Eaton (1988) suggested that particular consideration should be given to oxidant sensitivity of parasitized erythrocytes in the design of antimalarial drugs. The speculation that there are 'iron pools' in the parasitized erythrocytes (Jacobs, 1977) strengthens the idea that this iron can be used as a redox active transition metal maintaining oxidant stress. As in mammalian cells, such a class of bioavailable iron is probably vital for the parasite and is most likely accessible to iron chelators. However, depending on their chemical structure and the chelator complex formed, they might act not only as iron scavengers but also as free radical inducers. However, the propensity for free

Table 9.4. The effect of tempol on *P. falciparum* development.

Tempol (μM)	% Inhibition
200	1
400	15
800	34

P. falciparum at 5%Hct, 8% parasitaemia, was exposed to tempol for 4 h. Control values were 40,077 cpm.

radical formation and potential cytotoxicity, precluded their clinical application. We therefore shall focus on hydroxamate-based chelators, due to their therapeutic value in the treatment of iron overload diseases and since some members of this class have been shown to inhibit parasite growth *in vitro* and *in vivo* (rodent and primates) (Fritsch *et al.*, 1985; Pollack *et al.*, 1987), and some are even clinically effective as antimalarials. Inhibition of parasite growth with iron chelators was usually detected while *in vitro* cultures were being treated with the drug, but recovery of growth was often observed after discontinuation of the treatment, particularly at given stages of parasite growth. Similar results were obtained with malaria patients treated with desferrioxamine (DFO), since most patients suffered from recrudescences 7–10 days after termination of therapy (Gordeuk *et al.*, 1994).

DFO is a hydrophilic agent which permeates very slowly into parasitized cells, and consequently has a rather limited window of action on growing parasites. The nature of the selective parasite cytotoxicity of DFO and other hydroxamate iron chelators has only been partially resolved. The prevailing view is that its action involves sequestration of intracellular iron *en route* to integration into macromolecules. The metabolic iron utilized by the parasite is apparently acquired from the host during a restricted period of the parasite's life cycle. Therefore, chelators acting outside that period could irreversibly affect parasite development, as iron is essential throughout the entire parasite life cycle as an integral component of a variety of enzymes, cofactors and even transcription factors (Cabantchik, 1995).

A possible means of improving the therapeutic potential of iron chelators as antimalarial agents is by increasing their accessibility to parasites, and thus leading to an increase in their speed of action. Another means is by augmenting the retention of the drug by the parasites, to ensure a more persistent cytotoxic effect. These seemingly contradictory properties of potential drugs, which on the one hand act rapidly and on the other hand cause a long-lasting inhibition, could be accomplished either by application of permeant prodrugs which produce intracellular impermeant drugs or application of a combina-

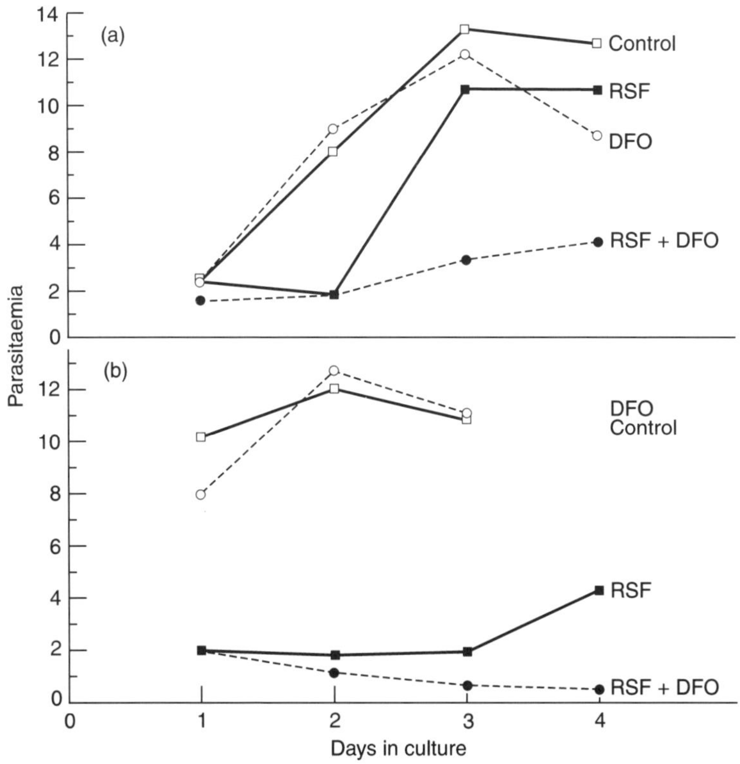

Fig. 9.3. Parasitaemias of *P. falciparum* treated with either DFO (16 μg ml^{-1}), RSF (12 μg ml^{-1}) or their combinations. Cultures of rings (a) or tropozoites (b) were treated with drugs for 8 h, washed and reincubated in drug-free medium.

tion of drugs with different action profiles, as shown in the present study. For the case of malaria parasites, the combined action of such drugs was anticipated to be greater than the sum of the individual effects. Such an effect was shown (Golenser *et al.*, 1995; Tsafack *et al.*, 1995) for pairs of iron chelators of the hydroxamate type, whose action on malaria parasites is associated with their capacity to permeate into parasites and scavenge essential iron. The distinctly different permeation features of the two chelators confer them with markedly different accessibility to parasite compartments and determined their speed of action and stage specificity. Since they are assumed to affect similar parasite targets, by scavenging iron(III) they should theoretically inhibit in a classically additive manner. However, the faster permeating drugs the reversed siderophores (RSFs)), although showing faster inhibitory action on parasites, also allowed parasites to recover from drug treatment upon drug removal. Therefore, the combined action of slow- and fast-permeating drugs can be expected to inhibit beyond the theoretical additive level and cause more persistent inhibitory effects on parasites exposed for a given period to drugs. The experimental results were in accord with the above hypothesis. They also showed that the combined *in vitro* antimalarial action of RSF and DFO was neither additive nor independent but demonstrably over-additive (Fig. 9.3). The potentiation of one drug effect by another drug, both in the continuous and discontinuous modes of drug treatment, might be indicative of a synergetic mechanism. In view of results obtained in the *Plasmodium vinckei*–mouse model, in which RSF outperformed DFO in terms of speed of action but not in terms of cure after discontinuation of treatment, we propose that such a type of drug combination should lead to improved antimalarial performance of DFO in *in vivo* animal studies and also in humans.

Conclusion

It is well established that malaria erythrocytes are highly susceptible to oxidant-mediated stress. This is especially significant in erythrocytes with abnormalities such as G6PD deficiency, where susceptibility to oxidant-mediated stress is highly emphasized. It is also likely that endogenous factors such those produced by effector cells, and exogenous physiological and toxicological factors, can induce production of deleterious OH radicals, particularly in the presence of iron-containing structures which are released during parasite development.

Future design of antimalarial drugs should take into account the unique features of iron metabolism and increased oxidant stress which prevail in parasitized erythrocytes. The use of combinations of drugs whose action spans the entire life cycle of the parasite and whose *modus operandii* show synergistic features, could lead to the development of potent therapeutic tools.

Acknowledgements

This work was supported in part by a grant from the National Institutes of Health AI-20342.

References

Allison, A.C. (1954) Protection afforded by sickle-cell trait against malaria infection. *British Medical Journal* 1, 290–291.

Arese, P., Turrini, F. and Ginsburg, H. (1991) Erythrophagocytosis in malaria: host defence or menance to the macrophages? *Parasitology Today* 7, 25–28.

Atamna, H. and Ginsburg, H. (1993) Origin of reactive oxygen species in erythrocytes infected with *P. falciparum*. *Molecular and Biochemical Parasitology* 61, 231–242.

Atamna, H., Pascarmona, G.P. and Ginsburg, H. (1994) Hexose-monophosphate shunt activity in intact *P. falciparum*-infected erythrocytes and in free parasites. *Proceedings of ICOPA VIII*, Istanbul.

Cabantchik, Z.I. (1990) Properties of permation pathways induced in the human red blood cell membrane by malaria parasites. *Blood Cells* 16, 421–432.

Cabantchik, Z.I. (1995) The mode of action of iron (III) chelators as antimalarials. The biochemical basis of selective cytotoxicity. *Parasitology Today* 11, 74–78.

Clark, I.A., Chaudhri, G. and Cowden, W.B. (1989) Some roles of free radicals in malaria. *Free Radical Biology and Medicine* 6, 315–321.

Cranston, H.A., Boylan, C.W., Carroll, G.L., Sutera, S.P., Williamson, J.R., Gluzman, I.Y. and Krogstad, D.J. (1984) *P. falciparum* maturation abolishes physiologic red cell deformability. *Science* 223, 400–403.

Elford, B.C. and Pinches, R.A. (1992) Inducible transport systems in the regulation of parasite growth in malaria infected red blood cells. *Biochemical Society Transactions* 20, 790–796.

Friedman, M.J. (1979) Oxidant damage mediates variant red cell resistance to malaria. *Nature* 280, 245–247.

Fritsch, G., Treumer, J., Spira, D.T. and Jung, A. (1985) *P. vinckei*: suppression of mouse infections with desferrioxamine B. *Experimental Parasitology* 60, 171–174.

Golenser, J., Miller, J., Spira, D.T., Kosower, N.S., Vande Waa, J.A. and Jensen, J.B. (1988) Inhibition of the intraerythrocytic development of *P. falciparum* in G6PD deficient erythrocytes is enhanced by oxidants and crisis form factor. *Tropical Medicine and Parasitology* 39, 273–276.

Golenser, J., Marva, E. and Chevion, M. (1991) The survival of *Plasmodium* under oxidant stress. *Parasitology Today* 7, 142–146.

Golenser, J., Kamyl, M., Tsafack, A., Marva, E., Cohen, A., Kitrossky, N. and Chevion, M. (1992) Correlation between destruction of malarial parasites by polymorphonuclear leucocytes and oxidative stress. *Radical Research Communications* 17, 249–262.

Golenser, J., Tsafack, A., Amichai, Y., Libman, J., Shanzer, A. and Cabantchik, Z.I. (1995) The antimalarial action of hydroxamate-based iron chelators. Effects of drug combinations. *Antimicrobial Agents and Chemotherapy* 39, 61–65.

Gordeuk, V.R., Brittenham, G. and Thuma, P.E. (1994) In Bergeson, B.J. and Brittenham, G.M. (eds) *The Development of Iron Chelators for Clinical Use*. CRC Press, Boca Raton, London, pp. 11–130.

Halliwell, B. (1990) How to characterize a biological antioxidant. *Free Radical Research Communications* 9, 1–32.

Halliwell, B. and Gutteridge, J.M.C. (1984) Oxygen toxicity, oxygen radicals, transition metals and disease. *Biochemical Journal* 219, 1–14.

Har-El, R., Marva, E., Chevion, M. and Golenser, J. (1993) Malaria derived redox active iron containing structures play an important role in inducing oxidant stress. Hemin is a possible candidate. *Free Radical Research Communications* 18, 279–290.

Hunt, N.H. and Stocker, R. (1990) Oxidative stress and the redox status of malaria-infected erythrocytes. *Blood Cells* 16, 499–526.

Jacobs, A. (1977) Low molecular weight intracellular iron transport compartments. *Blood* 50, 433–439.

Kutner, S., Baruch, D., Ginsburg, H. and Cabantchik, Z.I. (1982) Alterations in membrane permeability of malaria-infected human erythrocytes are related to the growth stage of the parasite. *Biochimie et Biophysica Acta* 687, 113–117.

Luzzatto, L. (1979) Genetics of red cells and susceptibility to malaria. *Blood* 54, 961–976.

Marva, E., Golenser, J., Cohen, A., Kitrossky, N., Har-El, R. and Chevion, M. (1992) The effects of ascorbate-induced free radicals on *P. falciparum*. *Tropical Medicine and Parasitology* 43, 17–23.

Nagel, R.N. and Roth, E.F. (1989) Malaria and red cell genetic defects. *Blood* 74, 1213–1221.

Pollack, S., Rossan, R.N., Davidson, D.E. and Escjadilo, A. (1987) Desferrioxamine suppresses *P. falciparum* in *Aeotus* monkeys. *Proceedings of the Society of Experimental Biology and Medicine* 184, 162–167.

Saltman, P. (1989) Oxidative stress: a radical view. *Seminars in Hematology* 26, 249–256.

Schirmer, R.H., Schollhammer, T., Eisenbrand, G. and Krauth-Siegel, R.L. (1987) Oxidative stress as a defence mechanism against parasitic infections. *Free Radical Research Communications* 3, 3–12.

Struchler, D. (1989) How much malaria is there worldwide. *Parasitology Today* 5, 39–40.

Tsafack, A., Golenser, J., Libman, J., Shanzer, A. and Cabantchik, Z.I. (1995) Natural and synthetic hydroxamate-based siderophores show synergistic inhibition of *P. falciparum* growth *in vitro*. *Molecular Pharmacology* (in press).

Vennerstrom, J.L. and Eaton, J.W. (1988) Oxidants, oxidant drugs and malaria. *Journal of Medicinal Chemistry* 31, 1269–1277.

Yuthavong, Y. and Wilairat, P. (1993) Protection against malaria by thalassaemia and haemoglobin variants. *Parasitology Today* 9, 241–245.

10

Immunogenicity of Multiple Antigen Peptides Containing T- and B-cell Epitopes of *Plasmodium falciparum* CS Protein

E.H. NARDIN,[1] G.A. OLIVEIRA,[1] J.M. CALVO-CALLE,[1] Y. MUNESINGHE,[1] P. CLAVIJO,[1] R. FARFAN,[1] R. RODRIGUEZ,[2] A. MORENO[2] AND R. NUSSENZWEIG[1]

[1]*Department of Medical and Molecular Parasitology, New York University School of Medicine, 341 East 25th Street, New York, NY 10016, USA and* [2]*Instituto de Inmunologia, Hospital San Juan de Dios, Bogota, Colombia*

Introduction

The clinical trials of a malaria vaccine candidate $(NANP)_3$-TT, carried out in 1986–1989 (Herrington *et al.*, 1987, 1990), were the first test of a synthetic peptide vaccine against a communicable disease in man. These Phase I and II trials demonstrated that the synthetic vaccine could induce antibody reactive with *Plasmodium falciparum* sporozoites in the majority of the vaccinees receiving the highest dose of vaccine. Although the overall antibody titres were suboptimal, there was a good correlation between the anti-repeat antibody response and reactivity with *P. falciparum* sporozoites. When challenged by the bite of *P. falciparum*-infected mosquitoes, one of three volunteers with the highest anti-sporozoite antibody titre was completely protected, i.e. did not develop a patent infection. The remaining two volunteers demonstrated a delayed prepatent period suggesting that there had been neutralization of the majority of infectious sporozoites. The magnitude of the antibody response was dose dependent, however, administration of larger doses of the vaccine, which consisted of 20% peptide by weight, were precluded by toxicity limitations due to the tetanus toxoid carrier.

The problems associated with the first generation synthetic peptide vaccines based on peptides conjugated to foreign proteins, i.e. carrier toxicity, low epitope density, possible epitopic suppression, and lack of parasite-derived T-cell epitopes have been overcome by multiple antigen peptides (MAPs). Synthetic MAP constructs, originally developed by J. Tam at Rockefeller University (Tam, 1988; Tam and Lu, 1989), replace the foreign protein carrier with a peptide core matrix that utilizes the amino groups of lysine to construct multiple dendritic branches on which peptide antigens can be synthesized.

MAPs are currently under development as vaccines for protozoan, helminthic and viral pathogens (Wolowczuk *et al.*, 1991; Darcy *et al.*, 1992; Keller *et al.*, 1994; Reynolds *et al.*, 1994; Nardin *et al.*, 1995). The first demonstration that MAPs could induce a protective immune response against an infectious disease was obtained in the *P. berghei* rodent malaria model (Tam *et al.*, 1990; Zavala and Chai, 1990; Chai *et al.*, 1992). Mice were immunized with MAPs containing various ratios and configurations of a T-cell epitope from the C terminus and a B-cell epitope from the repeat region of the circumsporozoite (CS) protein of *Plasmodium berghei*. The optimal immunogen, $(BT)_4$, induced high levels of anti-repeat and anti-sporozoite antibody as measured by enzyme-linked immunosorbent assay (ELISA) or the indirect fluorescent antibody test (IFAT). Antibody titres correlated with protection against challenge with viable sporozoites, as determined by the percentage of mice that did not have detectable blood stage parasitaemias on day 12 post-sporozoite challenge.

Design of a *P. falciparum* MAP Vaccine

MAP technology has been used to design a synthetic *P. falciparum* vaccine containing T- and B-cell epitopes of the *P. falciparum* CS protein that not only can induce high levels of anti-sporozoite antibodies in naive recipients, but can also elicit an anamnestic response in sporozoite-primed hosts.

Selection of T-helper cell epitopes for inclusion in MAP vaccine

The choice of *P. falciparum* T-cell epitope for inclusion in the MAP malaria vaccine was determined empirically by testing the immunogenicity, in mice, of MAP constructs containing the $(NANP)_3$ B-cell epitope in combination with different T-helper epitopes that have been identified in the repeat and non-repeat regions of the *P. falciparum* CS protein (Fig. 10.1). A total of four T-cell epitopes were tested. T1 represents an epitope within the repeat region (Nardin *et al.*, 1989) while T2 and T3 represent Th2R and Th3R, respectively, two epitopes identified by Good and colleagues based on murine and human T-cell

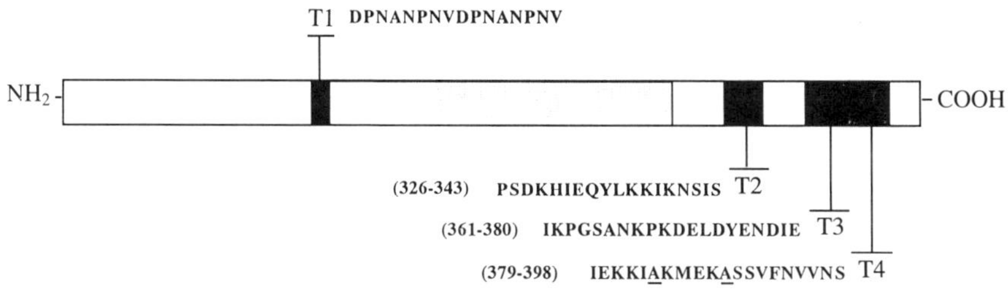

Fig. 10.1. T-helper epitopes of the *P. falciparum* CS protein.

responses (Good *et al.*, 1987, 1988). T4 represents the 'universal' T-helper cell epitope identified by Sinigaglia *et al.* (1988) using cells of naive individuals.

The optimal immunogen defined by these studies was the MAP construct containing the T-cell epitope T1 from the repeat region of the *P. falciparum* CS protein. The T1 epitope was originally identified by human $CD4^+$ T-cell clones, derived from a volunteer immunized with *P. falciparum* sporozoites (Nardin *et al.*, 1989). Mapping of the epitope recognized by these $CD4^+$ T-cell clones was carried out using a series of truncated recombinant CS proteins. Deletion of the 5′ repeat region from the recombinant CS proteins abrogated the T-cell proliferative response demonstrating that the T-cells recognized an epitope within the alternating NVDPNANP sequences. In all *P. falciparum* strains sequenced to date, two to three alternating NVDP and NANP sequences are found in the 5′ repeat region while the 3′ repeat regions consists of 33–43 NANP repeats (Lockyer and Schwartz, 1987; McCutchan *et al.*, 1988).

Synthetic peptides containing varying combinations of the NANP and NVDP repeats were used to confirm the identity of the epitope within the 5′ repeats. The 16mer peptide DPNANPNVDPNANPNV induced optimal stimulation of the repeat specific T-cell clones and was thus chosen for incorporation into the *P. falciparum* MAP construct.

Synthesis and immunogenicity of *P. falciparum* MAPs

MAPs containing the T1 epitope, in combination with the B-cell epitope $(NANP)_3$, were constructed by stepwise solid-phase peptide synthesis. The $(T1B)_4$ MAP is illustrated in Fig. 10.2 showing the T1 epitope (hatched bars) distal to the lysine core while the B-cell epitope (closed bars) are proximal to the core. Additional constructs reversing the orientation, i.e. BT, or varying the molar ratios of the T- and B-cell epitopes, were also tested.

All of the MAP constructs containing the T1 epitope were antigenic and stimulated proliferation and lymphokine secretion by the

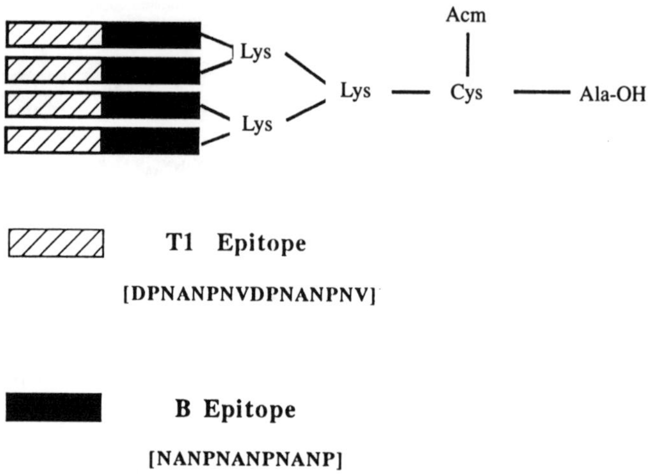

Fig. 10.2. Schematic diagram of $(T1B)_4$ MAP containing T- and B-cell epitopes of the repeat region of *P. falciparum* CS protein.

human $CD4^+$ T-cell clones specific for the T1 epitope. On a molar basis, the $(T1B)_4$ MAP construct was more efficient than the linear 16mer peptide in inducing proliferation and interferon-gamma production by the human T-cell clone (Table 10.1). The response was specific for the T1 epitope as the MAP construct containing only the B-cell epitope, $(B)_4$, did not stimulate the T-cells.

The immunogenicity of these MAP constructs was determined in four different strains of mice immunized with 50 μg MAPs administered intraperitoneally in Freund's adjuvant. Anti-repeat IgG antibody titres, measured by $(NANP)_{50}$ ELISA, were determined after each immunization on days 0, 21 and 42 (Fig. 10.3a). The $(T1B)_4$ construct was found to be optimal in that it induced high titre antibody responses in three

Table 10.1. Recognition of *P. falciparum* repeat MAPs by $CD4^+$ T-cell clone derived from a sporozoite-immunized volunteer.

Antigen	Concentration (M)	Proliferation SI	Interferon (U ml^{-1})
T1 16 mer	10^{-8}	33.2	24.0
	10^{-9}	3.7	nd
$(T1B)_4$	10^{-8}	33.4	22.7
	10^{-9}	23.5	13.8
	10^{-10}	4.0	nd
$(B)_4$	10^{-8}	1.2	<0.1

nd, Not determined; SI, stimulation index.

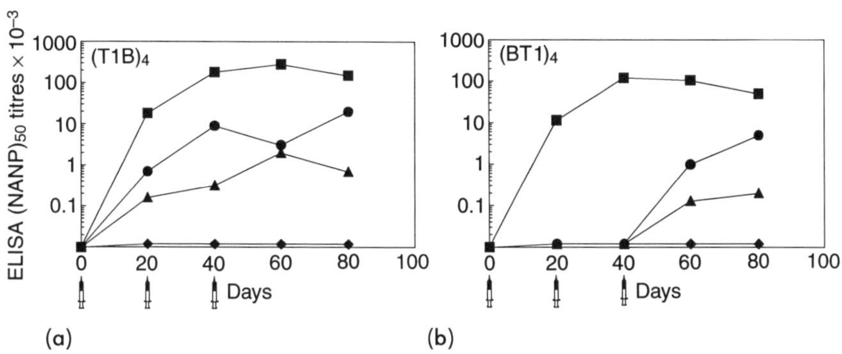

Fig. 10.3. Antibody response in different strains of mice immunized with (a) $(T1B)_4$ MAP or (b) $(BT1)_4$ MAP. Results are shown as geometric mean titres for each groups of five mice, as determined by $(NANP)_{50}$ ELISA. ■, C57Bl; ●, A/J; ▲, C3H; ♦, BALB/c.

of four strains of mice. The high responder mice, C57Bl/10^5, gave peak antibody titres of 10^5 while intermediate responders, A/J and C3H, had titres of 10^3–10^4. BALB/c mice did not respond to immunization with any of the MAP constructs in this formulation.

A MAP of the reverse orientation $(BT1)_4$, in which the B-cell epitope precedes the T-cell epitope, was less immunogenic (Fig. 10.3b). Intermediate responder strains required three doses of $(BT1)_4$ MAPs before significant anti-repeat antibody titres could be detected. Additional constructs containing altered ratios of T- and B-cell epitopes, i.e. one copy of the B-cell epitope and four copies of the T-cell epitope, or vice versa, were the least immunogenic.

The orientation of the T- and B-cell epitopes in the optimal $(T1B)_4$ MAP is the same as in the native *P. falciparum* CS protein, i.e. the T1 epitope in the 5′ repeat region precedes the B-cell epitope contained in the 3′ repeat region. Similarly, in the *P. berghei* MAPs which contain a T-cell epitope derived from the C terminus, the $(BT)_4$ MAP construct was optimal and this orientation also reflects the organization of the B- and T-cell epitopes in the *P. berghei* CS protein. However, thus far no pattern of T/B orientation has been defined that can predict the immunogenic properties of MAP antigens and the optimal stoichiometry and configuration of each MAP must be determined empirically.

Fine specificity of anti-$(T1B)_4$ MAP antibody responses. A previous drawback associated with synthetic immunogens has been the failure of anti-peptide antibodies to react with native protein. However, the $(T1B)_4$ MAP induced anti-peptide antibodies that reacted at high titres with the native CS protein on the *P. falciparum* sporozoites (Munesinghe *et al.*, 1991). In fact, IFAT titres of over 10^6 were obtained in the sera of some of the C57Bl/10 mice, levels that had not previously been induced by immunization with any synthetic or recombinant *falciparum* vaccine candidate. CS protein titres in excess of 1 : 100 were also observed in the serum of the MAP immunized C57Bl/10 mice.

It is also noteworthy that the anti-MAP antibodies reacted not only with the 3′ NANP repeats but also the 5′ NANPNVDP repeats of the *P. falciparum* CS protein (Munesinghe *et al.*, 1991). Therefore antibodies induced by either the B- or the T-cell epitope effectively reacted with the repeat region of the *P. falciparum* CS protein and contributed to a parasite-specific antibody response. In the high responder C57Bl/10 mice, the fact that each of the epitopes contained in the $(T1B)_4$ MAP could function as both B- and T-cell epitopes may also have played a role in enhancing the magnitude of the immune response.

Persistence of anti-$(T1B)_4$ MAP antibody responses. The immune response induced by $(T1B)_4$ MAP immunization was measured over a period of 6 months following the last injection of antigen (Fig. 10.4). As measured by both ELISA and IFAT, the anti-repeat and anti-sporozoite antibodies induced by immunization with $(T1B)_4$ MAP remained at high titres for a prolonged period of time. Approximately 6 months after the last dose of MAP, ELISA titres of 81 and 920, and IFAT titres of 10^4, were still detectable in the sera of the MAP immunized mice.

MAPs containing T-helper cell epitopes from the C terminus of the *P. falciparum* CS protein

The correlation between the anti-peptide and anti-sporozoite antibody titres induced by the $(T1B)_4$ MAP was particularly noteworthy in light of the responses that were obtained using MAPs containing T-helper epitopes from the C terminus of the *P. falciparum* CS protein. A series of MAP constructs containing the T2, T3 or T4 epitopes from the C terminus of the *P. falciparum* CS protein (Fig. 10.1), combined with the $(NANP)_3$ B-cell epitope, were tested in configurations in which the T-epitope preceded the B-epitope or the reverse orientation (Calvo-Calle *et al.*, 1993).

As previously noted, the anti-MAP response in the $(T1B)_4$ immunized mice directly correlated with the anti-sporozoite IFAT titres, indicating that the anti-peptide antibodies effectively recognized the native CS protein. In contrast, while the MAP constructs containing the T-helper epitopes from the C terminus of the CS protein, $(BT2)_4$, $(T3B)_4$ and $(BT4)_4$, induced very high anti-peptide antibody titres, the anti-sporozoite antibody titres were orders of magnitude lower (Table 10.2). For example, anti-peptide ELISA titres of over 10^6 were detected in mice immunized with $(BT4)_4$ MAP while an IFAT titre of 5200 was obtained when tested with *P. falciparum* sporozoites.

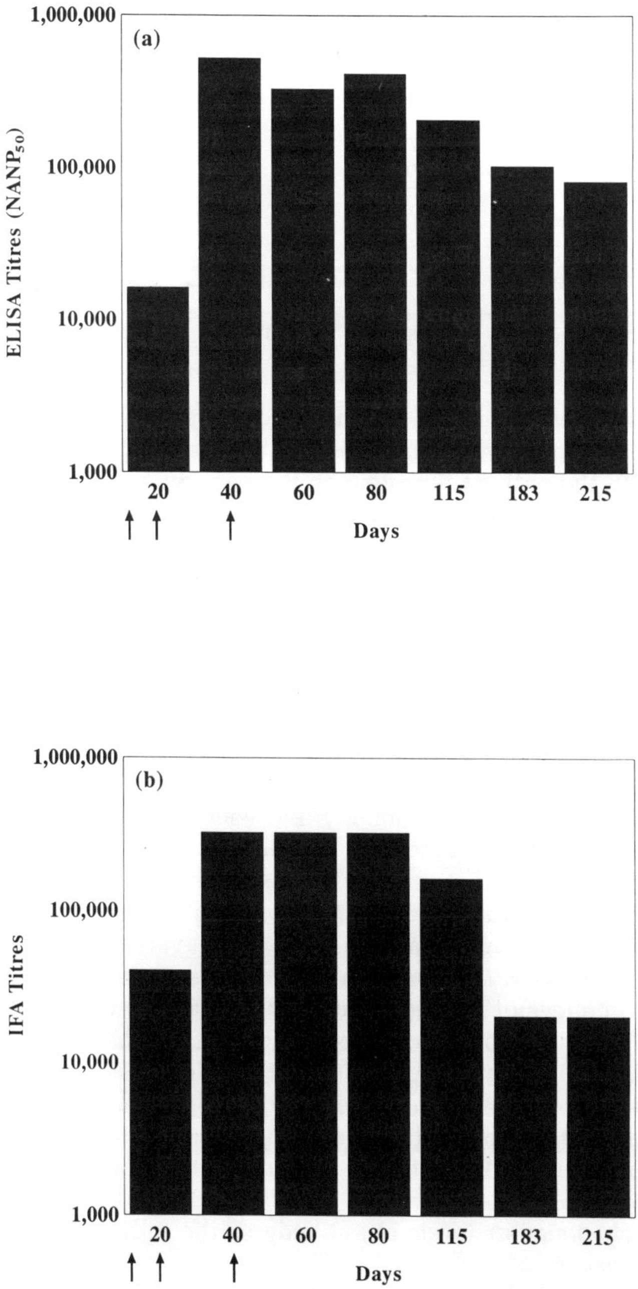

Fig. 10.4. Persistence of murine antibody response induced by immunization with $(T1B)_4$ MAP as measured by (a) $(NANP)_{50}$ ELISA and (b) IFAT against *P. falciparum* sporozoites.

Table 10.2. Immunogenicity of MAPs containing T-helper epitopes from the *P. falciparum* CS protein.

MAPs	Anti-sporozoite IFAT titre	Anti-peptide antibody ELISA titre	
		MAP	$(NANP)_{50}$
$(T1B)_4$	825,700	825,700	260,000
$(BT2)_4$	10,000	300,000	160,000
$(T3B)_4$	20,000	1,200,000	20,000
$(BT4)_4$	5,200	1,200,000	160,000

Source: Munesinghe *et al.* (1991); Calvo-Calle *et al.* (1992).

These findings suggested that a large proportion of the anti-peptide antibodies induced by immunization with these MAP constructs failed to react with the native CS protein on the parasite surface. The analysis of the fine specificity of the anti-MAP antibody response, in fact, demonstrated significant levels of antibody specific for the T-cell epitopes contained in these MAP constructs. Therefore, in contrast to the immune response to the $(T1B)_4$ MAP, in which antibodies generated against the T-cell epitope reacted with the sporozoite and thus provided a more effective immune response, the antibody generated against non-repeat T-cell epitopes did not enhance the anti-parasite response.

Immunogenicity of $(T1B)_4$ MAP in primate hosts, *Aotus* monkeys

To determine the immunogenicity and genetic restriction of the antibody response in a primate host, two different species of owl monkeys, *Aotus nancymai* and *Aotus vociferans*, were immunized with $(T1B)_4$ MAP in Freund's adjuvant (A. Moreno, unpublished). The two monkey species differed in phenotype (red neck versus grey neck) and karyotype (KI vs KV). Significant differences in the immune response of the *A. nancymai* versus *A. vociferans* could be observed following the first subcutaneous (s.c.) injection of MAPs. Three out of three *A. nancymai* had IgG anti-repeat ELISA titres in excess of 10^3 while the two *A. vociferans* did not have detectable responses. However, the first dose of MAPs had primed the immune system of the *A. vociferans*, as shown by a significant antibody response after the second injection of $(T1B)_4$ MAP. Following the third dose of $(T1B)_4$ MAP, the *A. vociferans* had reached similar levels of antibody as the *A. nancymai* with IFAT titres in excess of 10^5.

The magnitude and specificity of the anti-sporozoite antibody responses in the MAPs immunized *Aotus* was similar to that observed in the MAPs immunized mice. High titres of anti-MAPs antibody correlated with high titres of reactivity with *P. falciparum* sporozoites, as

Table 10.3. Immunogenicity of $(T1B)_4$ MAP using alum versus Freund's adjuvant.

Adjuvant	Anti-sporozoite IFAT titre	Anti-MAPs ELISA titre
$(T1B)_4$/alum	327,680	655,360
$(T1B)_4$/Freund's	163,840	463,409

Source: de Oliveira *et al.* (1994).

determined by indirect immunofluorescence. IFAT titres in excess of 1 million could be observed in the individual *Aotus* sera. As was also noted in the sera of the MAPs immunized mice, the anti-repeat antibodies induced in the MAPs immunized *Aotus* monkeys reacted with both the 5′ and 3′ repeats of *P. falciparum* CS protein.

$(T1B)_4$ MAP vaccine formulations

Based on its ability to induce high titres of anti-sporozoite antibodies in both murine and primate hosts, the $(T1B)_4$ MAP was chosen as the *falciparum* MAP vaccine candidate for Phase I and II trials in human volunteers. However, these high antibody titres in experimental animals were obtained with MAPs emulsified in Freund's adjuvant, a formulation that is not acceptable for human vaccines.

Studies of the immunogenicity of $(T1B)_4$ MAP in alum (aluminium hydroxide), a formulation that is acceptable for human use, were carried out in mice (de Oliveira *et al.*, 1994). After a single immunizing dose of MAPs, significant differences in the levels of anti-repeat antibodies were observed in the recipients of vaccine adjuvanted with Freund's versus alum. However, after three immunizing doses, the anti-repeat, and more importantly the anti-sporozoite, antibody response of the recipients of the MAPs/alum formulation were comparable to those obtained with Freund's adjuvant. Peak IFAT titres obtained were 3.2×10^5 versus 1.6×10^5 in the alum versus Freund's adjuvant groups respectively (Table 10.3).

Since intraperitoneal (i.p.) injection is not a practical route of immunization in man, the immunogenicity of the MAPs/alum preparation was compared when each vaccine formulation was administered s.c. The kinetics of the antibody response following multiple s.c. injections of $(T1B)_4$/alum was similar to i.p. immunization (de Oliveira *et al.*, 1994). A minimal response to the first injection of the vaccine was followed by a rapid rise of antibody titre in response to the second and third dose of $(T1B)_4$ MAP/alum s.c. reaching peak $(NANP)_{50}$ ELISA titres in excess of 10^5, similar to titres obtained with Freund's adjuvant. The corresponding IFAT titres after the third s.c. dose were

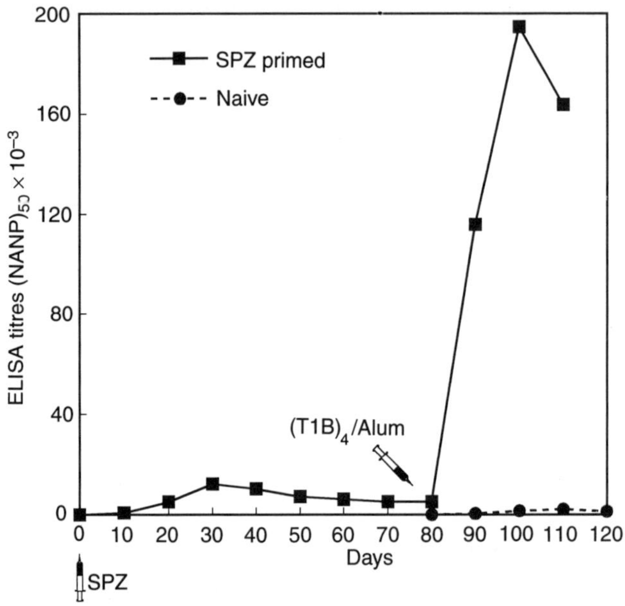

Fig. 10.5. $(T1B)_4$ MAP/alum elicits an anamnestic response in sporozoite(SPZ)-primed mice. Source: de Oliveira *et al.* (1994).

6.4×10^5 in pooled sera from the MAP/alum immunized mice versus 3.2×10^5 in the Freund's adjuvant group.

$(T1B)_4$ MAP elicit anamnestic antibody response in sporozoite-primed mice

The administration of the $(T1B)_4$ MAP/alum preparation was found to elicit a strong anamnestic response in sporozoite-primed mice (Fig. 10.5). Approximately 3 months after a single intravenous (i.v.) injection of *P. falciparum* sporozoites, mice were injected with 30 μg of MAP/alum s.c. Anti-repeat antibody titres reached 1.9×10^5, and IFAT titres 3.2×10^5, in the sporozoite-primed mice following a single injection of $(T1B)_4$ MAP. Titres in naive mice injected only with $(T1B)_4$ MAPs were less than 10^3.

These results suggest that the administration of $(T1B)_4$ MAP/alum vaccine would be likely to increase the antibody levels, and concomittantly, resistance to reinfection, of individuals living in malaria endemic areas, most of whom present low or suboptimal anti-sporozoite antibody titres. Since the 5′ repeat sequence is conserved in the CS protein of all *P. falciparum* isolates, the administration of the $(T1B)_4$ MAP should elicit a secondary response from individuals previously exposed to different *falciparum* strains. In addition, the presence of a conserved parasite-derived T-cell epitope in the vaccine should enable the natural exposure to infected mosquitoes in an endemic area to assure the persistance of MAP-induced antibody responses.

Conclusion

MAPs vaccines have been developed to eliminate the requirement for protein carrier and to obtain a highly immunogenic synthetic molecule containing T- and B-cell epitopes of the *P. falciparum* CS protein. In contrast to the low epitope density of the first generation peptide–protein conjugate vaccines (<20% peptide by weight), greater than 90% of the total molecular mass of the $(T1B)_4$ MAP consists of peptide epitopes relevant to a specific anti-sporozoite immune response. The high density of T- and B-cell epitopes in the $(T1B)_4$ MAP induced murine anti-sporozoite antibody titres that were one to two orders of magnitude higher than those obtained with previous *P. falciparum* CS protein subunit vaccines.

When compared with MAPs containing T-helper epitopes derived from the C terminus of the *P. falciparum* CS protein (T2, T3 or T4), only the $(T1B)_4$ MAP induced high titres of antibodies reactive with the native CS protein on the sporozoite surface. Antibodies to the T1 epitope, but not anti-T2, T3 or T4 antibodies, recognize *P. falciparum* sporozoites and thus contribute to a specific humoral anti-parasite response.

The presence of a parasite-derived T-helper epitope enables the $(T1B)_4$ MAP to elicit a strong anamnestic antibody response in *P. falciparum* sporozoite-primed mice. The inclusion of the CS T-helper epitope should also allow natural exposure to the bites of malaria-infected mosquitoes to enhance the persistence of the MAPs induced antibody response. Since the T1 epitope is conserved in all isolates of *P. falciparum*, administration of $(T1B)_4$ MAP vaccine should elicit and/or booster the immune response to all strains of *P. falciparum* and thus circumvent the problems associated with polymorphic T-helper cell epitopes in the C terminus of the CS protein.

The $(T1B)_4$ MAP was immunogenic in three of four mouse strains tested and induced high titres of anti-sporozoite antibodies in two different species of *Aotus* monkeys. The studies in rodent and primate hosts encourage the hope that the $(T1B)_4$ MAP vaccine will be immunogenic in individuals of diverse genetic backgrounds. The immunogenicity of MAPs in alum is comparable to Freund's adjuvant and thus the $(T1B)_4$ MAP vaccine formulation is compatible with human use. Phase I and II trials are planned to investigate the genetic restriction and immunogenicity of $(T1B)_4$ MAP/alum in human volunteers.

Acknowledgements

This work was supported by grants from the National Institutes of Health (AI-25085), the Agency for International Development (DPE-5979-A-00-006-00) and The Irma Hirschl Charitable Trust. G.A. Oliveira is a recipient of a Brazilian Council for Scientific and Technological Development fellowship.

References

Calvo-Calle, J.M., de Oliveira, G.A., Clavijo, P., Maracic, M., Tam, J.P., Lu, Y.A., Nardin, E.H., Nussenzweig, R.S. and Cochrane, A.H. (1993) Immunogenicity of multiple antigen peptides containing B and non-repeat T cell epitopes of the circumsporozoite protein of *Plasmodium falciparum*. *Journal of Immunology* 150, 1403–1412.

Chai, S.K., Clavijo, P., Tam, J.P. and Zavala, F. (1992) Immunogenic properties of multiple antigen peptide systems containing defined T and B epitopes. *Journal of Immunology* 149, 2385–2390.

Darcy, F., Maes, P., Gras-Masse, H., Auriault, C., Bossus, M., Deslee, D., Godard, I., Cesbron, M.F., Tartar, A. and Capron, A. (1992) Protection of mice and nude rats against toxoplasmosis by a multiple antigenic peptide construction derived from *Toxoplasma gondii* p30 antigen. *Journal of Immunology* 149, 3636–3641.

de Oliveira, G.A., Clavijo, P., Nussenzweig, R.S. and Nardin, E.H. (1994) Immunogenicity of an alum-adsorbed synthetic multiple-antigen peptide based on B- and T-cell epitopes of the *P. falciparum* CS protein: possible vaccine application. *Vaccine* 12, 1012–1017.

Good, M.F., Maloy, W.L., Lunde, M.N., Margalit, H., Cornette, J.L., Smith, G.L., Moss, B., Miller, L.H. and Berzofsky, J.A. (1987) Construction of synthetic immunogen: use of new T-helper epitope of malaria circumsporozoite protein. *Science* 235, 1059–1062.

Good, M.F., Pombo, D., Quakyi, I.A., Riley, E.M., Houghten, R.A., Menon, A., Alling, D.W., Berzofsky, J.A. and Miller, L.H. (1988) Human T-cell recognition of the circumsporozoite protein of *Plasmodium falciparum*: immunodominant T-cell domains map to the polymorphic regions of the molecule. *Proceedings of the National Academy of Sciences of the United States of America* 85, 1199–1203.

Herrington, D.A., Clyde, D.F., Losonsky, G., Cortesia, M., Murphy, J.R., Davis, J., Baqar, S., Felix, A.M., Heimer, E.P. and Gillessen, D. (1987) Safety and immunogenicity in man of a synthetic peptide malaria vaccine against *Plasmodium falciparum* sporozoites. *Nature* 328, 257–259.

Herrington, D.A., Clyde, D.F., Davis, J.R., Baqar, S., Murphy, J.R., Cortese, J.F., Bank, R.S., Nardin, E., DiJohn, D. and Nussenzweig, R.S. (1990) Human studies with synthetic peptide sporozoite vaccine (NANP)3-TT and immunization with irradiated sporozoites. *Bulletin of the World Health Organization* 68 (Suppl)., 33–37.

Kelker, H.C., Schlesinger, D. and Valentine, F.T. (1994) Immunogenic and antigenic properties of an HIV-1 gp120-derived multiple chain peptide. *Journal of Immunology* 152, 4139–4148.

Lockyer, M.J. and Schwarz, R.T. (1987) Strain variation in the circumsporozoite protein gene of *Plasmodium falciparum*. *Molecular Biochemistry and Parasitology* 22, 101–108.

McCutchan, T.F., De La Cruz, V.F., Good, M.F. and Wellems, T.E. (1988) Antigenic diversity in *Plasmodium falciparum* [Review]. *Progress in Allergy* 41, 173–192.

Munesinghe, D.Y., Clavijo, P., Calvo-Calle, J.M., Nussenzweig, R.S. and Nardin, E. (1991) Immunogenicity of multiple antigen peptides (MAP) containing T and B cell epitopes of the repeat region of the *P. falciparum* circumsporozoite protein. *European Journal of Immunology* 21, 3015–3020.

Nardin, E.H., Herrington, D.A., Davis, J., Levine, M., Stuber, D., Takacs, B., Caspers, P., Barr, P., Altszuler, R., Clavijo, P. and Nussenzweig, R.S. (1989) Conserved repetitive epitope recognized by $CD4^+$ clones from a malaria-immunized volunteer. *Science* 246, 1603–1606.

Nardin, E.H., Oliveira, G.A., Calvo-Calle, J.M. and Nussenzweig, R.S. (1995) The use of multiple antigen peptides in the analysis and induction of protective immunoresponses against infectious diseases. *Advances in Immunology* 60, 105–149.

Reynolds, S.R., Dahl, C.E. and Harn, D.A. (1994) T and B epitope determination and analysis of multiple antigenic peptides for the *Schistosoma mansoni* experimental vaccine triose-phosphate isomerase. *Journal of Immunology* 152, 193–200.

Sinigaglia, F., Guttinger, M., Kilgus, J., Doran, D.M., Matile, H., Etlinger, H., Trzeciak, A., Gillessen, D. and Pink, J.R. (1988) A malaria T-cell epitope recognized in association with most mouse and human MHC class II molecules. *Nature* 336, 778–780.

Tam, J.P. (1988) Synthetic peptide vaccine design: synthesis and properties of a high-density multiple antigenic peptide system. *Proceedings of the National Academy of Sciences of the United States of America* 85, 5409–5413.

Tam, J.P. and Lu, Y.A. (1989) Vaccine engineering: enhancement of immunogenicity of synthetic peptide vaccines related to hepatitis in chemically defined models consisting of T- and B-cell epitopes. *Proceedings of the National Academy of Sciences of the United States of America* 86, 9084–9088.

Tam, J.P., Clavijo, P., Lu, Y.A., Nussenzweig, V., Nussenzweig, R. and Zavala, F. (1990) Incorporation of T and B epitopes of the circumsporozoite protein in a chemically defined synthetic vaccine against malaria. *Journal of Experimental Medicine* 171, 299–306.

Wolowczuk, I., Auriault, C., Bossus, M., Boulanger, D., Gras-Masse, H., Mazingue, C., Pierce, R.J., Grezel, D., Reid, G.D., Tartar, A. and Capron,

A. (1991) Antigenicity and immunogenicity of a multiple peptidic construction of the *Schistosoma mansoni* Sm-28 GST antigen in rat, mouse, and monkey. 1. Partial protection of Fischer rat after active immunization. *Journal of Immunology* 146, 1987–1995.

Zavala, F. and Chai, S. (1990) Protective anti-sporozoite antibodies induced by a chemically defined synthetic vaccine. *Immunology Letters* 25, 271–274.

11

Cell Biology of the *Giardia lamblia* Life Cycle

F.D. GILLIN, J.M. MCCAFFERY, D.S. REINER, S.B. ALEY, T.-C. MENG AND M.L. HETSKO

University of California, San Diego, CA 92103–8416, USA

The initial identification of *Giardia lamblia* as a very primitive or early-branching eukaryote was based on evolutionary trees that were derived from ribosomal RNA sequences (Sogin *et al.*, 1989). The analyses of increasing numbers of protein-encoding genes has buttressed this hypothesis. In addition, both morphological (Friend, 1966) and biochemical studies have demonstrated the absence of mitochondria in Diplomonads. On the other hand, the commonly reported idea that *Giardia* has little endoplasmic reticulum (ER) and lacks a Golgi apparatus appears to be based largely on ultrastructural observations (Brugerolle, 1975). This contradicts one of the fundamental propositions of cell biology: that the complex endomembrane-mediated system of protein transport evolved with the true nucleus and is present in all eukaryotic cells (Alberts *et al.*, 1989).

We first became intrigued with this issue when we (Reiner *et al.*, 1989) and others (Ward *et al.*, 1990) found that the giardial cyst wall contained proteins and glycoproteins that were specifically synthesized during encystation. Moreover, we completed the *G. lamblia* life cycle *in vitro* for the first time. Using light microscopy, we found initial evidence of vesicular transport of regulated proteins to the nascent cyst wall (Reiner *et al.*, 1989). This stimulated us to ask how *Giardia* accomplishes the sorting of encystation-specific proteins destined for the cyst wall, and whether this organism has distinct pathways for recognition and transport of regulated versus constitutively expressed proteins. In higher organisms, these functions are accomplished during protein passage through successive compartments or cisternae of the ER and Golgi. It is generally agreed that proteins are carried from one compartment or cisterna to the next by small (~50 nm) transport vesicles (Alberts *et al.*, 1989).

In extensive ultrastructural analyses, we have shown that *G. lamblia* trophozoites are capable of sorting proteins to the three endomembrane-mediated protein transport pathways of higher eukaryotic cells (Reiner *et al.*, 1990, 1993; McCaffery *et al.*, 1994): a regulated pathway of cyst wall protein transport and secretion, as well as constitutive pathways of protein transport to the plasma membrane and to the lysosomal compartment.

We first focused on encysting cells because we hypothesized that elements of the giardial endomembrane transport machinery might be more active or more evident in cells that were actively synthesizing, sorting and transporting specific proteins to the cyst wall (Reiner *et al.*, 1990). In an extreme case, a Golgi apparatus might be assembled specifically during encystation. Interestingly, we found both numerous endomembrane elements whose presence was *independent* of the stage of the life cycle, as well as novel regulated secretory vesicles, whose assembly and secretion are *strictly limited* to encystation (Reiner *et al.*, 1989, 1990; McCaffery *et al.*, 1994).

A Regulated Pathway of Cyst Wall Protein Transport

We found a novel regulated secretory pathway in *G. lamblia* that is defined by polyclonal (Reiner *et al.*, 1989, 1990) or monoclonal (Reiner *et al.*, 1993) antibodies (GCSA-1, Ward *et al.*, 1990; and 8C5, Campbell and Faubert, 1994) against cyst wall proteins. Using frozen section immunoelectron microscopy (immunoEM), we found that early in encystation (~5–24 h), these cyst wall proteins localize to the interior, but not the membrane, of these encystation secretory vesicles or ESV. The ESV are large (>1 μm diameter), membrane-bound osmiophilic secretory vesicles, which transport specific proteins to the nascent cyst wall. The idea that this constitutes a *regulated* secretory pathway is based on our findings that ESV formation (Faubert *et al.*, 1991), like cyst antigen expression (Reiner *et al.*, 1989), is induced by exposure to alkaline pH and bile, potent stimuli of encystation that trophozoites encounter in the host small intestine. In extensive studies, we have never seen ESV or cyst wall structure that did not stain with anti-cyst antibodies (Reiner *et al.*, 1989; McCaffery *et al.*, 1994). Under our conditions, ESV are first observed at ~5 h of encystation, are maximal at ~16–24 h, and decline thereafter (Faubert *et al.*, 1990; McCaffery *et al.*, 1994). We do not yet know whether the release of the ESV contents is also regulated by external stimuli, like regulated secretion in many higher cells (Alberts *et al.*, 1989), or only by the progression of encystation. ESV are not observed in mature cysts (Schupp *et al.*, 1988) suggesting that they are no longer needed once the cyst wall is complete.

A Constitutive Pathway of Protein Transport to the Plasma Membrane

The cysteine-rich variable surface proteins (VSPs) make an ideal marker to trace protein transport to the trophozoite plasma membrane (Reiner *et al.*, 1990) because we found both the protein and its messenger (m) RNA to be present during both growth and encystation (Aley and Gillin, 1993; Meng *et al.*, 1993). Moreover, we have cloned and expressed the first *complete* VSP gene and made antibodies against the recombinant protein, called TSA 417 (Gillin *et al.*, 1990). In higher cells, constitutively expressed membrane proteins are transported to the plasmalemma by a so-called 'default pathway'. Although nascent proteins destined for secretion are directed into the lumen of the ER by their amino-terminal signal peptide, proteins that lack any signal to address them to other pathways are eventually transported to the cell membrane, defining a default pathway (Alberts *et al.*, 1989).

On the basis of its sequence (Gillin *et al.*, 1990) and hydropathy plot (Aley and Gillin, 1993), we predicted TSA 417 to be a Type 1 integral membrane protein with a single membrane spanning region (Gillin *et al.*, 1990), meaning that it crossed the plasma membrane only once. Moreover, the signal peptide predicted from the amino acid sequence was found to be cleaved from TSA 417 isolated from trophozoites (Aley and Gillin, 1993). As predicted, in contrast to the cyst antigens, we found that TSA 417, which is the major trophozoite outer surface protein, did *not* localize to the membrane or interior of the ESV during encystation (McCaffery *et al.*, 1994). Since TSA 417 protein and mRNA are present during growth and encystation (Aley and Gillin, 1993), it defines a *constitutive* pathway of transport to the plasma membrane.

A Constitutive Pathway of Acid Phosphatase Transport to the Lysosome-like Peripheral Vesicles

The ~300 nm peripheral vesicles (PV) that underlie the plasmalemma of much of the cell during both growth and encystation are thought to be the giardial equivalent of lysosomes because they contain acid phosphatase activity (Feely and Dyer, 1987). Therefore, we used acid phosphatase activity as a tracer of the constitutive pathway of transport to the lysosomal compartment. We found that during encystation, acid phosphatase activity did not localize to the ESV (Reiner *et al.*, 1990), but to the ER, Golgi-like membrane stacks and cytoplasmic tubules and cisternae, and, therefore, defined a distinct lysosomal transport pathway.

Complexity of Protein Transport Pathways in Growth and Encystation

In single (Reiner *et al.*, 1990) and double label (McCaffery *et al.*, 1994) immunoEM, we found that both the cyst wall antigens and TSA 417 followed the respective transport pathways that would be predicted for regulated and constitutive proteins. Both were observed in the nuclear envelope cisternae of the ER, in cytoplasmic ER cisternae and in parallel paranuclear membrane stacks, resembling a Golgi apparatus. At times, these two markers were found in the same ER or cisternal structure. In contrast, only TSA 417 was ever found on the plasma membrane, while only the cyst wall epitope 8C5 was found in ESV or on the cyst wall.

Interestingly, we also found both TSA 417 and 8C5 in some of the lysosome-like PV. PV were heterogeneous as they contained one, both, or neither antigen. TSA 417 was observed in PV in both vegetatively growing cells and throughout encystation. We observed 8C5 in these vesicles in small amounts in pre-encysting bile-starved cells and in increasing amounts later in encystation. Thus, 8C5 appeared in some PV prior to formation of the ESV, as well as later when most of this antigen was in ESV or the cyst wall. At present, we do not know whether the presence of these antigens in the PV is due to transport into or out of the cells. However, it suggests that the pathways of protein processing and transport are complex.

Ultrastructural Analyses of Antigenic Switching of TSA 417 During Encystation

Although TSA 417, the VSP expressed by our isolate (Gillin *et al.*, 1990), was present in Western blots throughout growth and encystation, the amounts in water-resistant cysts appeared to be lower. Moreover, TSA 417 + populations examined by light microscope immunostaining after they were excysted and grown for 48 h or longer, had switched to TSA 417 – (Meng *et al.*, 1993; T.-C. Meng *et al.*, unpublished). After switching, TSA 417 was replaced by other cysteine-rich VSP(s) that were not recognized by anti-TSA 417. Because intact cysts excluded the antibodies (Meng *et al.*, 1993), we used immunoEM to evaluate when the antigenic switching occurred.

We found that TSA 417 covered the entire outer surface of the plasma membrane, including the adhesive disc and flagella during both growth (Gillin *et al.*, 1990) and encystation (McCaffery *et al.*, 1994). In younger cysts, TSA 417 was often in the plasma membrane that underlies the cyst wall, although it was sparser in the plasmalemma of mature cysts, and frequently was not seen beneath the cyst wall, but only in internal vesicles. This may be a sequence of cells in the process of switching their VSP. At the same time, TSA 417 continued to be in the PV, showing that these cells were TSA 417+. Since the PV are lysosomal and our antibody recognizes TSA 417 proteolytic fragments (Gillin *et al.*, 1990; Aley and Gillin, 1993), these vesicles may contain partially digested protein. Therefore, switching could entail internalization and possibly degradation of TSA 417. The methods used in these studies would not detect release of TSA 417 from the cell surface (Nash *et al.*, 1983), and the cyst wall might interfere with release. Therefore, these studies support the idea that switching began late in encystation or in the cyst stage.

We also do not know whether other VSPs undergo switching under the same conditions. This may be a form of immune evasion in which newly excysted populations present different surface antigens to a new host. Further studies are needed to establish the time sequence and mechanism of the switch process.

Ultrastructural Bases for Protein Transport during Growth and Encystation

In higher cells, sorting of proteins to constitutive, lysosomal, and regulated pathways is a function of the trans-Golgi network, where small (~50–80 nm) vesicles for transport to these targets arise (Farquhar, 1985; Pryer *et al.*, 1992). However, the structural bases for these activities in *Giardia* have not been clear. Most ultrastructural analyses have focused on the adhesive disc and cytoskeleton (e.g. Brugerolle, 1975; Holberton *et al.*, 1986; Peattie *et al.*, 1989). By using several complementary membrane-enhancing ultrastructural techniques to visualize the giardial endomembrane system, we focused on the relationships between endomembrane structures, as well as on the formation and traffic of the ESV during encystation. Certain membrane structures, including ER (including the perinuclear cisternae), PV, and plasmalemma have been well-described previously.

With the advent of additional functional markers and mutants, new compartments in the endomembrane transport system are being identified in higher cells. In order to look for other giardial endomembrane structures and to reveal their interrelationships, we used a fixation procedure for transmission EM that tends to minimize the contrast of cytoplasmic ribosomes that appear to obscure many giardial endomembranes, and found many smooth membrane stacks (Reiner *et al.*, 1990). In addition, more recently (McCaffery *et al.*, 1994), stained giardial cells with reduced osmium-KFeCN (Karnovsky, 1971), which tends to stain endomembranes, or with tannic acid (Simionescu and Simionescu, 1976), which also enhances coated pits and vesicles (Rothman and Orci, 1990).

These techniques revealed not only Golgi-like smooth membrane stacks, but also abundant rough ER, which is covered with ribosomes. Frequently, long rough ER membranes traversed much of the cell body both parallel to the dorsal surface, underlying the PV and in the perpendicular direction. In addition, we observed transitional elements with one smooth and one rough, or ribosome-studded, face. In higher cells, these are a post-ER element. These techniques also revealed numerous smooth membrane cisternae. Some of the nests or clusters of smooth cisternae and vesicles resemble tubular-vesicular elements that are an intermediate or pre-Golgi compartment in higher cells (Balch *et al.*, 1994). Finally, we observed numerous small (50–80 nm) cytoplasmic vesicles, many of which were coated, that resemble the small transport vesicles that carry proteins between successive compartments of the ER and Golgi. Vesicles of similar size were previously reported at the cell periphery (Owen, 1980). Importantly, many of these membrane elements were captured in the process of budding (or fusing). It is important to note that all these components of the endomembrane system were present during both trophozoite growth and encystation, showing that they were not assembled only for cyst wall biosynthesis.

In contrast, the novel large ESV that transport cyst antigens to the growing wall, are assembled only in encystation. Our ultrastructural analyses suggest that the ESV may have unusual pathway(s) of formation and traffic. For example, early (~5 h) in encystation, sometimes smooth membrane cisternae appeared to be dilated at an edge, and may have been giving rise to larger vesicles, possibly ESV or their precursors. By 15 h of encystation, most cells had ESV, which were often extensive and filled much of the cytoplasm. Clearly, more work is needed to elucidate the formation and traffic of this novel organelle and transport pathway.

Biochemical support for an endomembrane transport system includes the recent isolation of a giardial ARF gene homologue (Lee *et al.*, 1992; Murtagh *et al.*, 1992), encoding an active GTP-binding ADP-ribosylation factor which is thought to be involved in vesicular protein transport between Golgi cisternae. Moreover, Gupta *et al.* (1994) recently isolated the gene for a giardial HSP 70 homologue (BiP), which in higher cells is a resident ER protein. Interestingly, the giardial BiP has the classic C terminal KDEL signal for ER retention. Taken together, these findings support the idea that *Giardia*, a primitive parasite, has complex organellar components and pathways for a highly developed secretory pathway.

We propose that the smooth membranes seen here are a giardial Golgi complex, although it may be simpler or have fewer structural compartments or functions than those of higher cells. Our studies suggest that *Giardia* has at least rough ER transitional elements, tubular-vesicular clusters, smooth Golgi membrane stacks, as well as trans-Golgi network functions (Reiner *et al.*, 1990; McCaffery *et al.*, 1994). Both the smooth membrane stacks and clusters may be manifestations of a dynamic and/or polymorphic Golgi complex. We also observed, for the first time, cytoplasmic 50–80 nm vesicles that resembled the 50 nm intercisternal Golgi transport vesicles of higher organisms (Pryer *et al.*, 1992). Importantly, we also observed vesicular structures budding from (or fusing with) others. Therefore, *Giardia* appears to have the morphological structures of the protein transport system of higher eukaryotes. Further biochemical, ultrastructural and molecular studies are needed to define its composition and complexity.

Acknowledgements

This work was supported by grants Al 24285, Al 19863, AM35108 and HD29275 from the National Institutes of Health. We are most grateful to Dr Marilyn G. Farquhar for support and for ongoing critique of this work, to Drs W. Balch, T. Hobman and C. Davis for critique of the manuscript, and to Tammie McQuistan for photographic assistance.

References

Alberts, B., Bray, D., Lewis, J., Raff, M., Roberts, K. and Watson, J.D. (1989) Intracellular sorting and maintenance of cellular compartments. *Molecular Biology of the Cell*, 2nd edn. Garland Publishing, New York, pp. 405–479.

Aley, S. B. and Gillin, F. D. (1993) *Giardia lamblia*: post-translational processing and status of exposed cysteine residues in TSA 417, a variable surface antigen. *Experimental Parasitology* 77, 295–305.

Balch, W.E., McCaffery, M., Plutner, H. and Farquhar, M.G. (1994) Vesicular stomatitis virus glycoprotein is sorted and concentrated during export from the endoplasmic reticulum. *Cell* 76, 841–852.

Brugerolle, G. (1975) Contribution a L'etude cytologique et phyletique des diplozoaires (zoomastigophorea, diplozoa, dangeard 1910). V. Nouvelle interpretation de l'organisation cellulaire de *Giardia*. *Protistologica* XI, 99–109.

Campbell, J.D. and Faubert, G.M. (1994) Recognition of *Giardia lamblia* cyst-specific antigens by monoclonal antibodies. *Parasite Immunology* 16, 211–219.

Farquhar, M.G. (1985) Progress in unraveling pathways of Golgi traffic. *Annual Review of Cell Biology* 1, 447–488.

Faubert, G., Reiner, D.S. and Gillin, F.D. (1991) *Giardia lamblia*: Regulation of secretory vesicle formation and loss of ability to reattach during encystation *in vitro*. *Experimental Parasitology* 72, 345–354.

Feely, D.E. and Dyer, J.K. (1987) Localization of acid phosphatase activity in *Giardia lamblia* and *Giardia muris* trophozoites. *Journal of Protozoology* 34, 80–83.

Friend, D.S. (1966) The fine structure of *Giardia muris*. *Journal of Cell Biology* 29, 317–332.

Gillin, F.D., Hagblom, P., Harwood, J., Aley, S.B., Reiner, D.S., McCaffery, M., So, M. and Guiney, D.G. (1990) Isolation and expression of the gene for a major surface protein of *Giardia lamblia*. *Proceedings of the National Academy of Sciences of the United States of America* 87, 4463–4467.

Gupta, R.S., Aitken, K., Falah, M. and Singh, B. (1994) Cloning of *Giardia lamblia* heat shock protein HSP70 homologs: Implications regarding origin of eukaryotic cells and of endoplasmic reticulum. *Proceedings of the National Academy of Sciences of the United States of America* 91, 2895–2899.

Holberton, D.V., Crossley, R.C., Marshall, J. and Clark, J.T. (1986) Immunocytochemical differentiation of microtubules in the cytoskeleton of *Giardia lamblia* using monoclonal antibodies to a-tubulin and polyclonal antibodies to associated low molecular weight proteins. *Journal of Cell Science* 80, 233–252.

Karnovsky, M.J. (1971) Use of ferrocyanide-reduced osmium tetroxide in electron microscopy. *Proceedings of the 14th Annual Meeting American Society of Cell Biology*, pp. 146–147.

Lee, F.J., Moss, J. and Vaughan, M. (1992) Human and *Giardia* ADP-ribosylation factors (ARFs) complement ARF function in *Saccharomyces cerevisiae*. *Journal of Biological Chemistry* 267, 24441–24445.

McCaffery, J.M. and Gillin, F.D. (1994) *Giardia lamblia*: Ultrastructural bases of protein transport during growth and encystation. *Experimental Parasitology* 79, 226–235.

McCaffery, J.M., Faubert, G.M. and Gillin, F.D. (1994) *Giardia lamblia*: traffic of a trophozoite variant surface protein and a major cyst wall epitope during growth, encystation, and antigenic switching. *Experimental Parasitology* 79, 236–249.

Meng, T.-C., Hetsko, M.L. and Gillin, F.D. (1993) Antigenic switching of TSA 417, a trophozoite variable surface protein, following completion of the life cycle of *Giardia lamblia*. *Infection and Immunity* 61, 5394–5397.

Murtagh, J.J. Jr, Mowatt, M.R., Lee, C.-M., Lee, F.-J.S., Mishima, K., Nash, T.E., Moss, J. and Vaughan, M. (1992) Guanine nucleotide-binding proteins in the intestinal parasite *Giardia lamblia*. *Journal of Biological Chemistry* 267, 9654–9662.

Nash, T.E., Gillin, F.D. and Smith, P.D. (1983) Excretory-secretory products of *Giardia lamblia*. *Journal of Immunology* 131, 2004–2010.

Owen, R.L. (1980) The ultrastructural basis of *Giardia* function. *Transactions of the Royal Society of Tropical Medicine and Hygiene* 74, 429–433.

Peattie, D.A., Alonso, R.A., Hein, A. and Caulfield, J.P. (1989) Ultrastructural localization of giardins to the edges of disk microribbons of *Giardia lamblia* and the nucleotide and deduced protein sequence of alpha giardin. *Journal of Cell Biology* 109, 2323–2335.

Pryer, N.K., Wuestehube, L.J. and Schekman, R. (1992) Vesicle-mediated protein sorting. *Annual Reviews of Biochemistry* 61, 471–516.

Reiner, D.S., Douglas, D. and Gillin, F.D. (1989) Identification and localization of cyst-specific antigens of *Giardia lamblia*. *Infection and Immunity* 57, 963–968.

Reiner, D.S., McCaffery, M. and Gillin, F.D. (1990) Sorting of cyst wall proteins to a regulated secretory pathway during differentiation of the primitive eukaryote, *Giardia lamblia*. *European Journal of Cell Biology* 53, 142–153.

Reiner, D.S., Hetsko, M.L., Das, S., Ward, H.D., McCaffery, M. and Gillin, F.D. (1993) *Giardia lamblia*: Absence of cyst antigens and reduced secretory vesicle formation and bile salt uptake in an encystation-deficient subline. *Experimental Parasitology* 77, 461–472.

Rothman, J.E. and Orci, L. (1990) Movement of proteins through the Golgi stack: a molecular dissection of vesicular transport. *Federated American Societies of Experimental Biology Journal* 4, 1460–1468.

Schupp, D.G., Januschka, M.M., Sherlock, L.A.F., Stibbs, H.H., Meyer, E.A., Bemrick, W.J. and Erlandsen, S.L. (1988) Production of viable *Giardia* cysts *in vitro*: Determination by fluorogenic dye staining, excystation, and animal infectivity in the mouse and the mongolian gerbil. *Gastroenterology* 95, 1–10.

Simionescu, N. and Simionescu, M. (1976) Galloglucoses of low molecular weight as mordant in electron microscopy. I. Procedure and evidence for mordanting effect. *Journal of Cell Biology* 70, 606–621.

Sogin, M.L., Gunderson, J.H., Elwood, H.J., Alonso, R.A. and Peattle, D.A. (1989) Phylogenetic meaning of the kingdom concept: an unusual ribosomal RNA from *Giardia lamblia*. *Science* 243, 75–77.

Ward, H.D., Kane, A.V., Ortega-Barrier, E., Keusch, G.T. and Pereira, M.E.A. (1990) Identification of developmentally regulated *Giardia lamblia* cyst antigens using GCSA-1, a cyst-specific monoclonal antibody. *Molecular Microbiology* 4, 2095–2102.

12

Metabolic Disturbances in Children with Chronic Giardiasis

N.N. TANELI

Dokuz Eylül University, Faculty of Medicine, Department of Pediatrics, Izmir, Turkey

In a letter to the London Society of Medicine, dated 4 November 1681 A. van Leeuwenhoek wrote what seems to be the first account of *Giardia* (*G*) as a human intestinal protozoan. However, *G* was first fully described by Lambl in 1859, and thus named *Giardia lamblia* (Dobell, 1920). The protozoan was initially thought to be a non-pathogenic inhabitant of the intestinal tract. Its true pathogenicity had not been established even as late as the 1960s (Mata *et al.*, 1985). Henceforth, *G* drew attention as being just another intestinal parasite causing unimpressive outbreaks in closed communities (Pickering and Engelkirk, 1988). Nevertheless, during the last two decades *G* became a menace for all underprivileged societies where sanitation is poor, water supplies are contaminated, and waste disposal, food handling and personal hygiene inappropriate. Mata *et al.* (1985) reported that in Latin America more than ten million children have two or more attacks of *G* infection per year. Nowadays, it is considered to be the most common human enteric protozoan parasite in temperate and tropical regions of the world (Petersen, 1972). Lately, some sophisticated methods, such as analyses of DNA endonuclease restriction patterns, evaluations of surface variant antigens, measurements of homologous and heterologous antibody responses, scanning electron microscopic studies of upper bowel biopsy specimens, antigen identification and isozyme evaluation techniques, are used in the identification of this formerly 'innocent' protozoan.

Where such a delicate ecological balance is disturbed, cysts are food- or water-borne, a priori, and *G* transmission is by person-to-person contact in the paediatric age group, especially via their caretakers (Keystone *et al.*, 1978). Although some unsuspected and untreated cases of chronic giardiasis last until adolescence, *G* infection is predominantly a childhood disease. Incidence is lower in the older age group and in long-term residents of endemic areas due to an acquired immune response to the protozoan (Rendtorff, 1954; Pickering and Woodward, 1982). Morbidity is related to several factors: (i) the number of cysts ingested; (ii) its specific strain; (iii) the non-immunologic protective factors in the gastrointestinal tract; and (iv) the immune response of the host to the organism. It is unfortunate that even as few as ten cysts are sufficient to initiate giardiasis (Rendtorff, 1954). Diversity among surface antigens of isolates can result in variable infectivity characteristics and immunological response of the host and therefore can present with varied clinical manifestations (Pickering and Engelkirk, 1988). Partially protective immunity to *G* develops as a host immune response and has both cellular and humoral components. Intestinal fluid immunoglobulin (Ig) increases in *G* infection; since *G* is predominantly a luminal parasite, gastrointestinal secretory (Sec) antibodies gain importance. Cellular immune response is produced by anti-*Giardia* Sec IgA and the mononuclear phagocytes ingest and kill *G* and produce *G* antigen in the Peyer's patches (Jones and Brown, 1976; Owen *et al.*, 1981).

In the gastrointestinal tract *G* is exposed to an ever-changing milieu of foods, digestive enzymes and luminal products toxic to *G*. Trophozoites are killed by *cis*-unsaturated free fatty acids, bacteria-induced deconjugated bile acids, and by the killing action of bile-stimulated lipase of human milk in exclusively breast-fed children. That clears the puzzle why *G* infection is not seen in exclusively breast-fed infants. However, in most underprivileged societies, where working mothers cannot sustain breast-feeding, giardiasis begins at weaning and is most prevalent during the first 5 years of age (Mata *et al.*, 1985); whereas it was interesting to see that all the cases with chronic giardiasis in our study groups were of primary school age or older. Non-mucine fraction of mucus and bile promote trophozoite colonization. While excystation occurs in an acidic environment, optimum colonization is possible only at 6.4–7.0 pH; which explains why they are found in profuse amounts in duodenal aspirates (Farthing, 1985). Besides, *G* flourish by competing with the host for ingested nutrients, especially carbohydrates. Reduction of carbohydrate content of the diet has detrimental effects on the parasite (Gillin and Diamond, 1988). *G* has an absolute requirement for a low oxidation/reduction potential which is provided by cysteine. Cysteine has an additional role on trophozoites as a growth promoter due to cysteine-dependent zinc binding to its protein membrane (Zhang *et al.*, 1993). Therefore, trophozoites also depend on nutrient zinc for flourishing (Schneider, 1961).

To review previous observations, Poley's (1985) scanning electron microscopic studies in normal children and in children with chronic giardiasis is recommended.

In normal children, the upper bowel villi pattern appears like tongues and ridges, uniform in height, free of mucus, perhaps with occasional creamy tufts scattered here and there. In giardiasis, at the onset, the whole mucosa is covered by a tenacious-looking mucus. Nevertheless, ridges and enterocytes stay intact under the heavy veil of mucus. Initially *G* trophozoites could be trapped in the tenacious net; however, that obstacle is easily abolished, since for easy anchorage trophozoites are adorned with ventral sucking discs and sawtooth-like flanges skirting peripherally. Previously, the vigorous beating of the ventrally situated pair of flagellae were presumed to establish the sucking force (Holberton, 1974). Later, tubulin of the disc microtubular structure, the contractile proteins and the lectin-like activity of the surface membrane were introduced as contributors to the adherence mechanism (Farthing *et al.*, 1982; Feely *et al.*, 1982). When anchorage is maintained, trophozoites may cover the mucosal surface in such great numbers that physically they block the absorptive function of the villi and actually compete for host nutrients; occasionally they may even invade villi crypt cells and the lamina propria.

As infection becomes chronic villi appear blunted and more widely spaced, with mucus still covering the surface like a crust or pseudomembrane, a phenomenon not observed in any other condition, except perhaps to lesser degree in cystic fibrosis. Excessive mucus, entrapped debris, constituents from extruded enterocytes and dead trophozoites are said to form this heavy, mucoid membrane (Poley, 1982), and if trophozoites are covered by that heavy blanket after anchorage, they would leave the imprints of their sucking discs long after they, themselves, were eliminated. Previously, such imprints had not been detected in humans by light microscopy techniques (Brunser and Araya, 1985).

Depending on the severity of *G* infection the absorptive, secretory and reabsorptive capacities for water and electrolytes, carbohydrates, fat and proteins, vitamins and trace elements diminish, resulting in a generalized malabsorptive state (Hoskins *et al.*, 1967), leading to malnutrition and failure to thrive. The sequential evolution of the metabolic disturbances in children with giardiasis are illustrated in Box 12.1.

When Box 12.1 is studied, it is easily seen that the main culprit is excessive mucus. When mucus totally covers the mucosal surface, it not only inhibits nutrient perfusion it also inhibits contact between carbohydrates and the brush-border enzymes. Thus, even in view of the normal disaccharidase assay the child becomes intolerant to mostly lactose and sucrose. High luminal lactose initiates osmotic diarrhoea and ends with profuse loss of luminal contents, e.g. water, nutrients and bile.

Box 12.1. Chronic giardiasis-triggered metabolic disturbances in children leading to malnutrition and failure to thrive.

Mucus↑ → CHO contact↓ → Luminal lactose↑ → **Osmotic diarrhoea** → Water, electrolytes, nutrients and bile↓

Bacteria↑ → Glucosidase activity↑ → Upper gut breakdown of mucin → **Mucosal injury**
Bile deconjugation↑
H_2O_2 production↑

Blunted villi → Disaccharidase↓ → **Lactase**↓

Permeability↑ → Macromolecule absorption↑ → Hypersensitivity↑

Gut hormone release↓ → **Pancreas** → Amylase↓ → **Malabsorption**
Secretin↓ Protease↓
Pancreozymin↓ Lipase↓
CCK↓ Sec **Zn, Se**↓

Enterohepatic circulation↓ → Gall bladder↓ → Bile pool↓ → **Steatorrhoea**

Bold type indicates important metabolic disturbances.
↑, Increase; ↓, decrease; CCK, cholecystokinin; CHO, carbohydrate; H_2O_2, hydrogen peroxide; Se, selenium; Sec, secretory; Zn, zinc.

Mucus also paves the way to secondary bacterial adhesion. Bacterial glucosidases, the noxious detergent effect of the bacteria-induced deconjugation of bile acids and bacterial toxic by-products, e.g. hydrogen peroxide, break the mucin coverage of the upper intestine and produce the well-documented mucosal injury seen in chronic giardiasis (Tandon *et al.*, 1974).

Crypts contain stem cells which mature into enterocytes, goblet cells, entero-endocrine cells and Paneth cells in healthy subjects. Chronic giardiasis will eventually lead to blunting of villi by shedding of the above-mentioned mature and competent mucosa cells. Increased mytotic activity and epithelial turnover end up with surfacing of immature cells and reduced digestive-absorptive function leading to impairment of brush-border disaccharidase activity. The one which diminishes the first would be lactase due to its spatial position on the lipid bilayer of the mucosal membrane (Loovard *et al.*, 1975). Only then, depending on the severity of blunting, could a diminished lactase activity be assayed. However, blunting is usually mild in giardiasis; flat mucosa is only rarely seen during parasitic invasion in an immuno-compromised host (Lebenthal, 1985). Breakdown of mucin reduces the physical barrier of the intestinal mucosa and enhances direct contact between food materials and the mucosal surface. Breaks in the microvilli surface coat function as entry points; increased permeability permits macromolecule absorption, acquisition of foreign antigens and thus provides the suggested link between giardiasis and hypersensitivity (Brunser and Araya, 1985).

Small bowel mucosal injury also impairs release of enteric hormones, e.g. secretin and pancreozymin-cholecystokinin (CCK). Such a deficiency in secretegogues leads to secondary pancreatic exocrine deficiency (Amini, 1963; Gupta and Mehta, 1973), and to amylase, protease and lipase deficiencies, accordingly; these, in turn, lead to carbohydrate, fat and protein malabsorptions. In addition to diminished absorption of, and competition for, zinc between parasite and host, endogenous secretions of trace elements also decrease in giardiasis (Yasso, 1983).

Excessive spillage of bile overcoming colonic salvage is lost alongside other luminal contents through osmotic diarrhoea. Reduction in the bile acid pool interrupts enterohepatic circulation. Sluggish emptying of the gall bladder complements the bile acid pool shortage, leading to decreased micellar lipid formation and ends up with overt steatorrhoea of giardiasis (Auricchio, 1984).

Thus, availability, transport and utilization of the principal nutrients are impaired in chronic giardiasis. When components of energy metabolism for the maintenance of tissue synthesis, stored energy and energetics of growth are insufficient, malnutrition becomes florid and failure to thrive is the ultimate inevitable outcome.

In our clinic, in different study groups of children with chronic giardiasis, the expected losses of carbohydrates, proteins, fats, some fat-soluble vitamins, e.g. vitamins A and E, trace elements, e.g. copper, zinc and selenium, and a trace-dependent enzyme, e.g. glutathione peroxidase, were studied to disclose the disturbances that tip the delicate balance of homeostasis. All our cases were studied on outpatient bases. Stool analyses revealed profuse discharges of cysts and trophozoites. After an oral load of glucose, 38.8% of cases showed lactose intolerance both during bouts of watery diarrhoea and during the intervals of defaecating foul-smelling, greasy stools (Taneli *et al.*, 1986). In contrast, the local lactose intolerance prevalence is only 8% (Taneli *et al.*, 1988a). Since clinical and laboratory manifestations showed normalizations within 15 to 20 days after beginning specific therapy for giardiasis (metronidazole 20 mg kg^{-1} for 10 days) duodenal biopsy and lactase assaying were not performed in our study cases. Faecal fat loss was found to be 30–90% by the steatocrite method (Erdem and Uysal, 1987). However, it was interesting to see that all our normal control groups who were consuming ordinary Turkish cuisine did not discharge any fat by the steatocrite method.

With so much fat lost through stools one expects quite a loss of fat-soluble vitamins too. With this idea in mind, serum vitamin E (Taneli *et al.*, 1989a) was assayed and found it to be significantly ($P < 0.020$) deficient. However, when vitamin A was taken into consideration, neither clinical examinations nor field surveys by (1% Rose Bengal-containing) eye drops (Kusin *et al.*, 1979) supplemented with biomicroscopic ophthalmologic studies disclosed any clinical deficiency manifestations in the eyes of the study group, e.g. xerophthalmia or the lesser finding, conjunctival rippling (Taneli *et al.*, 1989b). Even the serum carotene level studied in one of our most chronic cases was found to be around the lowest level, not below (Büyükgebiz *et al.*, 1993). This is not surprising a regional phenomenon in fact, attributed to ingestion of local foods rich in vitamin A. Nevertheles, in regions where fruits and vegetables are scarce, vitamin A loss through steatorrhoea would be expected to be hazardous.

Although peptide digestion and absorption of the small intestinal mucosa have a large reserve capacity, brush-border peptidases are reported to be reduced in giardiasis (Auricchio, 1984). In one study (Taneli *et al.*, 1988b) significant ($P < 0.02$) protein loss in chronic giardiasis was disclosed.

Zink loss in chronic giardiasis through diarrhoeal wash-out, results in reduced pancreatic exocrine secretions, desquamation of mucosal cells rich in zinc, competition between host and parasite, and diminished absorption due to colonization of the trophozoites at its absorptive site, the duodenum. Nutrient zinc, besides being a growth

Table 12.1. Predicted losses in children with chronic giardiasis.

	Control ± SEM	Giardiasis ± SEM	*t*	*P*
Plasma zinc (mg dl^{-1})	0.785 ± 0.048	0.578 ± 0.035	2.872	<0.01
Plasma selenium (μg l^{-1})	19.200 ± 1.245	12.000 ± 0.743	5.254	<0.001
Glutathione peroxidase (μg Hb^{-1})	63.980 ± 2.988	16.692 ± 0.593	14.593	<0.0001
Serum total α-tocopherol (mg dl^{-1})	1.17 ± 0.06	0.93 ± 0.06	2.55	<0.02

Control cases, 12; cases with chronic giardiasis, 10.

promoter for the trophozoites, is also important for their survival and colonization. In addition, as a metal cofactor, zinc plays a major role in the function of gustin in human beings in promoting the trophic effect on growth and development of taste buds. Therefore, in zinc-deficient states loss of the sense of taste ensues which leads to anorexia and inability to gain weight and is soon followed by overt malnutrition (Anal *et al.*, 1993a). Failure to thrive is more conspicuous in boys since zinc is important for testis maturation (Prasad *et al.*, 1963). With this concept in view, the predicted losses from giardiasis were studied in our department (Table 12.1).

When zinc status was studied, it was found to be significantly ($P < 0.01$) low; however, copper status was found to be unaltered ($P > 0.05$) (Özyürek *et al.*, 1993). In a consecutive study zinc was found to be depleted even in whole blood (Anal *et al.*, 1993b). Selenium deficiency was also expected since heavy wash-out and reduced absorption would have affected the selenium status. As seen in Table 12.1, significant ($P < 0.001$) selenium deficiency was disclosed in another of our studies (Anal *et al.*, 1993c). Since selenium is a cofactor to glutathione peroxidase, the enzyme, glutathione peroxidase, was studied simultaneously and its activity was also found to be significantly ($P < 0.0001$) lowered (Anal *et al.*, 1993c). In countries with selenium-poor provinces, e.g. Keshan and Beijing (Peking) in the People's Republic of China, one could easily comprehend the danger which chronic giardiasis might cause. Vegetables and staple food grown in such leached fields and the meat of cattlestock fed on such feeds would all be poor in selenium. However, some plants are natural sources of selenium, e.g. wheat. Selenomethionine-containing wheat is known to compensate for selenium deficiency (Anal *et al.*, 1993d) where people consume a lot of wheat bread (Yüregir, 1985) and omnivorous food as in the case in Turkey.

However, rice and maize are poor in view of selenium content.

Although serum selenium in the inhabitants of Central Anatolia (Hincal *et al.*, 1991) and the Aegean Province are normal (Anal *et al.*, 1993d), it is quite low in the Black Sea Province where people prefer mostly maize (Başaran, *et al.*, 1992). Thus, it is easy to see that people living in regions where fields are poor in selenium and where selenium-poor food is consumed should be alerted in view of grave selenium shortages where G is endemic.

When we look at the profile of the deficient elements in chronic giardiasis with scrutiny (Table 12.1), we see that all the losses can be grouped as anti-oxidants. Thus an overt anti-oxidant deficiency state is evident in children with chronic giardiasis. In an aerobic organism, free radicals (recently cited as reactive oxygen species) and anti-oxidants are in an actively balanced state; but increased formation of the former, and increased uptake or loss of the latter would tip this precarious balance easily. Besides, in a biological media, the presence of even a minute amount of peroxide and activation of peroxidase, accordingly, would stimulate lipoxygenase activity and induce production of the putative 'injury mediators'. The activated neutrophils, in turn, would lead to production of more free radicals such as superoxide ($O_2^- \cdot$) and hydrogen peroxide (H_2O_2), hypochlorous acid (HCPO) and nitrous oxide (NO) which would be noxious to the organism (Halliwell *et al.*, 1992). Such a triggering effect should be expected in giardiasis since increased radical formation is evident, e.g. H_2O_2 through the invading bacteria-induced toxic end-products, while glutathione peroxidase is known to be activated locally to compensate it and it is therefore consumed excessively (Anal *et al.*, 1993c). With this concept in view, Taneli, B. *et al.* (1994) showed significant increases of faecal leucotrienes, e.g. LTB4 and LTC4, and hydroxyeicosotetraenoic acid, e.g. 5-HETE, and phagocyte and neutrophil chemotaxis in children with chronic giardiasis. As 'injury mediators' LTC4 produces local ischaemia while LTB4 produces local inflammation. LTB4 and 5-HETE induce local leucocyte activation and degranulation which sustain this inflammation and activated leucocytes lead to further production of free radicals. Thus, excessive production of free radicals in giardiasis is important, a fact not noted until recently.

Zinc, by itself, is effective in controlling lipid peroxidation (Farthing, 1985). In addition, it is a metal cofactor to cytosolic superoxide dismutase (Zn/Cu SOD) which is an effective anti-oxidant, *in vivo*. Selenium is also a cofactor to both mitochondrial and cytosolic glutathione peroxidase and vitamins A, E and C are valuable anti-oxidant adjuncts. Anti-oxidants are more effective if they are present as essentials of the body, available at the site during radical generation. Supplementation after establishment of the pathology usually has little therapeutic value (Halliwell *et al.*, 1992).

Anti-oxidant adjuncts, such as vitamins A and E, are expected to be low in regions where fresh and dried fruits and vegetables are not readily obtainable. Also, watery diarrhoea of giardiasis might also enhance vitamin C loss, another complementary anti-oxidant. Our patients are doubly lucky since the Aegean Province of Turkey is rich in citrus foods (Taneli and Argun, 1973).

Superoxide and hydrogen peroxide are only potentially dangerous radicals; however, if there is even a minute amount of catalytic transient metal, e.g. iron, in the media, then the iron-catalysed Haber–Weiss reaction:

$$O_2^{-}\cdot + H_2O_2 \xrightarrow{Fe(II)} OH^{-} + O_2 + OH\cdot$$

would take place and these potentially dangerous radicals would gain kinetics and form the most potent one, the hydroxyl radical ($OH\cdot$) which accounts for damage to DNA and the phospholipid membranes of the cells (Halliwell *et al.*, 1992).

Iron deficiency anaemia is said to be a constant feature of *G* infection. Actually, in one of our study series (Taneli *et al.*, 1992), serum iron was found to be significantly ($P < 0.02$) reduced. However, it was amazing to see that the chronic giardiasis-induced iron deficiency anaemia was quickly abolished shortly after the institution of the specific therapy for *G* infection. Absence of overt gastrointestinal bleeding in giardiasis, ingestion of iron-containing foods sparingly by the anorexic host and the minute amount of catalytic iron released by the shed mucosa cell membranes during villi blunting are poor sources of 'low-molecular-mass-pool of iron'. Therefore the accumulated $O_2^{-}\cdot$ and H_2O_2 ions might falsely be assumed to have a limited damaging effect. However, treating chronic giardiasis-induced iron-deficient patients with ferrous or ferric salts would be hazardous, since iron salts would be the trigger mechanism; indeed, in activating the above-mentioned iron-catalysed Haber–Weiss reaction, *in vivo*. Thus, iron salts should never be instituted in giardiasis. Since zinc, which protects against iron-catalysed radical damage, has been shown to be diminished in giardiasis, vitamin E, if present at the site, would at least delay the damage (Halliwell *et al.*, 1992).

$$\text{Lipid-}O_2^{-}\cdot + \text{tocopherol-OH} \longrightarrow \text{Lipid-}O_2H + \text{tocopherol}\cdot$$

Unfortunately, tocopherol, itself, becomes activated during such a reaction. Nevertheless, ascorbic acid, being the most important anti-oxidant in human plasma, removes the activated tocopherol but becomes quickly depleted if not routinely supplemented in the diet.

$$\text{Tocopherol}\cdot + \text{ascorbate} \longrightarrow \text{Semidehydroascorbate} + \text{tocopherol-OH}$$

It is important to remember that ascorbate should also be available as a nutritional essential at the site of lipid peroxidation. Extra administration of vitamin C has little therapeutic effect as is the case in vitamin E (Halliwell *et al.*, 1992).

Although our study cases with chronic giardiasis are fortunate enough to live in a region abundant in foods rich in vitamins A and C, it is easy to comprehend that in regions where fresh fruits and vegetables are not adequate, compensating vitamins should be taken into consideration in food-donation programmes.

The majority of carcinogens have been found either to be free radicals themselves or are involved in or produce pathological free radical reactions. However, there are several mechanisms functioning against the effects of carcinogens. The most important one being the protective actions of the anti-oxidants. Many chemical carcinogens in the ecosystem need nitrosifying agents, amine and amide substrates, to reach to their terminal forms. Anti-oxidants compete for these nitrosifying agents or, at least, prevent formation of *N*-nitroso carcinogens (Fehér *et al.*, 1987). If *G* is capable of inducing the enhancement of free radical formation assumed by the above-mentioned anti-oxidant shortage (Taneli, N.N. *et al.*, 1994), one must be aware that chronic giardiasis might also be a potential carcinogen. Thus, giardiasis as a condition inducing the decrease of, or accelerated consumption of, anti-oxidants must be handled with this new concept.

Our studies in chronic giardiasis still continue. Since malonedialdehyde, which is known to have mutagenic and carcinogenic effects, is one of the terminal products of lipid peroxidation an ongoing study in our department is studying malonedialdehyde and Cu/Zn SOD in chronic giardiasis. Catalase activity was found to be normal in giardiasis (N.N. Taneli, unpublished data, 1994), nevertheless, its activity was expected to be low in giardiasis; because catalase is activated only when free radicals have accumulated in excessive amounts (Fehér *et al.*, 1987).

There is a tendency in not treating silent cyst shedders in older age groups and lately there is a trend in sustaining this policy in the paediatric age group too. However, a child with soft systemic manifestations such as anorexia, failure to gain weight and few bouts of watery diarrhoea could easily be overlooked by inattentive parents and these complaints would not be included at history giving. If *G* cysts are encountered in a routine stool examination in such children, we should consider them important. The cysts may be evidence of a well-harboured stock of trophozoites. Therefore, detailed history taking and a thorough physical examination are mandatory in prevalent areas. The aim of the present study on metabolic alterations in children with chronic giardiasis is to provide a new insight into the field of free radicals in medicine. My purpose is to highlight the importance of chronic giardiasis in paediatric and community practice. My approach to chronic giardiasis might appear theoretical; but I hope it will open new horizons for eventual and more comprehensive studies in giardiasis.

Conclusion

Since cysts and trophozoites are more commonly absent in the stool than present, absence of *G* trophozoites should not deter us in pursuing giardiasis. Even during active infection some patients might present with mild gastrointestinal manifestions but the predominant metabolic disturbances, e.g. anorexia, inability to gain weight, failure to thrive, even disturbed taste sensation or intolerance to milk, may easily be overlooked at the onset and such patients may be considered as asymptomatic. Therefore, health supervisors in underprivileged societies, where *G* is prevalent, should be alerted to the above-mentioned metabolic disturbances, and be warned against iron supplementation in iron deficiency anaemia of chronic giardiasis.

References

Amini, F. (1963) Giardiasis and steatorrhea. *Journal of Tropical Medicine and Hygiene* 66, 190–192.

Anal, Ö., Taneli, N.N., Ören, B. and Kavukçu, S. (1993a) Self-monitored zinc supplementation in mild, moderate and severe zinc deficiency states. Proceedings of the Fourth International Conference on Health and Disease: Effects of Essential and Toxic Trace Elements, 8–12 February 1993, New Delhi (abstract). *International Journal of Toxicology, Occupational and Environmental Health* 2, 110–111.

Anal, Ö., Erdem, N., Gezer, S., Şen, A. and Töre, I.R. (1993b) Zinc depletion in chronic giardiasis. Proceedings of the Fourth International Conference on Health and Disease: Effects of Essential and Toxic Trace Elements, 8–12 February 1993, New Delhi (abstract). *International Journal of Toxicology, Occupational and Environmental Health* 2, 150.

Anal, Ö., Gezer, S., Şen, A., Güner, G. and Taneli, N.N. (1993c) Selenium status in chronic giardiasis. Proceedings of the Fourth International Conference on Health and Disease: Effects of Essential and Toxic Trace Elements, 8–12 February 1993, New Delhi (abstract). *International Journal of Toxicology, Occupational and Environmental Health* 2, 82–83.

Anal, Ö., Yenigün, A., Gezer, S., Güner, G. and Taneli, N.N. (1993d) Regional selenium status of Turkey (the Aegean Province). Proceedings of the Fourth International Conference on Health and Disease: Effects of Essential and Toxic Trace Elements, 8–12 February 1993, New Delhi (abstract). *International Journal of Toxicology, Occupational and Environmental Health* 2, 39.

Auricchio, S. (1984) Peptid digestion and absorption in the small intestinal mucosa during acute and chronic diarrhea. In: Lebenthal, E. (ed.) *Chronic Diarrhea in Children*. Vevey/Raven Press, New York, pp. 179–191.

Başaran, N., Çeliker, A., Bilgiç, A., Hincal, F., Özbarlas, N. and Baysal, K. (1992) Selenium levels of children with dilated cardiomyopathy in Turkey. Proceedings of the Fourth International Conference on Health and Disease: Effects of Essential and Toxic Trace Elements, 8–12 February 1993, New Delhi (abstract). *Journal of Toxicology, Occupational and Environmental Health* 2, 86–87.

Brunser, O. and Araya, M. (1985) Damage and repair of small intestinal mucosa in acute and chronic diarrhea. In: Lebenthal, E. (ed.) *Chronic Diarrhea in Children*. Vevey/Raven Press, New York, pp. 31–56.

Büyükgebiz, B., Eroğlu, Y. and Özer, E. (1993) A case of chronic giardiasis which could only be diagnosed by upper intestinal biopsy (in Turkish). *Dokuz Eylül Universitesi Tip Fakültesi Dergisi* 7, 52–55.

Dobell, C. (1920) The discovery of the intestinal protozoan of man. *History of Medicine* 13, 1–15. In: Pickering L. and Engelkirk P.G. (1988) *Giardia lamblia. Pediatric Clinics of North America* 35, 565–577.

Erdem, N. and Uysal, I. (1987) Measurements of fat loss by steatocrite method in giardiasis. *Journal of Dokuz Eylül University Faculty of Medicine* 2, 103–107.

Farthing, M.J.G (1985) Giardiasis: pathogenesis of chronic diarrhea and impact on clinical growth and development. In: Lebenthal, L.E. (ed.) *Chronic Diarrhea in Children*. Vevey/Raven Press, New York, pp. 237–252.

Farthing, M.J.G., Pereira, M.E.A. and Keusch, G.T. (1982) Cell surface lectin of *Giardia lamblia* (abstract). *Clinical Research* 30, 496 A.

Feely, D.E., Schollmeyer, J.V. and Erlandsen, S.L. (1982) *Giardia* spp.: distribution of contractile proteins in the attachment organelle. *Experimental Parasitology* 53, 145–154.

Fehér, J., Csomós, G. and Vereckei, A. (1987) *Free Radical Reactions in Medicine*. Springer-Verlag, Berlin, pp. 7–151.

Gillin, F.D. and Diamond, L.C. (1988) *Entemoba histolytica* and *Giardia lamblia*: growth responses to reducing agents. *Experimental Parasitology* 51, 282–291.

Gupta, R.K. and Mehta, S. (1973) Giardisis in children: a study of pancreatic functions. *Indian Journal of Medical Research* 61, 743–748.

Halliwell, B., Gutteridge, J.M.C. and Cross, C.E. (1992) Free radicals, antioxidants and human disease: where are we now? *Journal of Laboratory and Clinical Medicine* 199, 598–818.

Hincal, F., Yetgin, S., Başaran, N. and Ataçeri, N. (1991) Selenium status of Turkey: serum selenium levels of children and adults living in Ankara. In: Yüregir, G.T., Donma, O. and Kayrin, I. (eds.) *Trace Elements in Health and Disease*. Çukurova University Medical Press, Adana, Turkey, pp. 183–187.

Holberton, D.V. (1974) Attachment of giardia – a hiydrodynamic model based on flagellar activity. *Journal of Experimental Biology* 60, 207–221.

Hoskins, I.C., Winaver, S.Y., Broitman, S.A., Gottlieb, L.S. and Zamcheck, N.

(1967) Clinical giardiasis and intestinal malabsorption. *Gastroenterology* 53, 265–279.

Jones, E.G. and Brown, W.R. (1976) Serum and intestinal fluid immunglobulin in patients with giardiasis. *Digestive Diseases and Science* 19, 791.

Keystone, J.S., Krajden, S. and Warren, M.R. (1978) Person to person transmission of *Giardia lamblia* in day-care nurseries. *Canadian Medical Association Journal* 119, 241–258.

Kusin, J A., Soewando, W. and Sinaga, H.S.R. (1979) Rose Bengal and Lissamine Green vital stains: useful diagnostic aids for early stages of xerophthalmia. *American Journal of Clinical Nutrition* 32, 1559–1569.

Lebenthal, E. (1985) Prolonged small intestinal injury as a primary cause of intractable diarrhea of infancy. In: Lebenthal, E. (ed.) *Chronic Diarrhea in Children*. Vevey/Raven Press, New York, pp. 5–29.

Loovard, D., Maroux, S., Vennier, C.H. and Desnvelle, P. (1975) Topological studies on the hydrolases bound to the intestinal brush border membranes. *Biochemistry and Biophysiology Acta* 375, 236–248.

Mata, L., Urrutial, J.J. and Simhon, A. (1985) Infectious agents in acute and chronic diarrhea of childhood. In: Lebenthal, E. (ed.) *Chronic Diarrhea in Children*. Vevey/Raven Press, New York, pp. 237–252.

Owen, R.L., Allen, C.L. and Stevens, D. (1981) Phagocytosis of *Giardia muris* by macrophages in Peyer's patch epithelium in mice. *Infection and Immunity* 33, 591.

Özyürek, R., Köse, G., Taneli, N.N. (1993) Serum zinc and copper levels in giardiasis. Proceedings of the Fourth International Conference on Health and Disease: Effects of Essential and Toxic Trace Elements, 8–12 February 1993, New Delhi (abstract). *Journal of Toxicology, Occupational and Environmental Health* 2, 107.

Petersen, H.C. (1972) Giardiasis (lambliasis). *Scandinavian Gastroenterology* 14 (suppl.), 7–44.

Pickering, L. and Engelkirk, P.G. (1988) *Giardia lamblia*. *Pediatric Clinics of North America* 35, 565–577.

Pickering, L.K. and Woodward, W.E. (1982) Diarrhea in day-care centers. *Pediatric Infectious Diseases* 1, 47–62.

Poley, J.R. (1982) Malabsorption in giardiasis: mucoid pseudomembrane. A scanning electron microscopic study. *Journal of Pediatric Gastroenterology and Nutrition* 1, 63–80.

Poley, J.R. (1985) Ultrastructural topography of small bowel mucosa in chronic diarrhea in infants and children: investigations with scanning electron microscope. In Lebenthal, E. (ed.) *Chronic Diarrhea in Children*. Vevey/Raven Press, New York, pp. 57–105.

Prasad, A.S., Miale, A., Farid, Z., Schulert, A.R. and Sandstead, H.H. (1963) Zinc metabolism in patients with the syndrome of iron deficiency anemia, dwarfism, and hypogonadism. *Journal of Laboratory and Clinical Medicine* 61, 535–537.

Rendtorff, R.C. (1954) The experimental transmission of human intestinal parasites: II. *Giardia lamblia* cysts given in capsules. *American Journal of Hygiene* 59, 209–220.

Schneider, C.C. (1961) *Giardia muris* II. The influence of nutrition and other factors on the course of infection. *Tropical Medicine and Parasitology* 12, 265–285.

Tandon, B.N., Puri, B.K., Ghandi, P.C. and Tevari, S.G. (1974) Mucosal surface injury of jejunal mucosa in patients with giardiasis: an electron microscopic study. *Indian Journal of Medical Research* 62, 1838–1844.

Taneli, B. and Argun, T. (1973) The blood and urine vitamin C levels in children living in the Aegean Region of Turkey (in Turkish). *Ege Üniversitesi Tip Fakültesi Dergisi* 12, 505–510.

Taneli, B., Aydoğdu, S., Altintaş, N., Kasirga, E., Yağci, R.V., Çoker, I. and Hüseyinov, A. (1994) The role of the new inflammatory lipid mediators in the pathogenesis of giardiosis (poster presentation). Proceedings of the VIIIth International Congress of Parasitology, October 10–14, 1994, Izmir, Turkey.

Taneli, N.N., Erdem, N., Özkan, H. and Karamizrak, T. (1986) Lactose intolerance in giardiasis (in Turkish). *Ege Üniversitesi Tip Fakültesi Dergisi* 25, 1093–1086.

Taneli, N.N., Erdem, N. and Ozkan, H. (1988a) Incidence of primary lactose intolerance (in Turkish). *Ege Üniversitesi Tip Fakültesi Dergisi* 26, 563–568.

Taneli, N.N., Erdem, N. and Kayahan, H. (1988b) Protein losses in giardiasis (in Turkish). *Ege Üniversitesi Tip Fakültesi Dergisi* 27, 519–523.

Taneli, N.N., Erdem, N., Töre, R., Güner, G. and Özkan, H. (1989a) Vitamin E status in giardiasis (in Turkish). *Ege Üniversitesi Tip Fakültesi Dergisi* 28, 57–60.

Taneli, N.N., Erdem, N., Kaya, Ş. and Kayahan, H. (1989b) Giardiasis related ophthalmologic disturbances in the Aegean Province of Turkey (in Turkish). *Dokuz Eylül Üniversitesi Tip Fakültesi Dergisi* 4, 84–90.

Taneli, N.N., Köse, G., Özyürek, R. and Özkan, H. (1992) Iron deficiency in giardiasis. *World Health and Child Care* 4, 161–163.

Taneli, N.N., Anal, Ö., Erdem, N., Töre, R. and Güner, G. (1994) Antioxidant deficiencies in chronic giardiasis (in Turkish). *Türk Parazitoloji Dergisi* 18, 6–12.

Yasso, J.G. (1983) Nutrition in cystic fibrosis. In: Dodge, J.A. (ed.) *Topics in Pediatric Nutrition*. Pittman Press, Bath, pp. 212–224.

Yüregir, G. (1985) Problems encountered in trace element studies in developing countries. In: Said, M., Rahman, M.A. and D'silva, I.A. (eds) *Elements in Health and Disease*. MAS Printers, Karachi, Pakistan, pp. 90–101.

Zhang, Y.Y., Alley, S.B., Stanley S.L. Jr and Gillin, F.D. (1993) Cysteine-dependent zinc binding by membrane proteins of *Giardia lamblia*. *Infection and Immunity* 61, 520–524.

13

Amoebae in Relationship with Bacteria in Their Environment

C. HARF

Laboratoire d'Hydrobiologie de l'Environnement, Faculté de Pharmacie, Université Louis Pasteur Strasbourg, B.P. 24–67401, Illkirch Cedex, France

Introduction

Free-living amoebae are widely distributed in water and soil throughout the world. The vegetative forms or trophozoites move with their pseudopodia and multiply by binary fission. Under hostile environmental conditions, the trophozoite is transformed into the cystic resistant form. There are many species of amoebae, some of them are described as amphizoic protozoa, capable of living as parasites, or as free-living. They live in close association with other microorganisms in the same ecological niches, feeding on bacteria and fungi.

Free-living amoebae interact with a variety of bacteria. This interaction ranges from bacteria that serve as food sources or are sequestered as endosymbionts within amoebae to bacterial species that are amplified by amoebae.

Endocytosis

The mechanisms of endocytosis and of digestion are similar to those observed in macrophages. In *Acanthamoeba* bacteria are internalized individually, then grouped in one vacuole; after fusion with lysosomes, creating a phagolysosome, acid pH and lysosomal enzymes lyse the phagocytosed bacteria (Drozanski and Chmielewski, 1979). This similarity suggests that amoebae might be useful as a model to elucidate the interaction of bacteria with human phagocytic cells. However, virulence determinants for infection of, and mutiplication in, amoebae and phagocytes do not appear to be the same (Dowling *et al.*, 1992).

Endocytosis is microfilament independent, in contrast to macrophages (King *et al.*, 1991), and according to Fields (1993) a new terminology is needed to describe bacterial uptake by a microfilament-independent process.

The efficiency of consumption is different, depending on protozoan type and growth state, temperature, bacterial abundance, and size and nature of bacterial prey (Gonzalez *et al.*, 1990). Many of the bacteria are digested, especially enterobacteria which are used as nutrients. Gram-positive bacteria seem to be digested more slowly, owing to their thick cell wall.

Escaping Digestion

Some bacteria may avoid digestion by being liberated from the phagosome before lysosomal fusion or by resisting the cellular mechanisms of digestion in phagosomes and escaping into the host cell cytoplasm where they can multiply (Hall and Voelz, 1985; Reisser *et al.*, 1985). According to Soldo and Fokin (1994), the membranes of non-fusible vesicles such as symbiosomes include symbiont-derived lipopolysaccharides which prevent lysosomes from fusing with them. Fields (1993), investigating a model of *Legionella pneumophila* endocytosis by *Hartmannella vermiformis*, described a fusion of the endosome containing the internalized *Legionella* with the host cell's endoplasmic reticulum (ER). The ER becomes the site of bacterial multiplication as it is rich in nutrients and would prevent exposure to lysosomes.

Escaping digestion can evolve in different types of interactions between amoebae and bacteria:

- Endosymbionts are permanently harboured bacteria. Some endosymbionts cause fatal infections in their amoebal hosts. Drozanski (1991) described an obligate intracellular bacterial parasite, *Sarcobium lyticum*, causing fatal infection in *Acanthamoeba castellanii*, which shows a close relationship to an intracellular *Legionella*. Fritsche *et al.* (1993) and Michel *et al.* (1994) gave another example for bacterial obligate parasites in *Acanthamoeba*. A prolonged symbiotic association may evolve in a functional integration between symbionts and the host cell. Soldo and Fokin (1994) suggested a bacterial endosymbiotic origin for mitochondria and plastids in eukaryotic cells.
- Amoebae can use bacterial metabolism to protect themselves against xenobiotic pollutants. Our group showed a non-obligate or transitory symbiosis. where amoebae acquired mercury resistance by using harboured resistant bacteria (Hagneré and Harf, 1993).
- On the other hand, bacteria use amoebae as hosts; bacterial multiplication in amoebae is dependent on an intracellular environment

and not on soluble nutrients secreted by amoebae (King *et al.*, 1991). Bacteria also use amoebae as dispersal vectors and as a protection system against fluctuating environmental parameters (Harf and Monteil, 1988). Several authors have shown that bacteria grown in free-living amoebae are less susceptible to chlorination and biocides (King *et al.*, 1988; Kilvington and Price, 1990; Barker *et al.*, 1988). According to King *et al.* (1988), resistance to digestion by predatory protozoa might be an evolutionary precursor of bacterial pathogenicity and a survival mechanism for bacteria in aquatic environments.

Pathogenic Bacteria

As the cell cycle of *Acanthamoeba* consists of growth and replication by binary fission under optimal growth conditions and encystment under environmental stress, harboured bacteria are thus disseminated and are protected from chemical inactivation in cysts. Jadin (1973), isolating *Mycobacteria* from free-living amoebae, foresaw their part as hosts for virus and bacteria. Free-living amoebae were incriminated in the ecology and epidemiology of legionellosis (Tyndall and Domingue, 1982; Rowbotham, 1984; Fields, 1991). They are also likely to be natural hosts for well-known pathogens like *Vibrio cholerae and Listeria monocytogenes*, and nosocomial new pathogens, i.e. *Pseudomonas* sp., *Xanthomonas maltophilia, Flavobacterium* sp.

Legionella pneumophila

Legionella appear to have the capacity to survive as intracellular parasites of free-living protozoa as well as intracellular pathogens in humans. In Legionnaire's disease, the pneumonic form of legionellosis, the bacteria multiply intracellularly in the alveolar macrophages of an infected often immunodepressed host. This ability to multiply intracellularly is considered the bacterium's primary virulence factor.

Inhalation of aerosols containing *Legionella* is presumed to be the primary means of acquiring legionellosis (Fields, 1991). Known sources for transmission of *Legionella* via aerosols have been identified: showers, respiratory care equipment, air-conditioning cooling-towers, whirlpool baths, ultrasonic mist machines, lakes and rivers (Breiman, 1993). *Legionella* are ubiquitous in freshwater environments and were isolated from a number of aquatic habitats. The contrast between the ubiquity of these bacteria and the fastidiousness of their recovery from environmental samples, added to the fact that their growth in the environment in the absence of protozoa has not been documented, led to the assumption that protozoa are their primary means of proliferation in nature. Rowbotham (1980) was the first to report phagocytosis of *L. pneumophila* by amoebae and suggested that a vacuole or an amoeba, full of legionellae might be the infective particle for man rather than free legionellae themselves. According to Cirillo *et al.* (1994), *L. pneumophila* grown in *A. castellanii* acquire the expression of new proteins and are more competent to enter eukaryotic cells by an increased ability to enter by coiling phagocytosis.

The occurrence of several cases of legionellosis after immersion in the Rhine river (Hasselmann *et al.*, 1987), in addition to frequent isolation of free-living amoebae in the same natural sites (Harf and Monteil, 1989), led us to investigate the interactions between *Legionella* and free-living amoebae. We demonstrated by immunofluorescence assay (IFA) and inoculation to guinea pigs that washed and lysed amoebae from river water and sediment contained *L. pneumophila* (Harf and Monteil, 1988). Direct culturing of *Legionella* from environmental samples proved to be very difficult, due to the abundance of the associated bacterial flora overgrowing the cultures. In order to improve the detection of *Legionella* spp., we developed a polymerase chain reaction (PCR) assay, which was useful in river waters, and in *Acanthamoeba* cysts lysed by shaking with glass beads (Jaulhac *et al.*, 1993).

Listeria monocytogenes

L. monocytogenes, a saprophytic soil bacteria is also a facultative intracellular bacteria and a common pathogen throughout the world. Its epidemiology is mostly food-borne. Particularly susceptible hosts are pregnant mothers, their fetuses, and immunocompromised individuals. The major symptomatic manifestations of listeriosis are septicaemia and meningitis. The virulence mechanism enabling *L. monocytogenes* to cause these invasive diseases is thought to be its ability to survive and replicate in cells of the infected host.

Ly and Muller (1990a,b) suggested that protozoa might be the missing link in the ecology and pathology of *Listeria*, and reported that different species of *Listeria* were ingested and multiplied intracellularly in *Acanthamoeba*. The authors reported that haemolytic species, *L. monocytogenes* and *Listeria seeligeri* ruptured their host. However, these reports are preliminary and lack confirmation; as the authors used non-axenic amoebae, there might be other environmental bacteria interacting.

We investigated the possibility of free-living amoebae and *Listeria* interactions by trying co-cultures with different species of *Listeria* and an environmental strain of axenic *Acanthamoeba* sp. *Listeria* strains were kindly provided by Prof. P. Berche (Hôpital Enfants Malades,

Necker, Paris). By observing *in vivo* thin-plate cultures and Giemsa stainings (Harf, 1994) we saw intracellularly dividing *Listeria*.

Vibrio cholerae

A recent investigation (Thom *et al.*, 1992) was undertaken to determine whether *Acanthamoeba polyphaga* and *Naegleria gruberi* could be hosts for various clinical and environmental strains of *V. cholerae*. While the existence of an aquatic reservoir of *V. cholerae* and its long survival under nutrient stress were proved by laboratory and field studies, the precise identity of the ecological niche favouring its survival in an aquatic environment remains obscure. The results of this investigation have demonstrated that *V. cholerae* is capable of greater survival in co-cultures with amoebae than in control cultures. Furthermore, it was shown that an environmental strain of *V. cholerae* could survive encystment within *N. gruberi* cysts, but not in *Acanthamoeba* cysts. *V. cholerae* had not previously been associated with an intracellular mode of existence, and its pathogenesis does not involve cellular invasion. The survival of *V. cholerae* within cysts would provide a protected niche under unfavourable conditions and a means of dispersal.

Multiresistant nosocomial bacteria

In addition to their function as vectors for highly pathogenic bacteria, we observed that environmental free-living amoebae harbour various bacteria susceptible to pathogens for immunocompromised patients in hospitals. Harboured bacteria were isolated in amoebae lysates.

Bacteria of the *Pseudomonas* group (*Xanthomonas maltophilia*, *Pseudomonas* spp., *Comamonas acidovorans*) and of the *Flavobacterium* group (*Flavobacterium* sp., *Sphingobacterium multivorum*) are the most frequent in environmental waters, associated with *Acanthamoeba* and *Naegleria* in natural sites (Harf and Monteil, 1989). Nosocomial infections caused by multiresistant *Xanthomonas maltophilia* and *Flavobacterium* sp. occurred in hospitals using non-chlorinated groundwater (Monteil *et al.*, 1992). Environmental amoebae harboured *X. maltophilia* and *Flavobacterium* sp. highly resistant to various antibiotics. Axenization for further study of amoebae strains was impossible with antibiotics normally used (penicillin, streptomycin); intracellular bacteria persisted after numerous subcultures. Amoebae strains were finally cured after successive transfers on discs with new β-lactams and 6-fluoroquinolones (Hagneré and Harf, 1993). These new antibiotics were chosen for their high activity against multiresistant opportunistic bacteria involved in nosocomial hospital infections. Other nosocomial pathogens, i.e. *Acinetobacter* sp., *Alcaligenes* sp., *Aeromonas* spp., and *Legionella pneumophila*, already mentioned above, were also harboured by free-living amoebae isolated from environmental waters (Harf and Monteil, 1988; Mirgain *et al.*, 1992; Monteil *et al.*, 1992; Hagneré, 1993; Hagneré and Harf, 1993).

Amoebae are likely to be natural hosts for resistant nosocomial bacteria. This potential reservoir of hydric opportunistic bacteria in non-chlorinated or insufficiently treated water used in multiple ways in hospitals, may represent a high risk for immunocompromised patients and in intensive care units. Nosocomial infections may be the result of our man-made environments which favour the growth and transmission of nosocomial bacteria to susceptible hosts.

Conclusion

A number of bacteria have been reported to survive as parasites or endosymbionts of free-living amoebae, and some of these bacteria are also potential pathogens to vertebrates. Free-living amoebae might be dispersal vectors and protection systems for known human pathogens as *Mycobacteria* and *V. cholerae*, or nosocomial new pathogens. *Legionella pneumophila* has the ability to survive as an intracellular parasite both in protozoan and in mammalian cells. Another mammalian intracellular parasite, *Listeria monocytogenes* was also reported to multiply in *Acanthamoeba* sp. This might be a survival mechanism for bacteria in aquatic environments. Their interaction with protozoa in biofilms associated with intracellular replication and inclusion of pathogen bacteria in cysts give these pathogens a possibility of survival and transmission in unfavourable conditions. Some intracellular pathogens may even have acquired their virulence by first adapting to intracellular life in predatory free-living protozoa.

Acknowledgements

This work was supported by PIRE-Eau Alsace, CNRS France and Région Alsace. The author would like to thank Prof. H. Monteil, Institute of Bacteriology, Faculty of Medecine, ULP Strasbourg, for helpful discussions.

References

Barker, J., Brown, M.R.W., Collier, P.J., Farrell, I., Gilbert, P. (1992) Relationship between *Legionella pneumophila* and *Acanthamoeba polyphaga*: physiological status and susceptibility to chemical inactivation. *Applied Environmental Microbiology* 58, 2420–2425.

Breiman, R.F. (1993) Modes of transmission in epidemic and nonepidemic *Legionella* infection: directions for further study. In: Barbaree, J.M., Breiman, R.F. and Dufour, A.P. (eds) *Legionella, Current Status and Emerging Perspectives*. American Society of Microbiology, Washington DC, pp. 30–35.

Cirillo, J.D., Falkow, S. and Tompkins, L.S. (1994) Growth of *Legionella pneumophila* and *Acanthamoeba castellanii* enhances invasion. *Infection and Immunity* 62, 3254–3261.

Dowling, J.N., Saha, A.K., Glew, R.H. (1992) Virulence factors of the family Legionellaceae. *Microbiological Reviews* 56, 32–60.

Drozanski, W.J. (1991) *Sarcobium lyticum* gen. nov. sp. nov., an obligate intracellular bacterial parasite of small free-living amoebae. *International Journal of Systematic Bacteriology* 41, 82–87.

Drozanski, W. and Chmielewski, T. (1979) Electron microscopic studies of *Acanthamoeba castellanii* infected with obligate intracellular bacterial parasite. *Acta Microbiologia Polonica* 28, 476–511.

Fields, B.S. (1991) The role of amoebae in legionellosis. *Clinical Microbiology Newsletter* 13, 92–93.

Fields, B.S. (1993) Legionella and protozoa: interaction of a pathogen and its natural host. In: Barbaree, J.M., Breiman, R.F. and Dufour, A.P. (eds) *Legionella, Current Status and Emerging Perspectives*. American Society of Microbiology, Washington DC, pp. 129–136.

Fritsche, T.R., Gautom, R.K., Seyedirashti, S., Bergeron, D.L. and Lindquist, T.D. (1993) Occurrence of bacterial endosymbionts in *Acanthamoeba* spp. isolated from corneal and environmental specimens and contact lenses. *Journal of Clinical Microbiology* 31, 1122–1126.

Gonzalez, J.M., Irriberri, J., Ebea, L., Barcina, I. (1990) Differential rates of digestion of bacteria by freshwater and marine phagotrophic protozoa. *Applied and Environmental Microbiology* 56, 1851–1857.

Hagneré, C. (1993) Symbioses d'amibes libres et de bactéries de l'environnement hydrique induisant une mercuri-résistance et une limitation de la prolifération de bactéries pathogènes. Thèse de Doctorat de l'Université Louis Pasteur, Strasbourg, France, 99 pp.

Hagneré, C. and Harf, C. (1993) Symbiotic interactions between free-living amoeba and harboured mercury-resistant bacteria. *European Journal of Protistology* 29, 155–159.

Hall, J. and Voelz, H. (1985) Bacterial endosymbionts of *Acanthamoeba* sp. *Journal of Parasitology* 71, 89–95.

Harf, C. (1994) Free-living amoeba: interactions with environmental pathogenic bacteria. *Endocytobiosis and Cell Research* 10, 167–183.

Harf, C. and Monteil, H. (1988) Interactions between free-living amoebae and *Legionella* in the environment. *Water Science and Technology* 20, 235–239.

Harf, C. and Monteil, H. (1989) Pathogen microorganisms in environmental waters: a potential risk for human health. *Water International* 14, 75–79.

Hasselmann, M., Lutun, Ph., Schneider, F., Harf, C., Kieffer, P., Monteil, H. and Tempe, J.D. (1987) Maladie des légionnaires après noyade dans le Rhin. *Médecine et Maladies Infectieuses* 12, 752.

Jadin, J.B. (1973) De la méningoencéphalite amibienne et du pouvoir pathogène des amibes 'limax'. *Année Biologique* 12, 305–342.

Jaulhac, B., Harf, C., Nowicki, N. and Monteil, H. (1993) Detection of *Legionella* spp. in environmental water samples and free-living amoeba by using DNA amplification. In: Barbaree, J.M., Breiman, R.F. and Dufour, A.P. (eds) *Legionella, Current Status and Emerging Perspectives.* American Society of Microbiology, Washington DC, pp. 151–153.

Kilvington, S. and Price, J. (1990) Survival of *Legionella pneumophila* within cysts of *Acanthamoeba polyphaga* following chlorine exposure. *Journal of Applied Bacteriology* 68, 519–525.

King, C.H., Shotts, E.B., Wooley, R.E. and Porter, K.G. (1988) Survival of coliforms and bacterial pathogens within protozoa during chlorination. *Applied Environmental Microbiology* 54, 3023–3033.

King, C.H., Fields, B.S., Shotts, E.M. Jr and White E.H. (1991) Effects of cytocholasin D and methylamine on intracellular growth of *Legionella pneumophila* in amoebae and human monocyte-like cells. *Infection and Immunity* 59, 758–763.

Ly, T.M.C. and Muller, H.E. (1990a) Ingested *Listeria monocytogenes* survive and multiply in protozoa. *Journal of Medical Microbiology* 33, 51–54.

Ly, T.M.C. and Muller, H.E. (1990b) Interactions of *Listeria monocytogenes, Listeria seeligeri*, and *Listeria innocua* with protozoans. *Journal of General Applied Microbiology* 36, 143–150.

Michel, R., Hauröder-Philippczyk, B., Müller, K.-D. and Weishaar, I. (1994) *Acanthamoeba* from human nasal mucosa infected with an obligate intracellular parasite. *European Journal of Protistology* 30, 104–110.

Mirgain, I., Hagneré, C., Green, G.A., Harf, C. and Monteil, H. (1992) Synthetic oligonucleotide probes for detection of mercury-resistance genes in environmental freshwater microbial communities in response to pollutants. *World Journal of Microbiology and Biotechnology* 8, 30–38.

Monteil, H., Harf, C., Meunier, O. and Jehl, F. (1992) *Xanthomonas maltophilia*, bactérie opportuniste. *Revue Française des Laboratoires* 236, 22–26.

Reisser, W., Meier, R., Görtz, H.D. and Jeon, K.W. (1985) Establishment, maintenance, and integration mechanisms of endosymbionts in protozoa. *Journal of Protozoology* 32, 383–390.

Rowbotham, T.J. (1980) Preliminary report on the pathogenicity of *Legionella pneumophila* for freshwater and soil amoeba. *Journal of Clinical Pathology* 33, 1179–1183.

Rowbotham, T.J. (1984) Legionellae and amoebae. In: American Society of Microbiology (ed.) *Legionella: Proceedings of the 2nd International Symposium.* American Society of Microbiology, Washington DC, pp. 325–327.

Soldo, A.T. and Fokin, S. (1994) Interactions in symbiosis. In: Hausmann, K. and Hülsmann, N. (eds) *Progress in Protozoology, Proceedings of the IX International Congress of Protozoology*, Berlin, 1993. Gustav Fischer Verlag, Stuttgart, pp. 113–120.

Thom, S., Warhurst, D. and Drasar, B.S. (1992) Association of *Vibrio cholerae* with fresh water amoebae. *Journal of Medical Microbiology* 36, 303–306.

Tyndall, R.L. and Domingue, E.L. (1982) Cocultivation of *Legionella pneumophila* and free-living amoebae. *Applied and Environmental Microbiology* 44, 954–959.

14

Molecular and Cellular Biology of Invasion by *Entamoeba histolytica*

S.L. REED,[1] I. GIGLI,[1] L. ECKMANN,[1] M. KAGNOFF[1] AND J.H. MCKERROW[2]

[1] *UCSD Medical Center, 200 W Arbor Drive, San Diego, CA 92103-8426, USA and* [2] *VA Medical Center, San Francisco, CA 94143, USA*

Introduction

Amoebiasis is a major cause of morbidity and mortality in tropical countries and a significant public health problem in the United States. Up to 10% of the world's population is infected with *Entamoeba histolytica*, resulting in a morbidity second only to malaria and schistosomiasis (Walsh, 1988). In developed countries, the incidence of amoebiasis is estimated at less than 5% (Healy, 1986), but surveys in homosexuals have documented isolation of *E. histolytica* from up to 30% of stool specimens (William *et al.*, 1978). Despite its worldwide distribution, the mechanisms of invasion by *E. histolytica* are poorly understood.

Infection is initiated by the ingestion of cysts, which release motile trophozoites in the small intestine. In most infected patients, the trophozoites exist as harmless commensals in the lumen of the bowel where they ingest bacteria and particulate material. In other patients, they invade either the mucosa to cause symptomatic colitis, or the bloodstream to cause distant abscesses of the liver, lungs or brain (reviewed in Reed, 1992). The factors that trigger invasion are an area of active research.

It has long been appreciated that *E. histolytica* infection can cause either asymptomatic infection or a wide spectrum of clinical disease. The relative roles of the host and parasite in determining the severity of infection have been a major area of debate. Epidemiologic and genetic studies now strongly support the hypothesis that the intrinsic virulence properties of the infecting strain are the major determinant of invasion. Key epidemiologic studies by Sargeaunt *et al.* (1978) compared the isoenzyme electrophoresis patterns of more than 6000 clinical isolates of *E. histolytica* and identified the patterns of two enzymes in particular, hexokinase and phosphoglucomutase, which differentiated between strains isolated from asymptomatic patients and those isolated from patients with colitis and liver abscesses. When isoenzyme analysis was extended to isolates of *Entamoeba* cultured from homosexual men and patients with acquired immune deficiency syndrome (AIDS), it was found that all harboured non-pathogenic zymodemes and remained free of invasive amoebiasis. These studies suggested that acute infection with *Entamoeba dispar* does not result in detectable morbidity, even in AIDS patients (Allason-Jones *et al.*, 1988; Reed *et al.*, 1991). Biological and biochemical confirmation of the strain differences comes from the finding that surface antigens of pathogenic and non-pathogenic amoebae differ and can be distinguished by monoclonal antibodies (Blakely *et al.*, 1990; Petri *et al.*, 1990; Tachibana *et al.*, 1990; Reed *et al.*, 1992). The first biological difference was our finding that isolates from patients with invasive disease are resistant to complement-mediated lysis, while strains from asymptomatic carriers are sensitive (Reed *et al.*, 1983). This observation suggests that non-pathogenic strains may be limited to the bowel lumen by complement-mediated killing in the bloodstream or tissues.

More recently, DNA probes based on rRNA sequences were shown to be specific for isolates of either pathogenic or non-pathogenic zymodemes (Garfinkel *et al.*, 1989; Clark and Diamond, 1991; Que and Reed, 1991) and restriction fragment length patterns for genes encoding actin (Edman *et al.*, 1987), surface antigens (Tannich *et al.*, 1989; Edman *et al.*, 1990) and cysteine proteinases differed (Tannich *et al.*, 1991). In addition, we have shown that pathogenic strains contain a unique gene (*acp 1*) encoding a cysteine proteinase that is not present in non-pathogenic strains (Reed *et al.*, 1993). Thus, these consistent differences support the suggestion, first made by Brumpt (1925), that *E. histolytica* should be separated into two species, *E. histolytica* and *E. dispar* (Clark and Diamond, 1991; Sargeaunt, 1992).

Successful invasion by *E. histolytica* trophozoites requires multiple steps including attachment, degradation of the extracellular matrix to allow bowel invasion, killing of host cells, and evasion of local and systemic immunity. Trophozoites must first attach to host cells, a process mediated by the galactose-specific lectin (Petri *et al.*, 1987). Penetration through the epithelial layer requires the digestion of extracellular matrix, one of the critical roles of cysteine proteinases (Keene *et al.*, 1986). Once the trophozoites come into contact with complement components in tissue or serum, the cysteine proteinase activates complement in the fluid phase by cleaving C3 (Reed *et al.*, 1989a), but rapidly degrades the anaphylatoxins, C3a and C5a, potentially limiting the influx of inflammatory cells (Reed *et al.*, 1994). Host cells appear to be killed by a pore-forming peptide (Leippe *et al.*, 1991), phospholipases (Long-Krug *et al.*, 1985) and a newly described haemolysin (Jansson *et al.*, 1994). This overview will focus on recent advances in our understanding of each of these steps in invasion.

Results and Discussion

Attachment and the Gal/GalNac lectin

To initiate symptomatic infection, the trophozoites within the bowel lumen must penetrate the mucus layer and adhere to the underlying mucosa. Attachment of trophozoites to colonic mucins, bacteria, and host epithelial and inflammatory cells is mediated by the Gal/GalNac lectin (Chadee *et al.*, 1987; Petri *et al.*, 1987, 1989). Although both non-pathogenic and pathogenic strains have related lectins, monoclonal antibodies have revealed epitope differences (Petri *et al.*, 1990). The lectin is composed by a 170 kDa heavy chain encoded by at least three genes (McCoy *et al.*, 1994). The Gal/GalNac lectin appears to have multiple functions. Contact-dependent cytotoxicity is dependent on adherence by the lectin (Chadee *et al.*, 1987), and it has recently been shown to play a role in resistance to lysis by the membrane attack complex of complement (Braga *et al.*, 1992). Immunization with purified lectin also provided protection against liver abscesses in gerbils (Petri and Ravdin, 1991).

Tissue invasion mediated by cysteine proteinases

The early pathologic changes of invasive amoebiasis have been best characterized in animal models (Chadee and Meerovitch, 1985). In rodents, epithelial cells underwent marked changes, including micro-villi shortening and separation at the apical portion of cells prior to direct contact (Takeuchi and Phillips, 1975). Trophozoites were then observed to penetrate the interglandular epithelium. These pathologic findings suggest that dissolution of the extracellular matrix may precede invasion and contact-dependent killing of host cells. The extra-cellular neutral cysteine proteinase appears to be the most likely candidate for this step in invasion because it can degrade collagen, and other extracellular matrix macromolecules *in vitro* (Keene *et al.*, 1986). Cysteine proteinases are responsible for lysis of tissue culture mono-layers, the most widely used assay for amoebic toxins and other viru-lence factors. Baby hamster kidney (BHK) cells are lysed by the purified proteinase from axenized *E. histolytica*, and lysis is completely blocked by specific cysteine proteinase inhibitors (Keene *et al.*, 1990). Mutant clones of axenic strain HM-1 that do not express cysteine proteinase are not cytopathic (Keene *et al.*, 1990) and, most recently, we have shown that the cytopathic effect on fibroblast monolayers by supernatants of pathogenic clinical strains is completely inhibited by Z-Phe-Arg-CH_2F, an irreversible cysteine proteinase inhibitor that is not toxic to host cells (Reed *et al.*, 1993). We found that clinical isolates from patients with invasive amoebiasis release significantly more cysteine proteinase activity, and these patients also develop an antibody response to the amoeba cysteine proteinase, demonstrating that it is released during the course of human infection (Reed *et al.*, 1989b).

Cysteine proteinases of *E. histolytica* appear to be encoded by at least three genes (Tannich *et al.*, 1991; Reed *et al.*, 1993). Two genes (*acp 2* and *3*) are present in both pathogenic and non-pathogenic *E. histolytica*, while only strains capable of invasion have a third gene, *acp 1*, which is only 35 and 45% identical to the other genes and located on a different chromosome (Mirelman *et al.*, 1983; Reed *et al.*, 1993). The presence of the *acp 1* gene was associated with increased levels of proteinase-specific mRNA as determined by Northern hybridizations and ribonuclease protection assays (Reed *et al.*, 1993). Strains with the *acp 1* gene also had elevated proteinase activity and caused an enhanced cytopathic effect on a fibroblast cell monolayer, an *in vitro* model of virulence. The nucleotide sequence of the unique *acp 1* gene has significant regions of homology with cysteine proteinases released by transformed macrophages and invasive tumour cells (Reed *et al.*, 1993). Both amoebae and tumour cells release extracellular cysteine proteinases which are active at neutral pH and may be present as active pro-enzymes (Reed *et al.*, 1993). These similarities suggest that cysteine proteinases may be an evolutionarily conserved mechanism of tissue invasion.

Mechanisms of cell killing

The exact mechanism by which host cells are killed by *E. histolytica* trophozoites has not been elucidated, but adherence via the Gal/GalNac lectin appears to be the first step. Cytotoxicity may be mediated by a pore-forming peptide with homology to melittin, the active peptide in bee venom (Leippe *et al.*, 1991), phospholipases (Long-Krug *et al.*, 1985) and a newly described haemolysin (Jansson *et al.*, 1994).

Interactions with local and systemic immunity

Amoebic trophozoites must successfully penetrate the bowel mucosa to cause invasive disease. It has recently been appreciated that intestinal epithelial cells have more complex functions than simple absorption of nutrients. A number of epithelial cell line monolayers have been shown to express soluble mediators, particularly cytokines, which may serve to signal inflammatory cells. Kagnoff's group has demonstrated that interleukin (IL)-8, IL-1β, IL-10, and tumour necrosis factor-alpha (TNF-α) are variably expressed by a number of epithelial cell lines (Eckmann *et al.*, 1993a). In addition, they have shown that IL-8 is secreted by intestinal and cervical epithelial cells in response to invasion by *Salmonella* and *Listeria*, but not in response to lipopolysaccharide (LPS) or non-invasive bacteria (Eckmann *et al.*, 1993b).

IL-8 is a potent chemoattractant and activator of neutrophils demonstrated by the accumulation of neutrophils at intradermal injection sites (Colditz *et al.*, 1989). The rationale for investigating the IL-8 response of epithelial cells in the intestine and liver is based on the pathologic observation that neutrophils are the first cells to respond to amoebic invasion (Takeuchi and Phillips, 1975). Although well-established amoebic liver abscesses are notable for the paucity of acute inflammation (Brandt and Perez-Tamayo, 1970), studies of animal models have shown that an early neutrophil infiltrate occurs in both the intestine and the liver (Chadee and Meerovitch, 1984, 1985; Tsutsumi and Martinez-Palomo, 1988). *In vitro* studies with Chang liver cells (Salata and Ravdin, 1986) and *in vivo* studies of leucopenic hamsters (Perez-Tamayo *et al.*, 1991) have suggested that the killing of neutrophils by amoebic trophozoites may be critical to the destruction of liver cells and tissue necrosis.

In collaboration with Dr Kagnoff's group, we designed studies to evaluate whether release of cytokines by the intestinal epithelium is part of the host immune response to invasion by pathogenic *E. histolytica*. Washed pathogenic, axenic HM-1 trophozoites were added to confluent cell monolayers in 24 well plates, and cultures were incubated for various periods. Supernatants were removed, filtered, and the IL-8 concentration was determined by ELISA (Eckmann *et al.*, 1993). Eight to ~300-fold induction of IL-8 secretion was detected in a number of cell lines, including T84 colon epithelial cells, HT29 colon epithelial cells, W138 fibroblasts, and HepG2 hepatoma cells, when the monolayer was more than 80% intact. These findings suggest that the chemokine IL-8 is produced by epithelial cells in response to damage by *E. histolytica* trophozoites which may act as chemoattractant for neutrophil influx characteristic of early amoebic infection.

Once amoebae penetrate the mucosa of the bowel, they encounter components of the complement system. To cause a liver abscess, amoebic trophozoites must survive invasion through the bloodstream. Previous studies of axenic *E. histolytica* strains, which were initially isolated from patients with invasive amoebiasis, but have lost their virulence to variable degrees during passage in culture (Phillips *et al.*, 1972), had shown marked serum sensitivity of all strains tested (Huldt *et al.*, 1979). Our studies in recent isolates of pathogenic strains isolated from patients with invasive disease, however, found that these strains were resistant to lysis by activated complement components (Reed *et al.*, 1983; Reed and Gigli, 1990).

Entamoeba trophozoites activate the alternative pathway of complement by a unique mechanism, specific cleavage of C3 by the extracellular proteinase. The resulting C3b fragment is haemolytically active (Reed *et al.*, 1989). Active complement components are generated in the fluid phase which result in passive lysis of *E. dispar* strains, preventing further host invasion, in contrast to pathogenic *E. histolytica* strains which are not lysed (Reed *et al.*, 1989). The remaining C3a portion of the molecule, which has anaphylatoxin activity, is very sensitive to proteolytic degradation. We directly assessed the effect of the proteinase on purified C3a by cellulose acetate electrophoresis of the cleavage products. These studies revealed that one-tenth of the proteinase concentration needed to cleave the α-chain of C3 between residues 78 and 79 rapidly cleaved C3a with formation of two smaller fragments (Reed *et al.*, 1994). We also showed that the bioactivity of C3a, as measured by guinea pig platelet aggregation, was also destroyed. These studies demonstrate that the amino-terminal portion of the α-chain which forms C3a is rapidly degraded by the extracellular amoebic proteinase, limiting the host response.

Thus, there has been significant progress in understanding the multiple important host and parasite interactions which determine the outcome of infection by *E. histolytica*.

References

Allason-Jones, E., Mindel, A., Sargeaunt, P.G. and Katz, D. (1988) Outcome of untreated infection with *Entamoeba histolytica* in homosexual men with and without HIV antibody. *British Medical Journal* 297, 654–657.

Blakely, P., Sargeaunt, P.G. and Reed, S.L. (1990) An immunogenic 30-kDa surface antigen of pathogenic clinical isolates of *Entamoeba histolytica*. *Journal of Infectious Diseases* 162, 949–954.

Braga, L., Ninomiya, H., McCoy, J.J., Eacker, S., Wiedmer, T., Pham, C., Wood, S., Sims, P.J. and Petri, W.A. (1992) Inhibition of the complement membrane attack complex by the galactose-specific adhesin of *Entamoeba histolytica*. *Journal of Clinical Investigation* 90, 1131–1137.

Brandt, H. and Perez-Tamayo, R. (1970) Pathology of human amebiasis. *Human Pathology* 1, 351–384.

Brumpt, E. (1925) Etude sommaire de l'*Entamoeba dispar* n. sp. amibe akkystes quadrinuclees, parasite de l'homme. *Bulletin of the Academy of Medicine (Paris)* 94, 942–952.

Chadee, K. and Meerovitch, E. (1984) The pathogenesis of experimentally induced amebic liver abscess in the gerbil (*Meriones unguiculatus*). *American Journal of Pathology* 117, 71–80.

Chadee, K. and Meerovitch, E. (1985) The pathology of experimentally induced cecal amebiasis in gerbils (*Meriones unguiculatus*). *American Journal of Pathology* 119(3), 485–494.

Chadee, K., Petri, W.A., Innes, D.J. and Ravdin, J.I. (1987) Rat and human colonic mucins bind to and inhibit adherence lectin of *Entamoeba histolytica. Journal of Clinical Investigation* 80, 1245–1254.

Clark, C.B. and Diamond, L.S. (1991) Ribosomal RNA genes of 'pathogenic' and 'nonpathogenic' *Entamoeba histolytica* are distinct. *Molecular and Biochemical Parasitology* 49, 297–302.

Colditz, I., Zwahlen, R., Dewald, B. and Baggiolini, M. (1989) *In vivo* inflammatory activity of neutrophil-activating factor, a novel chemotactic peptide derived from human monocytes. *American Journal of Pathology* 134, 755–760.

Eckmann, L., Jung, H.C., Schurer-Maly, C., Panja, A., Morzycka-Wroblewska, E. and Kagnoff, M.F. (1993a) Differential cytokine expression by human intestinal epithelial cells: regulated expression of interleukin-8. *Gastroenterology* 61, 4569–4574.

Eckmann, L., Kagnoff, M.F. and Fierer, J. (1993b) Epithelial cells secrete the chemokine interleukin-8 in response to bacterial entry. *Infection and Immunity* 61, 4569–4574.

Edman, U., Meza, I. and Agabian, N. (1987) Genomic and cDNA actin sequences from a virulent strain of *Entamoeba histolytica. Proceedings of the National Academy of Sciences of the United States of America* 84, 3024–3028.

Edman, U., Meraz, M.A., Rausser, S., Agabian, N. and Meza, I. (1990) Characterization of an immuno-dominant variable surface antigen from pathogenic and nonpathogenic *Entamoeba histolytica. Journal of Experimental Medicine* 172, 879–888.

Garfinkel, L.I., Giladi, M., Huber, M., Gitler, C., Mirelman, D., Revel, M. and Rozenblatt, S. (1989) DNA probes specific for *Entamoeba histolytica* possessing pathogenic and nonpathogenic zymodemes. *Infection and Immunity* 57, 926–931.

Healy, G.R. (1986) Immunologic tools in the diagnosis of amebiasis: epidemiology in the United States. *Reviews of Infectious Diseases* 8, 239–245.

Huldt, G., Davies, P., Allison, A.C. and Schorlemmer, H.U. (1979) Interactions between *Entamoeba histolytica* and complement. *Nature* 277, 214–216.

Jansson, A., Gillin, F., Kagardt, U. and Hagblom, P. (1994) Coding of hemolysins within the ribosomal RNA repeat on a plasmid in *Entamoeba histolytica. Science* 263, 1440–1443.

Keene, W.E., Pettit, M.G., Allen, S. and McKerrow, J.H. (1986) The major neutral proteinase of *Entamoeba histolytica. Journal of Experimental Medicine* 163, 536–549.

Keene, W.E., Hidalgo, M.E., Orozco, E. and McKerrow, J.H. (1990) *Entamoeba histolytica*: correlation of the cytopathic effect of virulent trophozoites with secretion of a cysteine proteinase. *Experimental Parasitology* 71, 199–206.

Leippe, M., Sebastian, E., Schoenberger, O.L., Horstmann, R.D. and Muller-Eberhard, H.J. (1991) Pore-forming peptide of pathogenic *Entamoeba histolytica. Proceedings of the National Academy of Sciences of the United States of America* 88, 7659–7663.

Long-Krug, S.A., Fischer, K.J., Hysmith, R.M. and Ravdin, J.I. (1985) Phospholipase A enzymes of *Entamoeba histolytica*: description and subcellular localization. *Journal of Infectious Diseases* 152, 536–541.

McCoy, J.J., Mann, B.J. and Petri, W.A. (1994) Adherence and cytotoxicity of *Entamoeba histolytica* or how lectins let parasites stick around. *Infection and Immunity* 62, 3045–3050.

Mirelman, D., Feingold, C. and Wexler, A. (1983) Interactions between *Entamoeba histolytica*, bacteria and intestinal cells. *CIBA Foundation Symposium* 99, 2–18.

Perez-Tamayo, R., Martinez, R.D., Montfort, I., Becker, I., Tello, E. and Perez-Montfort, R. (1991) Pathogenesis of acute experimental amebic liver abscess in hamsters. *Journal of Parasitology* 77, 982–988.

Petri, W.A. and Ravdin, J.I. (1991) Protection of gerbils from amebic liver abscess by immunization with the galactose-specific adherence lectin of *Entamoeba histolytica. Infection and Immunity* 59, 97–101.

Petri, W.A., Smith, R.D., Schlesinger, P.H., Murphy, C.F. and Ravdin, J.I. (1987) Isolation of the galactose-binding lectin that mediates the *in vitro* adherence of *Entamoeba histolytica. Journal of Clinical Investigation* 80, 1238–1244.

Petri, W.A., Chapman, M.D., Snodgrass, T., Mann, B.J., Broman, J. and Ravdin, J.I. (1989) Subunit structure of the galactose and *N*-acetyl-D-galactosamine-inhibitable adherence lectin of *Entamoeba histolytica. Journal of Biological Chemistry* 264, 3007–3012.

Petri, W.A., Jackson, T.F.H.G., Gathiram, V., Kress, K., Saffer, L.D., Snodgrass, T.L., Chapman, M.D., Keren, Z. and Mirelman, D. (1990) Pathogenic and nonpathogenic strains of *Entamoeba histolytica* can be differentiated by monoclonal antibodies to the galactose-specific adherence lectin. *Infection and Immunity* 58, 1802–1806.

Phillips, B.P., Diamond, L.S., Bartgis, I.L. and Stuppler, S.A. (1972) Results of intracecal innoculation of germ-free and conventional guinea pigs and germ-free rats with axenically cultivated *Entamoeba histolytica. Journal of Protozoology* 19, 498–499.

Que, X. and Reed, S.L. (1991) Nucleotide sequence of a small subunit ribosomal RNA (16s-like rRNA) gene from *Entamoeba histolytica*: differentiation of pathogenic from nonpathogenic isolates. *Nucleic Acids Research* 19, 5438.

Reed, S.L. (1992) Amebiasis: an update. *Clinical Infectious Diseases* 14, 385–393.

Reed, S.L. and Gigli, I. (1990) Lysis of complement-sensitive *Entamoeba histolytica* by activated terminal complement components. Initiation of complement activation by an extracellular neutral cysteine proteinase. *Journal of Clinical Investigation* 86, 1815–1822.

Reed, S.L., Sargeaunt, P.G. and Braude, A.I. (1983) Resistance to lysis by human serum of pathogenic *Entamoeba histolytica. Transactions of the Royal Society of Tropical Medicine and Hygiene* 77, 248–253.

Reed, S.L., Keene, W.E., McKerrow, J.H. and Gigli, I. (1989a) Cleavage of C3 by a neutral cysteine proteinase of *Entamoeba histolytica. Journal of Immunology* 143, 189–195.

Reed, S.L., Keene, W.E. and McKerrow, J.H. (1989b) Thiol proteinase expression correlates with pathogenicity of *Entamoeba histolytica. Journal of Clinical Microbiology* 27, 2772–2777.

Reed, S.L., Wessel, D.W. and Davis, C.E. (1991) *Entamoeba histolytica* infection and AIDS. *American Journal of Medicine* 90, 269–270.

Reed, S.L., Flores, B.M., Batzer, M.A., Stein, M.A., Stroeher, V.L., Carlton, J.E., Diedrich, D.L. and Torian, B.E. (1992) Molecular and cellular characterization of the 29-kDa peripheral membrane protein of *Entamoeba histolytica*: differentiation between pathogenic and nonpathogenic isolates. *Infection and Immunity* 60, 542–549.

Reed, S.L., Bouvier, J., Pollack, A.S., Engel, J.C., Brown, M., Hirata, K., Que, X., Eakin, A., Hagblom, P., Gillin, F.D. and McKerrow, J.H. (1993) Cloning of a virulence factor of *Entamoeba histolytica*: pathogenic strains possess a unique cysteine proteinase gene. *Journal of Clinical Investigation* 91, 1532–1540.

Reed, S.L., Ember, J.A., Herdman, D.S., Hugli, T.E. and Gigli, I. (1995) The extracellular neutral cysteine protein of *Entamoeba histolytica* degrades and anaphylatoxins C3a and C5a. *Journal of Immunology* (in press).

Salata, R.A. and Ravdin, J.I. (1986) The interaction of human neutrophils and *Entamoeba histolytica* increases cytopathogenicity for liver cell monolayers. *Journal of Infectious Diseases* 154, 19–26.

Sargeaunt, P.G. (1992) '*Entamoeba histolytica*' is a complex of two species. *Transactions of the Royal Society of Tropical Medicine and Hygiene* 86, 348.

Sargeaunt, P.G., Williams, J.E. and Greene, J.D. (1978) The differentiation of invasive and noninvasive *Entamoeba histolytica* by isoenzyme electrophoresis. *Transactions of the Royal Society of Tropical Medicine and Hygiene* 72, 519–521.

Tachibana, H., Kobayashi, S., Kato, Y., Nagakura, K., Kaneda, Y. and Takeuchi, T. (1990) Identification of a pathogenic isolate-specific 30,000-Mr antigen of *Entamoeba histolytica* by using a monoclonal antibody. *Infection and Immunity* 58, 955–960.

Takeuchi, A. and Phillips, B.P. (1975) Electron microscope studies of experimental *Entamoeba histolytica* infections in the guinea pig. I. Penetration of the intestinal epithelium by trophozoites. *American Journal of Tropical Medicine and Hygiene* 24, 34–48.

Tannich, E., Horstmann, R.D., Knobloch, J. and Arnold, H.H. (1989) Genomic DNA differences between pathogenic and nonpathogenic *Entamoeba histolytica*. *Proceedings of the National Academy of Sciences of the United States of America* 86, 5118–5122.

Tannich, E., Scholze, H., Nickel, R. and Horstmann, R.D. (1991) Homologous cysteine proteinases of pathogenic and nonpathogenic *Entamoeba histolytica*. *Journal of Biological Chemistry* 266, 4798–4803.

Tsutsumi, V. and Martinez-Palomo, A. (1988) Inflammatory reaction in experimental hepatic amebiasis, an ultrastructural study. *American Journal of Pathology* 130, 112–119.

Walsh, J.A. (1988) Prevelence of *Entamoeba histolytica* infection. In: Ravdin, J.I. (ed.) *Amebiasis. Human Infection* by Entamoeba histolytica. John Wiley and Sons, New York, pp. 93–105.

William, D.C., Shookhoff, H.B., Felman, Y.M. and DeRames, S.W. (1978) High rates of enteric protozoal infections in selected homosexual men attending a veneral disease clinic. *Sexually Transmitted Diseases* 5, 155–157.

15

Multidrug Resistance Gene Family in *Entamoeba histolytica*

P. Ayala,[2,3] M. del Consuelo Gómez,[1,2] D.G. Pérez,[1,2] J. Samuelson[4] and E. Orozco[1,2]

[1]*Program of Molecular Biomedicine and Departments of* [2]*Experimental Pathology and* [3]*Genetics and Molecular Biology, Centro de Investigacion y de Estudios Avanzados del I.P.N., México City, México and* [4]*Department of Tropical Public Health, Harvard School of Public Health, Boston, MA 02115, USA*

Introduction

Entamoeba histolytica is the protozoan parasite causative of human amoebiasis, which has infected 500 million people around the world and has killed about one hundred thousand persons in developing countries, where sanitation is poor (Walsh, 1986). Amoebiasis is primarily controlled by drug treatment (Norris and Radvin, 1988); however, sensitivity of trophozoites to amoebicidal drugs has not been routinely detected. In addition, little is known about the mechanisms of drug resistance in the parasite and their implication in the prevalence of amoebiasis in poor countries.

On the other hand, overexpression of a single protein called P-glycoprotein, confers multidrug resistance in cell lines and in cancers (Endicott and Ling, 1989). This increased expression can be the consequence of gene amplification, promoter mutations or rearrangements, or alterations in transcription regulation (Higgins, 1993).

To identify *EhPgp* genes we used P-glycoprotein ATP-binding site probes, synthesized by polymerase chain reaction (PCR), using *E. histolytica* DNA as a template (Samuelson *et al.*, 1990). However, the identity of the *E. histolytica* P-glycoprotein genes overexpressed in mutants was not determined, because ATP-binding site probes cross-reacted with each other. The present study was undertaken to detect the *EhPgp* genes overexpressed in drug-resistant mutants. We designed specific probes to be used in Southern and Northern assays, results showed that these genes are differentially expressed. In addition, the 5′-flanking

EhPgp1

5' ATTGGAATTATCACTATGATTTCTTTCTTTTGTTATGTTGGTTTGTTTATGGCAGCAGGA
GAAAAGATGATTGGAAGAATCCGAAGAAGATTTTA TTATTCTATTATGCACCAAAATGTA
TCATGGTTTGATCGAAGAGAAAATATGGTTGGTCGTGTAACAACAAAACTTACATCTGAC
CCAACGTCTTTACAAGGAATTTCTGCAGAACGTGTTGGTGATATAATAGAAATAATGTCA
ACAGTTGGCTTTGGATTCGGCATT 3'

EhPgp2

5' CGTGTTAGAAAAGATATGTATCATTCAATAATGCA TCAAAATATCTCATGGTTTGATAGA
AAAGAAAATATGGTAGGTTCATTAACAACAAGACTTGCATCTGATCCAACAACATTACAA
GGAATTTCAGGAGAACGTGTTGCAAATGTTATTCACATTATTTCAACAATTGGATTTGCC
TTAGGTATTGCGTTCTATTATGATTGGAAAGTTTC ATTAGCT 3'

EhPgp3

5' TTAAATTTATTGATTTGATAGTAATGATGATGGAA CTACAAGATGGAGTTGAACTAACAG
ACGAACAACAACATACATTAGTTAATACTATAATT TGGGTTATGGGAATTGCATTTGCTG
GACTTTTATCCACTTACTGTTATATTGGAATATTT GCATCTTCAGCAGAATATTTAATAG
GTAGTGTTAGAAGAAGAATATTTAAATCAATTGTC AAACAAGAAATAGGAT 3'

Fig. 15.1. Linker-probes specific for each *E. histolytica* P-glycoprotein genes (*EhPgp1*, *EhPgp2* and *EhPgp5*) used in the Southern and Northern blots. *EhPgp1* specific probe was 266 bp long (from amino acids 770 to 856 of the *EhPgp1* ORF), *EhPgp2* specific probe was 220 bp long (from amino acids 803 to 876 of the *EhPgp2* ORF) and *EhPgp5* specific probe was 235 bp long (from amino acids 733 to 809 of the *EhPgp5* ORF).

region of *EhPgp1* and *EhPgp5* genes was sequenced and showed 53% and 66% of positional identity to several promoter regions of *Dictyostelium discoideum* and some potential regulator elements.

Materials and Methods

Entamoeba histolytica cultures: Trophozoites of *E. histolytica* were axenically cultured in TYI-S-33 medium (Diamond *et al.*, 1978). Two clones were used as a source of nucleic acids: (i) wild-type, drug-sensitive amoeba (clone A) (Orozco *et al.*, 1985) and emetine-resistant mutant (C2(200)) which was cultured in 200 μM emetine (Ayala *et al.*, 1990).

Southern and Northern blots with specific probes for each of the three amoeba P-glycoprotein genes: Nucleic acid probes specific for each of the three amoeba P-glycoprotein genes (*EhPgp1*, *EhPgp2* or *EhPgp5*) were made from the linker region between the two ATP-binding sites of each gene, using three sets of primers and PCR amplification. The *EhPgp1* linker-probe was 200 base pairs (bp) long, *EhPgp2* linker-probe was 220 bp long and *EhPgp5* linker-probe was 235 bp long (Fig. 15.1). Phagemids containing the entire linker region of each of the three P-glycoprotein genes were cut with the appropriate restriction enzymes, separated in triplicate sets on agarose gels, transferred to nylon membranes, and probed with the specific linker-probes using a hybridization buffer (50% formamide, 6 × Denhardt's, 5 × SSC, 1% SDS) at 42°C and wash buffer (0.1 × SSC and 1% SDS) at 65°C. For Northern blots, 3 μg mRNAs from clones A and C2(200) were electrophoresed in parallel lanes in triplicate on formaldehyde gels containing 1% agarose, transferred to nylon membranes, and hybridized separately with the three linker-region probes and with an actin probe to demonstrate that similar amounts of RNA were added to each lane (Edman *et al.*, 1987).

Sequence analysis of the 5′ region of *EhPgp1* and *EhPgp5*: The upstream region of the *EhPgp* genes was cloned and sequenced by the Sanger method (Sanger, 1977). The computer analysis of the sequences was made by FASTA (Pearson and Lipman, 1988) and BLAST (Altschul *et al.*, 1990) research programmes versus EMBL update.

Results and Discussion

Detection of high divergent regions among *EhPgp1, EhPgp2* and *EhPgp5* genes

To determine which of the *E. histolytica* P-glycoprotein genes (*EhPgp1*, *EhPgp2* and *EhPgp5*) were transcribed in drug-resistant mutants, we aligned the three genes and synthesized probes from the highest divergent regions. Segments just distal to the N-terminal ATP-binding site were chosen, because in this region each gene is different from the two others in 48–59% (Fig. 15.1). The *EhPgp1* probe recognized a 4.3 kb band in Bluescript plasmid containing *EhPgp1* gene, but no hybridization to plasmids containing *EhPgp2* and *EhPgp5* genes was detected (Fig. 15.2A, lane 1; B, lane 1), *EhPgp2* probe bound to a 0.5 kb band corresponding to the *EhPgp2* gene but not to *EhPgp1* and *EhPgp5* genes (Fig. 2A, lane 2; B, lane 2), and the *EhPgp5* probe bound to a 4.3 kb band of the *EhPgp5* gene containing plasmid, but not to *EhPgp1* and *EhPgp2* genes (Fig. 2A, lane 3; B, lane 3). These results corroborated that *EhPgp1*, *EhPgp2* and *EhPgp5* genes have divergent regions that do not cross-hybridize each other. Therefore probes made of these regions are suitable to differentiate their expression.

Detection of *EhPgp* transcripts in multidrug-resistant and drug-sensitive *E. histolytica* mRNA

To further examine the relationship between *EhPgp* genes expression and emetine resistance in *E. histolytica*, we performed Northern blots

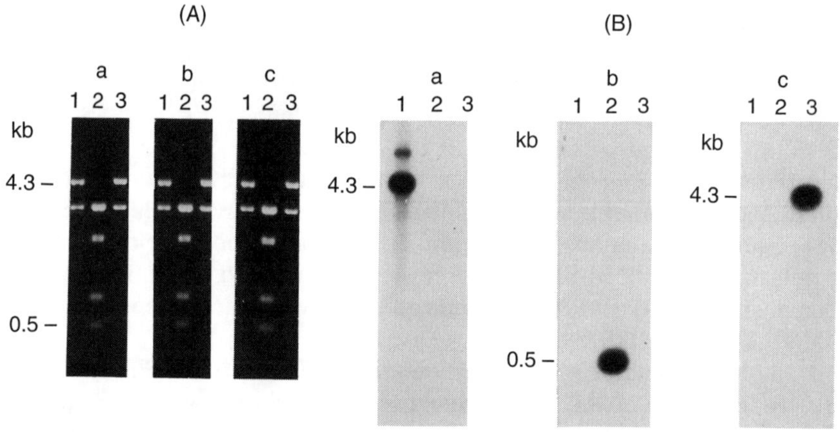

Fig. 15.2. Southern blot showing the specificity of the *E. histolytica* P-glycoprotein linker-probes. Cloned segments containing more than 50% of the coding regions of *EhPgp1* (lane 1), *EhPgp2* (lane 2) and *EhPgp5* (lane 3) were excised from the Bluescript vector and run in triplicate on agarose gels (A), transferred to nylon membranes and then (B) probed with *EhPgp1* (B,a), *EhPgp2* (B,b) and *EhPgp5* (B,c) linker-region probes. Each probe bound only to the gene from which the probe was made and did not cross-react with the other P-glycoprotein genes.

using mRNAs from wild-type clone A and mutant clone C2 that was growing in 200 μM emetine (Fig. 15.3). The *EhPgp1* transcript was only detected in mRNA from clone C2(200). On the other hand, the *EhPgp2* probe did not hybridize to any mRNAs, suggesting that this gene could not be involved in emetine resistance (Fig. 15.3). In contrast, the *EhPgp5* probe was detected in C2(200) mutant but the amount of transcript was higher than the *EhPgp1* transcript (Fig. 15.3). These results suggested that *EhPgp1* and *EhPgp5* genes could be participating in multidrug-resistance of *E. histolytica*, because they are expressed only in drug-resistant clone C2, but not in emetine-sensitive wild-type clone A. Mechanisms that are participating in the regulation of the expression of these genes are unknown. The study of regulatory regions in upstream sequences and DNA–protein interactions will help to elucidate this phenomenon.

Comparison of the 5′-flanking sequences of the *EhPgp1* and *EhPgp5* genes

To start the study of the mechanisms of regulation of the *EhPgp1* and *EhPgp5* genes expression, we sequenced their upstream flanking regions and searched their identity with other sequences in the EMBL Data Bank. *EhPgp1* and *EhPgp5* 5′-flanking sequences have between 53 and 66% of positional identity with promoter regions of several *D. discoideum*

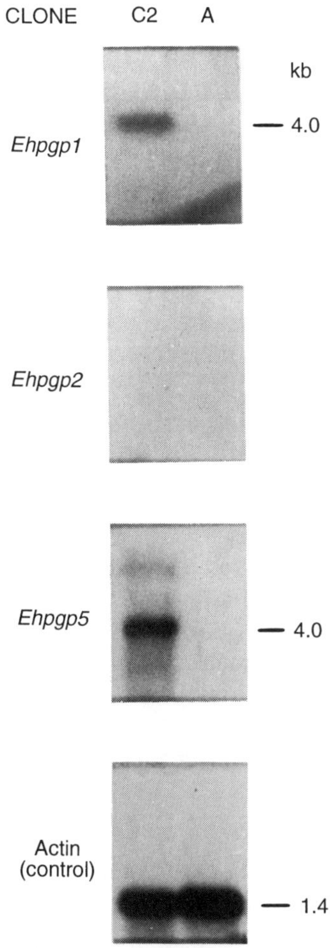

Fig. 15.3. Northern blot of mRNAs from drug-resistant clone C2 trophozoites, growing in 200 μM emetine, and from wild-type clone A trophozites growing without drug, hybridized with *EhPgp1*, *EhPgp2* and *EhPgp5* specific probes. The actin gene probe is shown as the control.

genes (Table 15.1), which encode proteins involved in the differentiation process and have a differential expression through this event. The sequence composition of A/T and G/C in the first 400 bases upstream was analysed. *EhPgp1* gene has 80% of A/T, similar to the promoter regions of *D. discoideum* that are characterized by an 85% A/T, whereas the *EhPgp5* gene has 55% of A/T. To localize the putative TATA boxes of *EhPgp1* and *EhPgp5* we aligned 100 bases of upstream regions of several *E. histolytica* reported genes (Fig. 15.4). We found some

Table 15.1. Percentage of positional identity between upstream regions of *EhPgp1* and *EhPgp5* with several *D. discoideum* promoters.[1]

Promoter region of:	Percentage of positional identity
UDPGP1 gene	66.0[2]
Dynein heavy chain gene	59.0[2]
ecmA gene	58.6[2]
DG17 gene	57.1[2]
Dd rasG gene	55.7[2]
D19 gene	54.1[3]
SP96 = spore-coat protein gene	53.1[3]

[1] Data were obtained from the EMBL Data Bank, using the FASTA and BLAST search programmes (Pearson and Lipman, 1988; Altschul *et al.*, 1990).
[2] *EhPgp1*.
[3] *EhPgp5*.

regions similar to that proposed as putative TATA boxes that could be involved in transcriptional control.

Finally, mechanistically, little is known about the control of P-glycoprotein expression. Consequently studies to identify proteins that interact with promoter or other regulator regions are required. Currently, we are studying the DNA–protein interactions and functional assays in order to find regions involved in the differential expression of multidrug-resistance genes in this parasite.

```
M17     TAGAAATATAAAAGAATGTTAAAAATGAAAACAAA CATAAAAAATAAGTG
ACTIN   AAAATAACGTAAGGAACTATGAAGTTCACCTTCAG TAAAAAAGAAGAAAG
ACTIN                                     G TTAACTCCAAACAAA
FERRED  TATTACACTAACTAAATTTCTTATTATTCTATTTA AATCACAAACAAACT
EhPgp5  TCCAATCAACAATGAATTTTTGGAATACCACAAAT AAGTTATGCCTCAAA
EhPgp1  AATTAAATTATTTAATATAGAATTTTAGATTTAAT GTGTTAGTGTTTGGT

M17     TATTTAAAGTGTTTTTAAAAAAACTAATTATTCAT AAATTAAAGTTATG
ACTIN   ACACTATTTAAAGACTGACAAAAACTGAATTGAAC ATTCAATAAATATG
ACTIN   AAGTATTTAAAGATCATAATGAACTGAATTAAATC ATTAATTAATTATG
FERRED  TGTCATTTAAATACAACAAATTAATCTTTTTATTC ATTCAAATCTAATG
EhPgp5  CTTTCTAAATTCTATTTAAAGAAAAAGATATTAAT GAGTCATTCGCATG
EhPgp1  AGTTCAAGCTGAACTAACTAACAGAACTAAAAATT ATCGTTCAAATATG
```

Fig. 15.4. Alignment of the 5′-flanking sequence from genes *M17*, *Actin*, *Ferredoxin*, *EhPgp1* and *EhPgp5*; the likely Goldberg–Hognes consensus sequences are double-underlined. The sequence similarities around the CAP site are underlined.

Acknowledgements

This work was supported by an International Research Scholars Award from the Howard Hughes Medical Institute (USA), by CONACyT (México), The MacArthur Foundation (USA) and the International Atomic Energy Agency.

References

Altschul, S.F., Gish, W., Miller, W., Myers, E.W. and Lipman, D.J. (1990) Basic local alignment search tool. *Journal of Molecular Biology* 215, 403–410.

Ayala, P., Samuelson, J., Wirth, D. and Orozco, E. (1990) *Entamoeba histolytica*: physiology of multidrug resistance. *Experimental Parasitology* 71, 169–175.

Diamond, L.S., Harlow, D.R. and Cunnick, C.C. (1978) A new medium for axenic cultivation of *Entamoeba histolytica* and other *Entamoeba*. *Transactions of the Royal Society of Tropical Medicine and Hygiene* 73, 431–432.

Edman, U., Meza, I. and Agabian, N. (1987) Genomic and cDNA sequences from a virulent strain of *Entamoeba histolytica*. *Proceedings of the National Academy of Sciences of the United States of America* 84, 3024–3028.

Endicott, J.A. and Ling, V. (1989) The biochemistry of P-glycoprotein-mediated multidrug resistance. *Annual Review of Biochemistry* 58, 137–171.

Higgins, C. (1993) The multidrug resistance P-glycoprotein. *Current Opinion in Cell Biology* 5, 687–689.

Norris, S.M. and Radvin, J.I. (1988) In: Ravdin, J.I. (ed.) *Amebiasis: Human Infection by Entamoeba histolytica*. John Wiley & Sons, New York, pp. 734–735.

Orozco, E., Hernández, F. and Rodríguez, M.A. (1985) Isolation and characterization of *Entamoeba histolytica* mutants resistant to emetine. *Molecular and Biochemical Parasitology* 15, 49–59.

Pearson, W.R. and Lipmman, D.J. (1988) Improved tools for biological sequence comparison. *Proceedings of the National Academy of Sciences of the United States of America* 85, 2444–2448.

Samuelson, J., Ayala, P., Orozco, E. and Wirth, D. (1990) Emetine-resistant mutants of *Entamoeba histolytica* overexpress mRNA for multidrug resistance. *Molecular and Biochemical Parasitology* 38, 281–290.

Sanger, F., Nicklen and Coulson, A.R. (1977) DNA sequencing with chain termination inhibitors. *Proceedings of the National Academy of Sciences of the United States of America* 74, 5463–5467.

Walsh, J.A. (1986) Problems in recognition and diagnosis of amebiasis: estimation of the global magnitude of morbidity and mortality. *Reviews in Infectious Diseases* 8, 228–238.

16

Cytoskeleton Activities in Pathogenic *Entamoeba histolytica*

N. GUILLÉN,[1] P. ARHETS,[1] P. GOUNON[2] AND P. SANSONETTI[1]

[1]*Unité de Pathogénie Microbienne Moléculaire, INSERM U389 and* [2]*Station Centrale de Microscopie Electronique, Institut Pasteur, 28 rue de Dr Roux, 75724 Paris Cédex 15, France*

Migration is a fundamental function in both single cell and multicellular organisms; it is necessary for development, immune system cell migration and metastatic spread. In biological systems, migrating cells move through a protein-rich medium and encounter other cells; cell–matrix and cell–cell contacts may be established. Different molecules account for cell movement: signalling cellular receptors that sense stimuli, extracellular matrix (ECM) receptors, cell-to-cell adhesion molecules and motion force-generating proteins. The wide range of phenotypes of cell migration, ranging from pseudopod extension, cytoplasmic streaming and tail retraction has been observed in a variety of cells such as leucocytes, macrophages, slime moulds, and parasitic amoebae. We are interested in the amoeboid movement of the parasitic protozoan *Entamoeba histolytica* which resides in the lumen of the human large intestine and has the capacity to invade the intestinal epithelium and to spread to other organs.

Amoebiasis accounts for between 40,000 and 100,000 deaths each year. The parasites reside in the lumen of the host's large intestine where they multiply, causing colitis and dysentery. It first depletes the mucus of the intestinal parenchyma and then interacts with human cells. Amoebae phagocytose human cells, and as they progress into the intestinal epithelium, tissue destruction occurs. The invasive parasite moves through the intestinal epithelium as a polarized cell, forming a large frontal pseudopod and a posterior appendix commonly referred to as a uroid (Guillén, 1993). During locomotion, surface-bound particles are translocated back-wards to the uroid (Bailey *et al.*, 1992) in which capped ligands accumulate. By an unknown mechanism the uroid is detached from the parasite without damaging it. A correlation

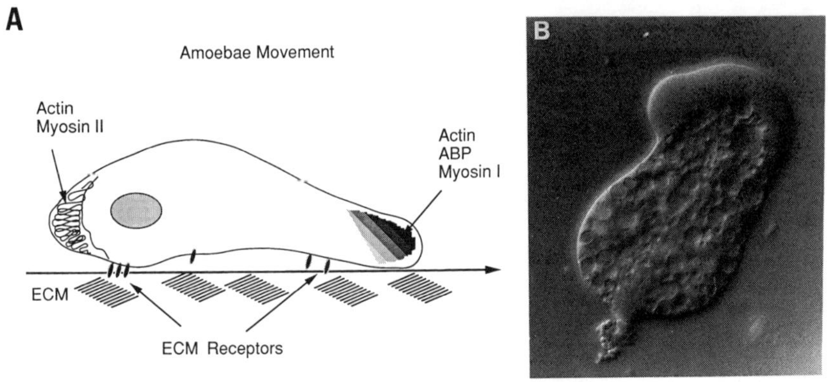

Fig. 16.1. Amoeboid movement. (A) Schematic representation of amoebae movement, see text for explanation. (B) *E. histolytica* moving as a polarized cell presenting a frontal pseudopod (bar = 6 μm). ABP, actin-binding protein; ECM, extracellular matrix.

between motility and capping of particles has been suggested for *E. histolytica* (Calderón *et al.*, 1980) and also for other types of motile cells (Bretscher, 1984).

During invasive amoebiasis and liver abscess formation, the parasite triggers the production of anti-amoeba antibodies and the activation of complement. A phenotype with resistance to complement-mediated lysis has been observed in trophozoites freshly isolated from a liver abscess (Hamelmann *et al.*, 1993). This phenotype is, in part, explained by the capping of proteins of the complement system. The fate of these proteins could be their degradation after endocytosis, or their translocation towards the uroid along the plasma membrane, followed by their elimination with the uropodial structure itself.

The mechanism of amoeboid movement involves three different processes: extension of pseudopods, transport of cytoskeleton compounds into pseudopods and retraction of the tails. Extension of pseudopods originates with a dynamic actin assembly. Actin is polymerized at the leading edge of migrating cells; myosin I and actin cross-linking proteins like ABP-120 have been found concentrated in the pseudopods of migrating *Dictyostelium discoideum*, a non-pathogenic amoeba (Fig. 16.1A) (Egelhoff and Spudich, 1991). Myosin II and actin are highly concentrated in the posterior regions of migrating cells and they are involved in tail contraction and in the related streaming activity of the cytoplasm. Thus, motion is considered as the consequence of the dynamic redistribution of the cortical actin-rich cytoskeleton. As observed by microscopy, the gliding movement of *E. histolytica* is similar to the amoeboid movement of *D. discoideum*

(Fig. 16.1B). Several lines of evidence suggest that actin-containing filaments exist in *E. histolytica* and it was proposed that the redistribution of attached surface molecules by capping stimulates parasite movement (Aust-Kettis and Uttler, 1984). However, few components of the *E. histolytica* cytoskeleton have been described (Guillén, 1993) and only genes encoding actin (Edman *et al.*, 1987; Huber *et al.*, 1989) or myosin II (Raymond-Denise *et al.*, 1993) have been isolated.

To understand the mechanism by which movement and spread of *E. histolytica* is accomplished during the development of the disease we analysed the cytoskeleton functions in this parasite. We have identified the *mhcA* gene encoding the myosin heavy chain A from *E. histolytica* which corresponds to a classical myosin II (Raymond-Dinise *et al.*, 1993). Using specific anti-myosin II serum, we localized this protein to the rear part of the moving cells, more specifically in the uroid and in the contractile ring involved in uroid extrudation, but not within the frontal pseudopod (Rahim *et al.*, 1993). It has been demonstrated that *E. histolytica* has the ability to accumulate large amounts of folded plasma membrane on the uroid, especially when cells are induced to cap antibodies or Concanavalin A (Con A) (Fig. 16.2). Thin sections observed during transmission electron microscopy revealed the folding of the plasma membrane and the presence of numerous vesicles at the uroid level. In addition, we demonstrated that myosin II co-localizes with capped Con A receptors into the uroid (Arhets *et al.*, 1995).

In *E. histolytica*, all surface-bound material moves to the uroid which spontaneously detaches from the trophozoite without cell damage, allowing the elimination of these molecules. To determine whether cytoskeleton compounds were present in the released uroid, biochemical analysis of uroid-released material was performed. After cap induction with Con A, the released material was isolated and immunoprecipitated with an anti-Con A antibody and proteins were resolved by electrophoresis (Fig. 16.3A). Electrophoretic profiles obtained by SDS-PAGE and by immunoblot analysis were observed (Fig. 16.3). In addition to Con A, several proteins of high molecular mass, ranging between 96 kDa and 200 kDa, were immunoprecipitated and visualized by silver staining of the gel. Immunoblot analysis using antibodies specific to cytoskeleton proteins was performed. The association of actin and myosin II with the Con A–receptor complex was investigated. It was determined that the *E. histolytica* anti-myosin II antibody specifically cross-reacted with a peptide of 250 kDa, indicating that myosin II is present in the released caps (Fig. 16.3C). Using an anti-actin monoclonal antibody, a peptide of 43 kDa was revealed (Fig. 16.3D), indicating that actin was also released with the caps.

To determine the influence of the cytoskeleton activities on capping and on the uroid formation, amoebae were induced to cap Con A in the

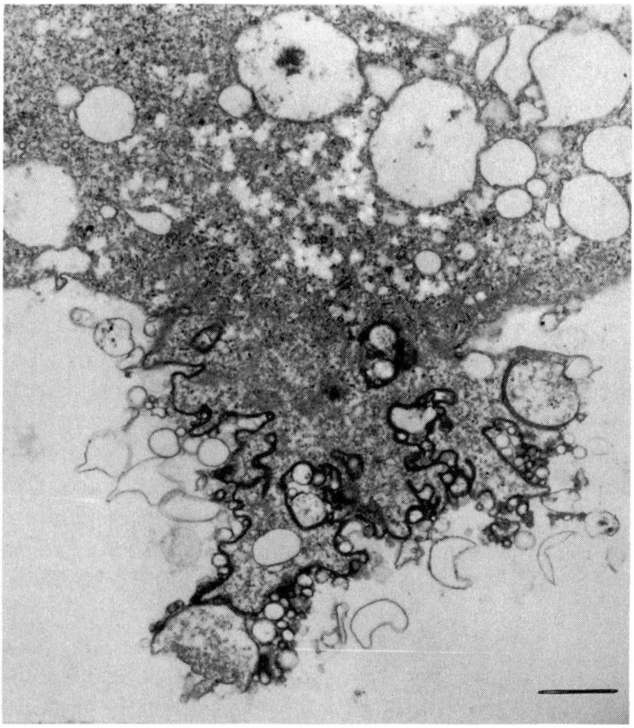

Fig. 16.2. *Entamoeba histolytica* undergoing capping of Con A. Growing amoebae were incubated at 4°C in the presence of Con A (20 μg ml^{-1}). Amoebae were harvested by centrifugation at 700 × ***g*** for 5 min, washed in phosphate-buffered saline (PBS) and incubated with 50 μg ml^{-1} of peroxidase. After washing cells were resuspended in 200 μl of PBS at a density of 5 × 10^6 amoebae in 100 μl. Trophozoites were fixed with 1.5% glutaraldehyde and post-fixed in 0.2% OsO_4. Thin sections were prepared and observed in a Jeol 1010 electron microscope operating at 80 kV. A detailed view of an induced uroid is shown. Bar = 0.5 μm.

presence of cytochalasin D, a drug that disrupts actin filamentation. Examination of cap formation with Con A–fluorescein isothiocyanate (FITC), in the presence of cytoskeleton inhibitors was carried out by microscopy. In control experiments, fluoresceinated Con A bound to the surface of the plasmic membrane and was polarly accumulated in the uroid region (Fig. 16.4B). In experiments where trophozoites were initially exposed to cytoskeleton inhibitory drugs, trophozoites bound Con A since their surface was fluorescent (Fig. 16.4C), indicating that Con A–receptor complexes were clustered in patches. Moreover, their translocation towards the uroid was inhibited. These experiments revelaed significant differences in the degree of capping, as well as

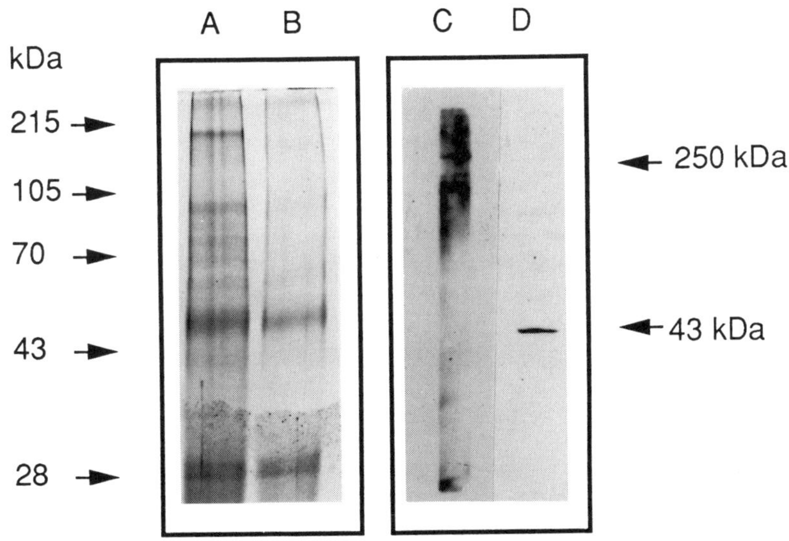

Fig. 16.3. Electrophoretic analysis of the immunoprecipitated uroid material after incubation in the presence of an anti-ConA antibody. (A) Uroid immunoprecipitated proteins. (B) Immunoprecipitation of commercial Con A. (C) Immunoblot of proteins from lane (A) incubated with an anti-myosin II antibody. (D) Immunoblot from lane A incubated with an anti-actin antibody. Immunoblots were treated with the ECL Western blotting detection reagent (Amersham).

dramatic changes in morphology of the amoebae. Pseudopods were always present on the opposite side of the vesiculated uroid, which indicates that motile amoebae were almost associated with the capping process. In the presence of cytochalasin D, trophozoites rounded up but they remained attached to the substratum indicating that the absence of uroid may be a consequence of the absence of capping.

These results indicate a fundamental role of the amoeba cytoskeleton, in particular the association of actin and myosin II, in the capping activity which could be associated with the general capping activity of other eukaryotic cells. A specific transmembrane interaction may occur betwen intracellular actomyosin filaments and surface receptors undergoing capping (Bourguignon and Bourguignon, 1984). Cytoskeleton could be directly or indirectly bound to an integral membrane protein (or class of proteins) following the binding of ligands.

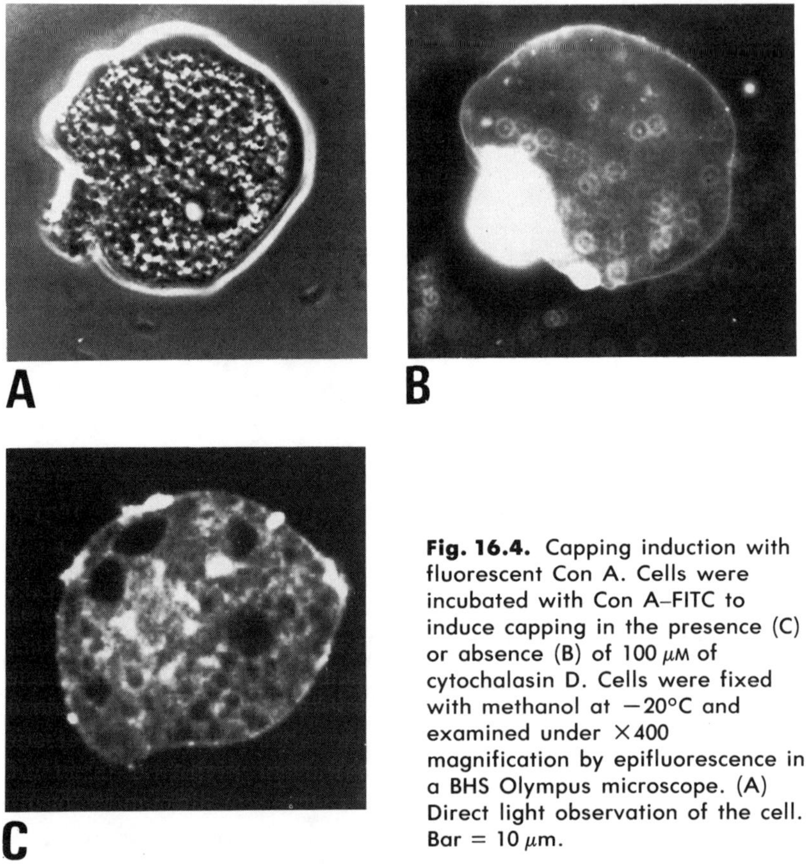

Fig. 16.4. Capping induction with fluorescent Con A. Cells were incubated with Con A–FITC to induce capping in the presence (C) or absence (B) of 100 μM of cytochalasin D. Cells were fixed with methanol at −20°C and examined under ×400 magnification by epifluorescence in a BHS Olympus microscope. (A) Direct light observation of the cell. Bar = 10 μm.

References

Arhets, P., Gounon, P., Sansonetti, P. and Guillén, N. (1995) Myosin II is involved in capping and uroid formation in the human pathogen *Entamoeba histolytica*. *Infection and Immunity* (in press).

Aust-Kettis, A. and Uttler, G. (1984) The influence of specific antibodies and complement on the mobility and the viability of *Entamoeba histolytica* trophozoites. *American Journal of Tropical Medicine and Hygiene* 33, 569–577.

Bailey, G.B., Day, D.B. and McCoomer, N.E. (1992) *Entamoeba* motility: Dynamics of cytoplasmic streaming, locomotion and translocation of surface-bound particles, and organization of the actin cytoskeleton in *Entamoeba invadens*. *Journal of Protozoology* 39, 267–272.

Bourguignon, L.Y.W. and Bourguignon, G.J. (1984) Capping and cytoskeleton. *International Review of Cytology* 87, 195–224.

Bretscher, M.S. (1984) Endocytosis: relation to capping and cell locomotions. *Science* 224, 681–686.

Calderón, J., Muñoz, M. and Acosta, H.M. (1980) Surface redistribution and release of antibody-induced caps in *Entamoebae*. *Journal of Experimental Medicine* 151, 184–193.

Edman, U., Meza, I. and Agabian, N. (1987) Genomic and cDNA actin sequences from a virulent strain of *Entamoeba histolytica*. *Proceedings of the National Academy of Sciences of the United States of American* 84, 3024–3028.

Egelhoff, T.T. and Spudich, J.A. (1991) Molecular genetics of cell migration: *Dictyostelium* as a model system. *Trends in Genetics* 7, 161–166.

Guillén, N. (1993) Cell signalling and motility in *Entamoeba histolytica*. *Parasitology Today* 9, 364–369.

Hamelmann, C., Urban, B., Foerster, B. and Horstmann, R.D. (1993) Complement resistance of pathogenic *Entamoeba histolytica* mediated by trypsin-sensitive surface component(s). *Infection and Immunity* 61, 1636–1640.

Huber, M., Koller, B., Gitler, C., Mirelman, D., Revel, M., Rozenblatt, S. and Garfinkel, L. (1989) *Entamoeba histolytica* ribosomal RNA genes are carried on palindromic circular DNA molecules. *Molecular and Biochemical Parasitology* 32, 285–296.

Rahim, Z., Raymond-Denise, A., Sansonetti, P. and Guillén, N. (1993) Localization of myosin heavy chain A in the human pathogen *Entamoeba histolytica*. *Infection and Immunity* 61, 1048–1054.

Raymond-Denise, A., Sansonetti, P. and Guillén, N. (1993) Identification and characterization of a myosin heavy chain gene (*mhcA*) from the human parasitic pathogen *Entamoeba histolytica*. *Molecular and Biochemical Parasitology* 59, 123–132.

17

Infection Sources, Reservoir and Transmission of Pneumocystosis

E. DEI-CAS,[1,2] E. MAZARS,[1] E.M. ALIOUAT,[1] C. ÖDBERG FERRAGUT,[1] I. DURAND,[1] C.M. DENIS[1] AND D. CAMUS[1,2]

[1]*INSERM (U.42), 369 rue J. Guesde, BP 39, 59650 Villeneuve d'Ascq, France and* [2]*Laboratory of Parasitology-Mycology, Faculty of Medicine and Hospital University Centre, 1 Place Verdun, 59045 Lille, France*

Pneumocystis carinii is an important agent of severe pneumonitis in immunosuppressed patients, especially in humans suffering from acquired immune deficiency syndrome (AIDS). Although *P. carinii* pneumonia (PCP) is a worldwide problem of public health, a number of important aspects of the basic biology of its agent have not been elucidated.

In recent years, significant advances have been made in the knowledge of the characteristics, genetic diversity and taxonomy of *P. carinii*. These advances are providing new bases to understand the transmission of PCP, the nature of the reservoir of *Pneumocystis* and the sources of host infection. These aspects are examined in this chapter.

Transmission of Pneumocystosis

Hughes (1987) has shown that animal-to-animal transmission of PCP takes place by air-borne routes. He showed, using germ-free rats housed in an isolator that were not carriers of *Pneumocystis*, that the rats were susceptible to the parasite and that they develop PCP after exposure to *Pneumocystis*-infected conventional rats. Moreover, he showed that the infection was transmitted by an air-borne route and not by feeding, drinking, or from infected lung tissue placed in the isolator.

It is now well established that *P. carinii*-free recipient hosts can be experimentally infected with *Pneumocystis*. Thus, *P. carinii*-free SCID mice developed PCP after exposure to outbreed mice developing corticosteroid-induced PCP (Soulez *et al.*, 1991). In these experiments we have shown that a single infected mouse placed in the same cage (or in another cage but under the same filter bonnet) was a sufficient source of *P. carinii* to infect at least 20 *P. carinii*-free SCID mice. What is more, even a short exposure (1 day) was sufficient for *P. carinii* transmission to SCID mice.

Moreover, we are able to obtain viable mouse-derived parasites from parasitized lungs to infect SCID mice by intranasal inoculation (Aliouat *et al.*, 1993). We are also able to infect corticosteroid-treated nude rats by intratracheal inoculation with rat-derived *Pneumocystis* (Aliouat *et al.*, 1994, 1995). Inoculated SCID mice or nude rats treated in these ways develop PCP within 40 days.

With regard to other routes of infection, the vertical transmission of *Pneumocystis* has been suggested in humans and rats (Bazaz *et al.*, 1970; Pifer *et al.*, 1984). In contrast, no infection of *P. carinii* seems to occur through the placenta in SCID mice (Ito *et al.*, 1991).

Although we are able to infect experimental hosts in the laboratory we do not known which life cycle stage(s) is involved in transmission nor how the transmission of *Pneumocystis* takes place in nature.

Infectious Life Cycle Stages of *Pneumocystis*

The infectious stage(s) of *Pneumocystis* that is involved in natural aerial transmission of the parasite is yet to be elucidated. In the PCP-inoculation models mentioned, all life cycle parasite stages coexist in the inoculum. Therefore, it is not possible to study comparatively the infectious power of different parasite stages. Recently a method to obtain either trophozoite or cyst samples, based on elutriation, was proposed (DeStefano *et al.*, 1994). This kind of method should lead to the comparative testing of the infectivity power of the different life cycle stages of *P. carinii*.

Moreover, various other approaches can be used. For instance, a given parasite stage can be selectively eliminated from a parasite sample. We are able to obtain *in vitro* a rat-derived parasite population composed almost exclusively of ultrastructurally unaltered trophic forms (Aliouat *et al.*, 1995; E.M. Aliouat *et al.*, unpublished). In this system, a few cystic or precystic forms are observed but they show ultrastructural alterations. The parasite samples obtained in this culture system were found to be more infectious to nude rats than freshly extracted rat *Pneumocystis* organisms. However, we can not be certain of the total absence of infectious non-trophic forms because they could be rare and, therefore, microscopically undetectable.

Schmatz *et al.* (1991) were able to selectively eliminate cysts in infected rats submitted to therapeutic doses of an echinocandin (L-671,329), an inhibitor of β-1,3-glucan synthesis. Using lower, prophylactic doses of the same molecule, the authors prevented the appearance of trophozoites. As L-671,329 does not directly affect trophozoites, the authors suggest that the cyst stage is required for trophozoite proliferation.

It is certain that during normal expiration, *P. carinii*-infected hosts eliminate highly infectious forms of the parasite in the air expelled from the lungs. As mentioned before a 1-day exposure to an infected mouse was sufficient for *P. carinii* transmission to SCID mice (Soulez *et al.*, 1991). The transmission of the parasite from one immunosuppressed patient to another could occur in this way. Highly infectious airborne *Pneumocystis* forms would be present in the air expelled from the patient's lungs. This could explain the recently reported nosocomial 'epidemics' of PCP (Santiago-Delpin *et al.*, 1988; Chave *et al.*, 1991; Jacobs *et al.*, 1991).

Environmental Sources of Infection

Susceptible hosts can also be infected by *Pneumocystis* from unidentified environmental sources. In fact, many experiments (critically summarized by Hughes, 1987, pp. 97–104) suggest that the environmental transmission of *P. carinii* occurs. Moreover, most authors working on this parasite obtain *Pneumocystis* organisms by administering corticosteroids to laboratory rodents and no exposure to infected animals is usually needed. To explain this spontaneous PCP 'revealed' by the corticosteroid treatment the following hypotheses can be proposed:

1. Laboratory animals are frequently healthy carriers of *Pneumocystis* and immunosuppressing treatment reveals latent infection allowing PCP to develop.
2. Laboratory animals are infected by widespread, highly infective forms of *Pneumocystis* present in the environment. These forms are able to infect mammals by airborne routes but only immunodeficient ones are susceptible and develop PCP. The existence of a free-living form of *Pneumocystis* able to produce an airborne spore-like infective stage was proposed (Wakefield *et al.*, 1992). However, the infective stages driven out of lungs of infected hosts could be resistant to environmental factors, remaining potentially infectious to new hosts for long periods.
3. PCP could develop from both latent forms of the parasite and infective airborne forms from exogenous sources.

We have observed a low number of *Pneumocystis* cysts in the lungs of apparently healthy, non-immunosuppressed rats and rabbits. This could support hypothesis **1**. However, we are unable to affirm that when these hosts are treated with corticosteroids PCP develops from latent forms: the *Pneumocystis* infection could also develop from infectious forms from an exogenous source.

The following recent observation supports hypothesis **2**: DNA sequences identical to those of *Pneumocystis* have been amplified by the polymerase chain reaction (PCR) from ambient air by two different teams working on different continents and using different DNA primers (Bartlett *et al.*, 1994; Wakefield, 1994).

Furthermore, in humans without PCP no *P. carinii* organisms have been detected using either specific monoclonal antibodies (Millard and Heryet, 1988) or PCR (Peters *et al.*, 1992). This suggests that immunosuppressed, susceptible humans develop *Pneumocystis* infections from widespread exogenous sources. This is consistent with many recent observations indicating that *Pneumocystis* is a fungus (Wakefield *et al.*, 1992; Stringer, 1993). This concern will be examined in the following section.

Pneumocystis Seems to be an Atypical Fungus

Pneumocystis carinii was recently assigned to the kingdom of Fungi on the basis of DNA homology studies. First, the 16S-like RNA and gene of *P. carinii* presents affinities with Ascomycota yeast rRNA (Edman *et al.*, 1988; Stringer *et al.*, 1989). The gene encoding for the *P. carinii* mitochondrial large subunit of rRNA (mlsrRNA) (Sinclair *et al.*, 1991) and seven other continuous mitochondrial genomic sequences (Pixley *et al.*, 1991), show significant homology with fungal DNA sequences. Other *Pneumocystis* genes encode proteins that are most similar to their homologous ones from fungi, including β-tubulin (Edlind *et al.*, 1992), transcription factor IID and a P-type cation-translocating ATPase (Meade and Stringer, 1991).

Second, the genes of two important enzymes, thymidylate synthase (TS), catalysing the conversion of deoxyuridylate (dUMP) to TMP (thymidylate), and dihydrofolate reductase (DHFR), catalysing the conversion of dihydrofolate to tetrahydrofolate, were found to be located on different chromosomes of *P. carinii* (Edman *et al.*, 1989; Lundgren *et al.*, 1990). Thus, in *P. carinii*, as in non-protozoan organisms, perhaps with the exception of plants, TS and DHFR reside in two polypeptide chains, clearly distinct and monofunctional (Edman *et al.*, 1989). This may be considered as evidence against *P. carinii* being a protozoan, in so far as these two enzymatic activities exist in one bifunctional protein in all protozoa studied to date (Kungkrai *et al.*, 1990).

Third, protein synthesis in eukaryotic cells requires two soluble proteins or elongation factors (EF-1 and EF-2) for the peptide elongation cycle. In contrast, protein synthesis in several genera of fungi requires three elongation factors (EF-1, EF-2 and EF-3). EF-3 could not be detected in *P. carinii* using immunological methods (Jackson *et al.*, 1991). However, a gene capable of encoding a protein that shows a marked structural similarity with EF-3 of *Saccharomyces cerevisae* and *Candida albicans*, was later detected in *P. carinii* (Ypma-Wong *et al.*, 1992).

Now, in which fungal group must *Pneumocystis* be included? In contrast to most results obtained by using nuclear gene sequences suggesting affinities with Ascomycetes, the sequence of the mlsrRNA gene of *Pneumocystis* has been shown to be similar to those of the ustomycetous red yeast fungi (Basidiomycetes) (Wakefield *et al.*, 1992). These discordant results emphasize the fact that *Pneumocystis* is a unique microorganism (Hughes, 1989), that can not be easily included in the usually recognized fungal groups (Stringer, 1993).

Indeed, the affirmation that '*P. carinii* is a fungus' can provide some intellectual comfort but the idea can be disturbing for those who really know fungi. The atypicalness of *P. carinii* can be considered regarding its fungal nature as well as its pathogenic role.

P. carinii has been assigned to the fungal kingdom on the basis of genomic studies. Now, when we want to 'switch' from the genomic level to the phenotypic one, we find a number of difficulties. Marked differences exist between the cystic wall of *Pneumocystis* and the cell wall of yeasts; they were analysed elsewhere (Palluault *et al.*, 1990, 1992a,b; Dei-Cas *et al.*, 1992). Strikingly, *Pneumocystis* shows an outer surface membrane, similar to that observed in prokaryotes, as was reported by DeStefano *et al.* (1990) for cyst forms, and by us for all stages (Palluault *et al.*, 1992a).

With regard to the trophozoite, its amoeboidal morphology, especially the presence of filopodia, is not typical of fungal cells. The attachment process of *Pneumocystis* is different from that of pathogenic yeast to mammals (Dei-Cas *et al.*, 1991, 1992; Aliouat *et al.*, 1993a). Likewise, most antifungal drugs are inefficacious against *P. carinii*. The presumed absence of ergosterol in its plasma membrane could explain the absence of response of this parasite to amphotericin B (Dei-Cas *et al.*, 1987). *Pneumocystis carinii* possesses a pathway for *de novo* folate synthesis (using pteridines and *p*-aminobenzoic acid) which is susceptible to sulphonamides, and DHFR (Kovacs *et al.*, 1989) which is the target for trimethoprim. Finally, *Pneumocystis* does not develop in typical fungal culture media although limited development was reported in media with neopeptona and *N*-acetylglucosamine at low pH (without feeder cells) (Cushion and Ebbets, 1990).

Table 17.1. *Pneumocystis carinii*: cross-infection experiments using *P. carinii*-free SCID mice.[1]

Inoculum from:	Microscopic detection	PCR	Specific mAbs	Source
Mouse	+	+	+	Gigliotti *et al.*, 1993
Ferret	−	−	−	
Mouse	+	+	ND	Aliouat *et al.*, 1993b
Rat	−	−	ND	
Rabbit	−	−	ND	
Monkey	−	ND	ND	Furuta *et al.*, 1993

[1] +, Pneumocystosis; −, *P. carinii* absent; mAbs, monoclonal antibodies; ND, not determined.

Is Pneumocystosis a Deep Mycosis-like Disease?

It seems reasonable to compare PCP with deep mycoses that can be acquired by airborne route transmission, such as fungal diseases caused by dimorphous fungi (filamentous in their saprophytic state and becoming yeast-like when invading the host). *Paracoccidioides brasiliensis*, *Blastomyces dermatitidis* and *Histoplasma capsulatum* are typical dimorphous exosaprophytic fungal agents (Dei-Cas and Vernes, 1986). The comparison of PCP with fungal diseases caused by these fungi, strengthens the atypicalness of *Pneumocystis* (Dei-Cas *et al.*, 1992).

Deep mycoses may be acquired from environmental sources through the respiratory tract, but unlike PCP they cannot be transmitted from one infected host to another. Furthermore, whereas *P. carinii* strains are host species specific (Table 17.1) (Aliouat *et al.*, 1993b, 1994; Furuta *et al.*, 1993; Gigliotti *et al.*, 1993) no evidence of host species specificity was reported in exosaprophytic pathogenic fungi. The reservoir of these pathogens is the soil, that is the infection source for hosts.

In the geographic regions where the environmental development of pathogenic dimorphous fungi is observed, subclinical lung fungal infections have high incidence, but severe, clinical manifestations are rare, usually occuring in weakened hosts (Dei-Cas and Vernes, 1986). This pattern seems to be similar in PCP. Serological evidence suggests that in European, North American and sub-Saharan African populations (see Hughes, 1987 for review; Wakefield *et al.*, 1990), subclinical *P. carinii* infection commonly occurs in young healthy children.

In histoplasmosis (*H. capsulatum*), paracoccidiodomycosis, blastomycosis and coccidioidomycosis, the infection is primarily pulmonary but secondary lesions occur in other organs (Binford and Dooley, 1976). *Pneumocystis* is also able to disseminate from infected lungs to other organs where it is occasionally found (Raviglione, 1990; Cohen and

Stoeckle, 1991). The real incidence of *P. carinii* extrapulmonary infection in humans remains unknown but this dissemination seems relatively frequent in deeply immunosuppressed laboratory animals with PCP (Denis *et al.*, 1994).

Deep mycoses and PCP appear to resemble one another less when one considers pathological data. Dimorphous fungi usually elicit a granulomatous reaction of histiocytes, epithelioid cells and giant cells of the foreign body or of Langhans' type (Binford and Dooley, 1976). Fibrocaseous pulmonary nodules may be caused by *Coccidioides immitis*. Abscess formation, where polymorphonuclear neutrophils are accumulated in large numbers may be seen in blastomycosis. Caseous necrosis may develop in histoplasmosis where calcification is a characteristic feature of old necrotic lesions (Binford and Dooley, 1976). In PCP, alveolar type II cell hypertrophy, macrophagic infiltrate in the alveolar septa and filling of the alveolar spaces with a foamy eosinophilic exudate are the most typical histopathological changes (Hughes, 1987; Dei-Cas *et al.*, 1990; Bartlett and Smith, 1991; Walzer, 1991). Eosinophilic exudate and necrosis were reported in extrapulmonary pneumocystosis (Cohen and Stoeckle, 1991).

In summary, pneumocystoses and deep mycoses share some features: (i) high frequency of subclinical forms; and (ii) severe disease in weakened hosts. The two groups of diseases also show marked differences: (i) unlike deep mycoses, pneumocystosis is transmissible from an infected host to another; (ii) *Pneumocystis* shows host species specificity whereas exosaprophytic fungi do not; and (iii) deep mycoses and pneumocystosis induce different histopathological changes in the host.

Reservoir and Infection Sources

Although *P. carinii* strains of different host species are usually considered morphologically indistinguishable (Walzer, 1993), a cluster of genomic (Sinclair *et al.*, 1991; Wakefield *et al.*, 1992; Stringer, 1993), karyotypic (Weinberg and Bartlett, 1991; Stringer, 1993) and antigenic differences (Furuta and Ueda, 1987; Linke *et al.*, 1989; Gigliotti, 1992; Bauer *et al.*, 1993), which seemed host species-linked, was reported for the last few years. In our laboratory, *Pneumocystis* isolates from four host species have been compared by using restriction mapping, gene sequence analysis, multilocus enzyme electrophoresis (MEE), ultrastructure and *in vitro* cultures. Host species specific differences were recorded using genomic (Mazars *et al.*, 1995) or MEE methods (Mazars *et al.*, 1994). Ultrastructural differences were observed between rabbit and mouse *Pneumocystis* organisms (Dei-Cas *et al.*, 1994). The different *in vitro* behaviour of rat and mouse *Pneumocystis* (Aliouat *et al.*, 1993a and unpublished), indicates that biological properties of parasite strains from different host species are different.

It is unclear to what extent these genotypic or phenotypic variations involved different infectious powers or pathogenicity. A crucial point is if animal-to-animal transmission takes place between mammals of different species. This is important in so far as neither the sources of the human *P. carinii* infection, nor how humans become infected by this agent, have been clearly established. Therefore, cross-infection experiments were performed in order to compare the infectious power and pathogenicity of *Pneumocystis* isolates of different mammals to experimental *P. carini*-free recipient hosts. Results obtained by three different teams suggests *Pneumocystis* is a stenoxenic parasite (Table 17.1) (Aliouat *et al.*, 1993b, 1994; Furuta *et al.*, 1993; Gigliotti *et al.*, 1993). Actually, only mouse-derived *Pneumocystis* isolates were infectious to SCID mice (Table 17.1) and only rat-derived parasites were infectious to corticosteroid-treated nu/nu rats used as recipient hosts (Aliouat *et al.*, 1994). Up-to-date, cross-infection experiments have shown that *Pneumocystis* from a given mammal cannot infect other host species.

From an epidemiological point of view, the host species specific genetic differences of *Pneumocystis* as well as its strong host species specificity suggest that the reservoir of the human PCP agent is man. Therefore, PCP should not be a zoonotic disease. On the taxonomic front, present results suggest that '*P. carinii*' represents several species which could be distinguished by genomic, biochemical or other markers, and through their biological properties such as their infectivity to a given host.

The results summarized in this chapter should be confirmed by increasing the number of tested parasite samples from different hosts and geographic origins. Thus, an extensive programme of study and identification of *Pneumocystis* isolates is being developed by 19 research teams from seven European countries in the framework of a European Concerted Action (Biomed-1, Ref No PL94-1118).

References

Aliouat, E.M., Dei-Cas, E., Ouaissi, M.A., Palluault, F., Soulez, B. and Camus, D. (1993a) *In vitro* attachment of *Pneumocystis carinii* from mouse and rat origin. *Biology of the Cell* 77, 209–217.

Aliouat, E.M., Mazars, E., Dei-Cas E., Cesbron, J.Y. and Camus, D. (1993b) Intranasal inoculation of mouse, rat or rabbit-derived *Pneumocystis* in SCID mice. *Journal of Protozoology Research* 3, 94–98.

Aliouat, E.M., Mazars, E., Dei-Cas, E., Delcourt, P., Billaut, P. and Camus, D.

(1994) *Pneumocystis* cross infection experiments using SCID mice and nude rats as recipient host, showed strong host-species specificity. *Journal of Eukaryotic Microbiology* 41, S71.

Aliouat, E.M., Dei-Cas, E., Dujardin, L., Billiaut, P. and Camus, D. (1995) *Pneumocystis carinii* organisms from *in vitro* culture were highly infectious to the nude rat. *Parasitology Research* 81, 82–85.

Bartlett, M.S. and Smith, J.W. (1991) *Pneumocystis carinii*, an opportunist in immunocompromised patients. *Clinical Microbiology Review* 4, 137–149.

Bartlett, M.S., Lee, C.H., Lu, J.J., Bauer, N.L., Brady, S.L., Beltz, J.F., McLaughlin, G.L. and Smith, J.W. (1994) *Pneumocystis carinii* detected in air. *Journal of Eukaryotic Microbiology* 41, S75.

Bauer, N.L., Paulsrud, J.R., Bartlett, M.S., Smith, J.W. and Wilde, C.E. (1993) *Pneumocystis carinii* organisms obtained from rats, ferrets and mice are antigenically different. *Infection and Immunity* 61, 1315–1319.

Bazaz, G.R., Manfredi, O.L., Howard, R.G. and Claps, A.A. (1970) *Pneumocystis carinii* pneumonia in three full-term siblings. *Journal of Pediatrics* 76, 767–769.

Binford, C.H. and Dooley, J.R. (1976) Diseases caused by fungi and actinomycetes. In: Binford, C.H. and Connor, D.H. (eds) *Pathology of Tropical and Extraordinary Diseases*, Vol 2. A.F.I.P., Washington, pp. 551–609.

Chave, J.P., David, S., Wauters, J.P., Van Melle, G. and Francioli, P. (1991) Transmission of *Pneumocystis carinii* from AIDS patients to other immunosuppressed patients: a cluster of *Pneumocystis carinii* pneumonia in renal transplant recipients. *AIDS* 5, 927–932.

Cohen, O.J. and Stoeckle, M.Y. (1991) Extrapulmonary *Pneumocystis carinii* infections in the acquired immunodeficiency syndrome. *Archives of Internal Medicine* 151, 1205–1214.

Cushion, M.T. and Ebbets, D. (1990) Growth and metabolism of *Pneumocystis carinii* in axenic culture. *Journal of Clinical Microbiology* 28, 1385–1394.

Dei-Cas, E. and Vernes, A. (1986) Parasitic adaptation of pathogenic fungi to mammalian hosts. *CRC Critical Reviews in Microbiology* 13, 173–218.

Dei-Cas, E., Ajana, F., Deldicque, P., Bonaly, R. and Vernes, A. (1987) Place actuelle de l'Amphotéricine B dans le traitement des mycoses profondes. *Annales de Biologie Clinique* 45, 673–679.

Dei-Cas, E., Soulez, B., Palluault, F., Charet, P. and Camus, D. (1990) *Pneumocystis carinii*, un défi pour le biologiste. *Médecine/Sciences* 6, 517–525.

Dei-Cas, E., Jackson, H., Palluault, F., Aliouat, E.M., Hancock, V., Soulez, B. and Camus, D. (1991) Ultrastructural observations on the attachment of *Pneumocystis carinii in vitro*. *Journal of Protozoology* 38, 205–207.

Dei-Cas, E., Cailliez, J.C., Palluault, F., Aliouat, E.M., Mazars, E., Soulez, B., Suppin, J. and Camus, D. (1992) Is *Pneumocystis carinii* a deep mycosis-like agent? *European Journal of Epidemiology* 8, 460–470.

Dei-Cas, E., Mazars, E., Ödberg-Ferragut, C., Durand, I., Aliouat, E.M., Dridba, M., Palluault, F., Cailliez, J.C., Séguy, N., Tibayrenc, M., Mullet, C., Creusy, C. and Camus, D. (1994) Ultrastructural, genomic, isoenzymatic and biological features make it possible to distinguish rabbit *Pneumocystis* from other mammal *Pneumocystis* strains. *Journal of Eukaryotic Microbiology* 41, S84.

Denis, C.M., Cailliez, J.C., Aliouat, E.M., Delcourt, P., Lepage, G., Hermant, N., Lamouret, M. and Dei-Cas, E. (1994) Does *Pneumocystis carinii* remain infectious in the bloodstream? *Journal of Eukaryotic Microbiology* 41, S86.

DeStefano, J.A., Cushion, M., Sleight, R. and Walzer, P.D. (1990) Analysis of *Pneumocystis carinii* cyst wall. I. Evidence for an outer surface membrane. *Journal of Protozoology* 37, 428–435.

DeStefano, J.A., Foy, J.M., Sullivan, D.W., Dawes, S.M. and Walzer, P.D. (1994) Fractionation of *Pneumocystis carinii* developmental stages by counterflow centrifugal elutriation and sequential filtrations. *Parasitology Research* 80, 1–9.

Edlind, T.D., Bartlett, M.S., Weinberg, G.A., Prah, G.N. and Smith, J.W. (1992) The β-tubulin gene from rat and human isolates of *Pneumocystis carinii*. *Molecular Microbiology* 6, 3365–3373.

Edman, J.C., Kovacs, J.A., Masur, H., Santi, D.V., Elwood, H.J. and Sogin, M.L. (1988) Ribosomal RNA sequence shows *Pneumocystis carinii* to be a member of the Fungi. *Nature* 334, 519–522.

Edman, U., Edman, J.C., Lundgren, B. and Santi, D.V. (1989) Isolation and expression of the *Pneumocystis carinii* thymidylate synthase gene. *Proceedings of the National Academy of Sciences of the United States of America* 86, 6503–6507.

Furuta, T. and Ueda, K. (1987) Intra- and inter-species transmission and antigenic difference of *Pneumocystis carinii* derived from rat and mouse. *Japanese Journal of Experimental Medicine* 57, 11–17.

Furuta, T., Fujita, M., Mukai, R., Sakakibara, I., Sata, T., Miki, K., Hayami, M., Kojima, S. and Yoshikawa, Y. (1993) Severe pulmonary pneumocystosis in simian acquired immunodeficiency syndrome induced by simian immunodeficiency virus – its characterization by the polymerase-chain-reaction method and failure of experimental transmission to immunodeficient animals. *Parasitology Research* 79, 624–628.

Gigliotti, F. (1992) Host species-specific antigenic variation of a mannosylated surface glycoprotein of *Pneumocystis carinii*. *Journal of Infectious Diseases* 159, 329–336.

Gigliotti, F., Harmsen, A.G., Haidaris, C.G. and Haidaris, P.J. (1993) *Pneumocystis carinii* is not universally transmissible between mammalian species. *Infection and Immunity* 61, 2886–2890.

Hughes, W.T. (1987) Pneumocystis carinii *Pneumonitis*. CRC Press, Boca Raton.

Hughes, W.T. (1989) *Pneumocystis carinii*: taxing taxonomy. *European Journal of Epidemiology* 5, 265–269.

Ito, M., Tsugane, T., Kobayashi, K., Kuramochi, T., Hioki, K., Furuta, T. and Nomura, T. (1991) Study on placental transmission of *Pneumocystis carinii* in mice using immunodeficient SCID mice as a new animal model. *Journal of Protozoology* 38, 218–219.

Jackson, H.C., Colthurst, D., Hancock, V., Marriot, M.S. and Tuite, M.F. (1991) No detection of characteristic fungal protein Elongation Factor EF-3 in *Pneumocystis carinii*. *Journal of Infectious Diseases* 163, 675–677.

Jacobs, J.L., Libby, D.M. and Winters, R.A. (1991) A cluster of *Pneumocystis carinii* pneumonia in adults without predisposing illnesses. *New England Journal of Medicine* 324, 246–250.

Kovacs, J.A., Allegra, C.J., Beaver, J., Boarman, D., Lewis, M., Parrillo, J.E., Chabner, B.A. and Masur, H. (1989) Characterisation of de novo folate synthesis in *Pneumocystis carinii* and *Toxoplasma gondii*: potential for screening therapeutic agents. *Journal of Infectious Diseases* 160, 312–320.

Kungkrai, J., Webster, H.K. and Yuthavong, Y. (1990) Folate and cobalamin metabolism in *Plasmodium falciparum*. *Parasitology Today* 6, 388–391.

Linke, M.J., Cushion, M.T. and Walzer, P.D. (1989) Properties of the major antigens of rat and human *Pneumocystis carinii*. *Infection and Immunity* 57, 1547–1555.

Lundgren, B., Cotton, R., Lundgren, J.D., Edman, J.C. and Kovacs, J.A. (1990) Identification of *Pneumocystis carinii* chromosomes and mapping of five genes. *Infection and Immunity* 58, 1705–1710.

Mazars, E., Ödberg-Ferragut, C., Durand, I., Tibayrenc, M., Dei-Cas, E. and Camus, D. (1994) Genomic and isoenzymatic markers of *Pneumocystis* from different host species. *Journal of Eukaryotic Microbiology* 41, S104.

Mazars, E., Ödberg-Ferragut, C., Dei-Cas, E., Fourmaux, M.N., Aliouat, E.M., Brun-Pascaud, M., Mougeot, G. and Camus, D. (1995) Polymorphism of *Pneumocystis carinii* from different host species. *Journal of Eukaryotic Microbiology* 42, 26–32.

Meade, J.C. and Stringer, J.R. (1991) PCR amplification of DNA sequences from the transcription factor IID and cation transport ATPase genes in *Pneumocystis carinii*. *Journal of Protozoology* 38, 66–68.

Millard, P.R. and Heryet, A.R. (1988) Observations favouring *Pneumocystis carinii* pneumonia as a primary infection: a monoclonal antibody study on paraffin sections. *Journal of Pathology* 154, 365–370.

Palluault, F., Dei-Cas, E., Slomianny C., Soulez, B. and Camus, D. (1990) Golgi complex and lysosomes in rabbit-derived *Pneumocystis carinii*. *Biology of the Cell* 70, 73–82.

Palluault, F., Slommianny, C., Dei-Cas, E., Soulez, B. and Camus, D. (1992a) High osmotic pressure enables fine ultrastructural and cytochemical studies on *Pneumocystis carinii*. I. Epon embedding. *Parasitology Research* 78, 437–444.

Palluault, F., Soulez, B., Slommianny, C., Dei-Cas, E. and Camus, D. (1992b) High osmotic pressure for *Pneumocystis carinii* LR White embedding allows fine ultrastructural and cytochemical studies. I. Golgi complex and cell wall synthesis. *Parasitology Research* 78, 482–488.

Peters, S.E., Wakefield, A.E., Sinclair, K., Millard, P.R. and Hopkin, J.M. (1992) A search for *Pneumocystis carinii* in post-mortem lungs by DNA amplification. *Journal of Pathology* 166, 195–198.

Pifer, L.L., Lattuada, C.P., Edwards, C.C., Woods, D.R. and Owen, D.R. (1984) *Pneumocystis carinii* infection in germ-free rats: Implications for human patients. *Diagnosis, Microbiology and Infectious Diseases* 2, 23–36.

Pixley, F.J., Wakefield, A.E., Banerji, S. and Hopkin, J.M. (1991) Mitochondrial gene sequences show fungal homology for *Pneumocystis carinii*. *Molecular Microbiology* 5, 1347–1351.

Raviglione, M.C. (1990) Extrapulmonary pneumocystosis: the first 50 cases. *Review of Infectious Diseases* 12, 1127–1138.

Santiago-Delpin, E.A., Mora, E., Gonzalez, Z.A., Morales-Otero, L.A. and Bermudez, R. (1988) Factors in an outbreak of *Pneumocystis carinii* in a transplant unit. *Transplantation Proceedings* 20, 462–465.

Schmatz, D., Powles, M., Mcfadden, D.C., Pittarelli, L.A., Liberator, P.A. and Anderson, J.W. (1991) Treatment and prevention of *Pneumocystis carinii* pneumonia and further elucidation of the *P. carinii* life cycle with 1,3-β-glucan synthesis inhibitor L-671,329. *Journal of Protozoology* 38, 151–153.

Sinclair, K., Wakefield, A.E., Banerji, S. and Hopkin, J.M. (1991) *Pneumocystis carinii* organisms derived from rat and human hosts are genetically distinct. *Molecular Biochemical Parasitology* 45, 183–184.

Soulez, B., Palluault, F., Cesbron, J.Y., Dei-Cas, E., Capron, A. and Camus, D. (1991) Introduction of *Pneumocystis carinii* in a colony of SCID mice. *Journal of Protozoology* 38, 123–125.

Stringer, J.R. (1993) The identity of *Pneumocystis carinii*: not a single protozoan, but a diverse group of exotic fungi. *Infectious Agents and Disease* 2, 109–117.

Stringer, S.L., Stringer, J.R., Blase, M.A., Walzer, P.D. and Cushion, M.T. (1989) *Pneumocystis carinii*: sequence from ribosomal RNA implies a close relationship with fungi. *Experimental Parasitology* 68, 450–461.

Wakefield, A.E. (1994) Detection of DNA sequences identical to *Pneumocystis carinii* in samples of ambient air. *Journal of Eukaryotic Microbiology* 41, S116.

Wakefield, A.E., Stewart, T.J., Moxon, E.R., Marsh, K. and Hopkin, J.M. (1990) Infection with *Pneumocystis carinii* is prevalent in healthy Gambian children. *Transactions of the Royal Society of Tropical Medicine and Hygiene* 84, 800–802.

Wakefield, A.E., Peters, S.E., Banerji, S., Bridge, P.D., Hall, G.S., Hawksworth, D.L., Guiver, L.A., Allen, A.G. and Hopkin, J.M. (1992) *Pneumocystis carinii* shows DNA homology with the Ustomycetous red yeast fungi. *Molecular Microbiology* 6, 1903–1911.

Walzer, P.D. (1991) Immunopathogenesis of *Pneumocystis carinii* infection. *Journal of Laboratory Clinical Medicine* 118, 206–216.

Weinberg, G.A. and Bartlett, M.S. (1991) Comparison of pulsed field gel electrophoresis karyotypes of *Pneumocystis carinii* derived from rat lung, cell culture and ferret lung. *Journal of Protozoology* 38, 64–65.

Ypma-Wong, M.F., Fonzi, W.A. and Sypherd, P.S. (1992) Fungus-specific translation elongation factor 3 gene present in *Pneumocystis carinii*. *Infection and Immunity* 60, 4140–4145.

18

Schistosomiasis and Food-borne Trematode Infections: Diseases of Social Dimensions

K.E. MOTT

Schistosomiasis Control Unit, Division of Control of Tropical Diseases, World Health Organization, Geneva 27, Switzerland

The World Health Organization (WHO) reflects its member states. We need their support and persistence to assist WHO to set the correct priorities and provide feasible solutions. The other chapters in this volume largely deal with new scientific data. This gives me the unique opportunity to raise questions for which the answers have been or will be provided by others.

Social Disease

'Les maladies de caracter sociale', or diseases with social characteristics, have special meaning in government budgets. Due to their social characteristics, leprosy and tuberculosis are two diseases for which the funds for the purchase of drugs for treatment derive from the central government budget rather than that of the Ministry of Health. Neither schistosomiasis nor food-borne trematode infections currently enjoy this status.

The social aspects of trematode infections are intuitively recognized. But to varying degrees our scientific and biological thinking interferes with our contributions to the pragmatic task of control. It is not a profound statement to say that these diseases are social diseases.

We are familiar in our circles with the biological statement that one egg can generate millions of cercariae. However, the concept that one person can infect many others, the public health or social statement, is only just beginning to take effect. If you have any doubts about the social nature of diseases, one only needs to go near any freshwater body in any endemic country to observe the social nature of water contact. Social interaction determines schistosomiasis distribution.

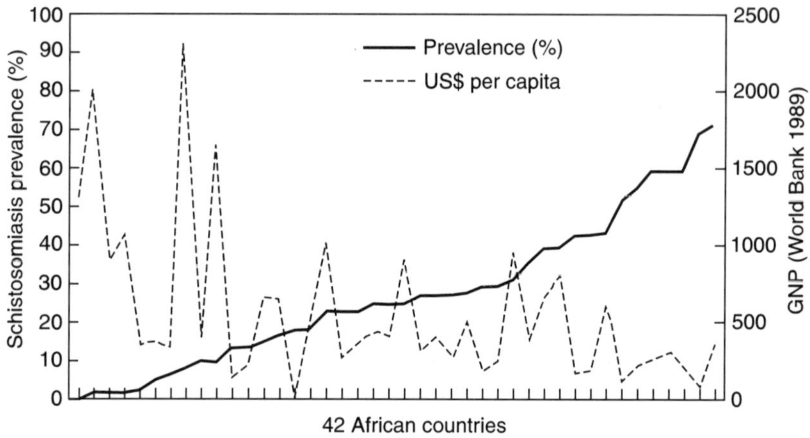

Fig. 18.1. Relationship between prevalence of schistosomiasis and GNP.

Schistosomiasis and gross national product (GNP)

More profoundly, however, schistosomiasis is a disease of poverty. The GNP of the endemic African countries is nearly inversely proportional to the prevalence of schistosomiasis (Fig. 18.1). Schistosomiasis is promoted by poverty and the lack of means to deal with it assures that it will remain a problem.

Schistosomiasis and primary schools

A similar correlation is shown between level of school enrolment and the prevalence of the disease (Fig. 18.2). Poverty and ignorance are the foundations of disease.

Disease as a Product

Product life cycle: a business concept

It is fashionable to go to the economic sector to look for explanations of social phenomena. The 'product life cycle' (Fig. 18.3) is a standard concept for all students of economy and business administration. If we change the parameters of this concept, this can be called a 'disease life cycle' (Fig. 18.4).

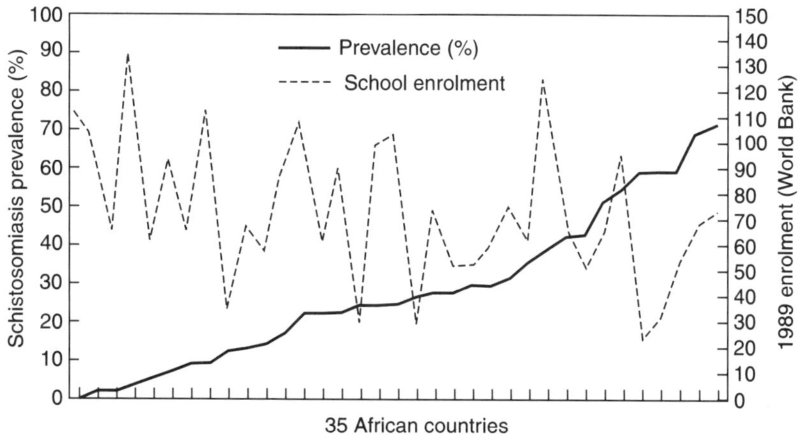

Fig. 18.2. Relationship between prevalence of schistosomiasis and enrolment in primary schools.

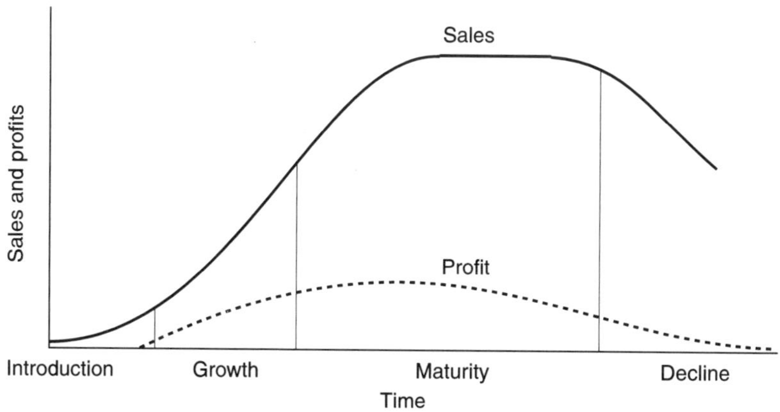

Fig. 18.3. Product life cycle: a business concept.

Disease life cycle: an epidemiological concept

In the endemic interval one can shift these curves in many directions. On the other hand, at the end of the time scale, when eradication approaches, the intensity of infection drops to such low levels that prevalence may be a weak measurement of the true situation, as in China or Morocco.

We can review the geographical dispersion of schistosomiasis over time. In 1972 it had been reported in 71 countries where 180–200 million persons were estimated to be infected. In Israel, the focus

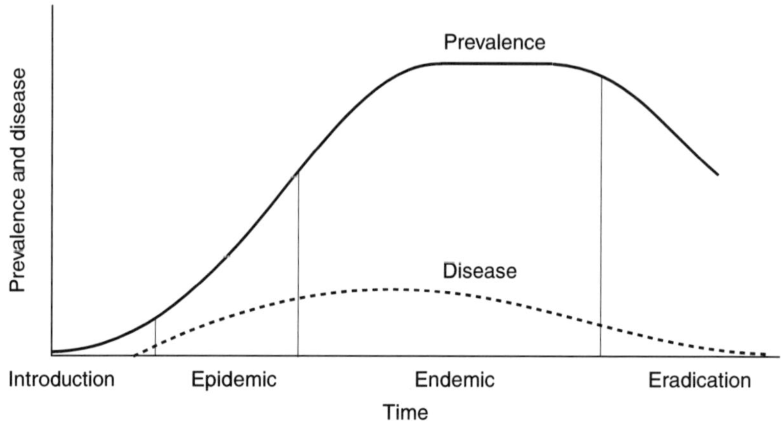

Fig. 18.4. Disease life cycle: an epidemiological concept.

of *Schistosoma mansoni* in the Yarkon River area and the focus of *Schistoma haematobium* in the Algarve, Portugal, had disappeared. At the time it was first noted that schistosomiasis was beginning to be reported in urban areas. With the spread of perennial irrigation in lower Egypt following the construction of the Aswan Dam it was suggested that the prevalence and intensity of schistosomiasis would be on the rise. A similar prediction was made regarding Lake Volta.

In 1994, there were 74 endemic countries, but transmission has ceased in Montserrat, Tunisia and Japan. There are still areas of change. Schistosomiasis is increasingly an urban disease in the northeast of Brazil and in West Africa. The extent of the problem in Laos and Cambodia continues to grow. Transmission has returned in some areas of central China where it had been eradicated. In the estuaries of rivers in Egypt and Ghana with large upstream dams (Aswan in Egypt and Akasombo and Kpong in Ghana) *S. mansoni* has displaced *S. haematobium* as the most prevalent infection. Most recently in Mauritania, *S. mansoni* is shown to be present for the first time and at over 30% prevalence in villages around Rosso.

Diseases of Development

As social diseases, schistosomiasis and food-borne trematode infections are also diseases of development. The recent WHO publication *Parasitic Diseases in Water Resources Development* (Hunter *et al.*, 1994) has comprehensively documented the link between schistosomiasis and water resources development as a basis for strengthening

Box 18.1. Risks in urinary schistosomiasis due to *S. haematobium.*

- 2 : 10 Bladder polyps
- 1 : 10 Hydronephrosis
- 1 : 1000 Bladder cancer
- 1 : 1000 Death

Box 18.2. Risks in intestinal schistosomiasis due to *S. mansoni* and *Schistosoma japonicum.*

- 1 : 10 Hepatosplenomegaly
- 1 : 20 Portal hypertension
- 1 : 1000 Haematemsis
- 1 : 10,000 Death

the role of ministries of health in the intersectoral dialogue. There are no simplistic answers as to how to achieve a greater financial input for mitigation and prevention of these diseases in development. Unfortunately, the people who are most affected have no political forum and few advocates who express how important these diseases might be to them.

Disabling Diseases

The World Development Report on Health of the World Bank (1993) has had two major impacts:

1. It sends a message to governments that health is a rational investment of national resources.
2. We need to improve how to determine the effects of parasitic diseases on morbidity, disability and mortality.

Since the time of the Han dynasty in Hunan Province of China, schistosomiasis has been recognized as a chronic infection which is cumulative in the lifetime of any person at risk of infection in the endemic area. At each succeeding year after the initial infection the probability of disability and disease increases over the previous year. This probability will be determined by the independent variables of: initial intensity of infection, the age of the victim when first infected and the rate of accumulation of infection due to exposure. I have purposely left out the immune side of this equation; there are other papers and publications which deal with that.

Box 18.3. Risks in all forms of schistosomiasis.

- 1 : 100,000 Partial CNS-related motor disability
- 1 : 200,000 Paraplegia

These are variables that we can do something about. The following estimates of risk will be adjusted with further studies and verification by comparison with real data. They are included here for critical review. The information in Boxes 18.1 and 18.2 will be published later in the WHO\World Bank Review.

For central nervous system (CNS) involvement, in many parts of the world, we have only cumulative case reports rather than systematic reporting (Box 18.3).

Most recently, we have encouraging news from the cancer specialists. Now in a bold move, the International Agency for Research on Cancer (IARC) has started to look at biological causes of cancer (Table 18.1). This is good news and means that parasitic causes of cancer will benefit from increased research effort.

Costs of Control

Earlier, I considered the epidemiological changes over 25 years. Now it is appropriate to look at changes in the cost of drugs, as these are the tools used to control the disease (Table 18.2).

The low cost of drugs to treat schistosomiasis has not resolved the problem of control. To prevent the application of drugs becoming part of the problem and to ensure that their use contributes to the solution, their role should be understood and they should be appropriately used. Questions are raised by their use, for example, after semiannual and annual treatment of school-age children in Pemba with drugs, the prevalence of haematuria initially declined, reaching a plateau of reduced infection after 2 years. This level was maintained for a further 3 years, but then started to rise. No other effective interventions were undertaken (L. Savioli and M. Albonico and the Ministry of Health of Zanzibar, personal communication).

Other Challenges

There are many problem areas associated with food-borne trematode infections (WHO, 1995). The freshwater fish involved in the life cycle of *Clonorchis* and *Opisthorchis* are closely linked to aquaculture, thus

Table 18.1. Carcinogenicity of biological agents (IARC, 1994).

Organism	Carcinogenicity
Schistosoma haematobium	A cause of bladder cancer
Opisthorchis viverrini	A cause of cholangiocarcinoma
Clonorchis sinensis	Probably a cause of cholangiocarcinoma
Schistosoma japonicum	Possibly a cause of colorectal cancer
Schistosoma mansoni	Not a cause of cancer
Opisthorchis felineus	Data not available to assess it as a cause of cholangiocarcinoma

Table 18.2. Cost of a single treatment of a 25 kg child by current antischistosomal drugs (based on US$ bulk purchase prices).

Drug	1984	1994
Praziquantel	1.39	0.14
Metrifonate	0.13	0.18
Oxamniquine	0.89	0.98
1US$ =	2.40 DM	1.65 DM

their economic potential in trade may be limited by the new GATT regulations.

Similarly, further research and epidemiological surveillance is needed with regard to those organisms that are responsible for swimmer's itch or paddy itch. Cercerial dermatitis or swimmer's itch due to a wide range of schistosomes (mostly avian parasites) appears to be increasing in central European countries. It is also present in many developing tropical countries.

Conclusion

Ten years ago, WHO attempted to reorient its health education activities in the schistosomiasis control programme by focusing on the person who is infected; this same person is the source of transmission. As we move into a global free market era governments are recognizing that public health legislation may have to be reinforced. Several endemic countries are reviewing older public health laws. If these are not reinforced, there may be dire consequences.

Ten years ago, WHO started with new goals. But, as Herakleitos said in Ephesus in the 5th century BC: 'No one can wash his feet twice in the same water because everything is changing; everything is turning'.

We can now say that these diseases have new dimensions. Social dimensions are not the least of these. There are major social changes under way: pushed by economic prerogatives of the free market, the effectiveness of new technology to bring information quickly and efficiently and upset by political and civil unrest as never before. On one hand we have the tools, on the other hand we have the same problems.

References

Hunter, J., Rey, L., Chu, K.Y., Adekolu-John, E.O. and Mott, K.E. (1993) *Parasitic Diseases in Water Resources Development: The Need for Intersectoral Negotiation*. World Health Organization, Geneva, 152 pp. (This publication is now available free of charge in French, English or Spanish due to a generous grant from the Rockefeller Foundation. It is available free of charge for developing countries in large quantities for conferences/seminars on this topic.)

International Agency for Research on Cancer (IARC) (1994) *Schistosomes, Liver Flukes and* Helicobacter pylori. IARC Monographs on the Evaluation of Carcinogenic Risks of Humans, Vol 61. IARC, Lyon, 270 pp.

World Bank (1993) *World Development Report 1993*. Washington, DC, 329 pp.

World Health Organization (1993) *The Control of Schistosomiasis*. Second Report of the WHO Expert Committee. Technical Report Series no. 830, 86 pp. WHO, Geneva.

World Health Organization (1995) *The Control of Foodborne Trematode Infections*. Report of a WHO Study Group. Technical Report Series no. 849, 157 pp. WHO, Geneva.

19

Genetic Variability in Parasitic Helminths

D.P. McManus

Molecular Parasitology Unit, Tropical Health Program, The Queensland Institute of Medical Research, 300 Herston Road, Brisbane 4029, Queensland, Australia

Introduction

The study of molecular variation in parasites is an area of increasing importance, as most, if not all, parasites so far investigated exhibit at least some genetic variability. Such parasite variability has profound implications for drug, diagnostics and vaccine development and a special thematic issue of *Acta Tropica* (Vol. 53, 1993, 165–348) was recently devoted to a series of reviews and papers on 'Molecular variation in parasites and its implications'. A discussion of molecular variation and the implications of such variability in three groups of parasitic helminths, *Echinococcus*, *Taenia* and *Schistosoma japonicum* is highlighted here.

Genetic Variability in *Echinococcus*

Of the four accepted species in the genus *Echinococcus*, namely *Echinococcus granulosus*, *Echinococcus multilocularis*, *Echinococcus vogeli* and *Echinococcus oligarthrus*, only *E. granulosus* and *E. multilocularis* are of public health significance. Until relatively recently, there had been limited attempts made to define differences between them at the DNA level because of the lack of appropriate molecular techniques. However, the application of a number of DNA approaches, especially the polymerase chain reaction (PCR), is dramatically expanding our understanding of variation and taxonomy within *Echinococcus*. A full discussion of newly acquired molecular information on genetic variation in *Echinococcus* has been provided by Bowles and McManus (1993a) and in several chapters of Thompson and Lymbery (1995).

Variation within certain species of *Echinococcus* is now a well-accepted phenomenon (Thompson and Lymbery, 1988) and a number of intraspecific variants or strains of *E. granulosus* have been characterized using a range of procedures. These studies have been carried out as genetic diversity may reflect differences in infectivity, especially to humans, with important implications for the epidemiology of hydatid disease. In addition, the phenomenon of strain variation is an important consideration in the future design and development of vaccines, diagnostic reagents and drugs effective against the *Echinococcus* organisms.

DNA techniques developed for assessing genetic variability in *Echinococcus*

A number of traditional approaches including geographical distribution, host range, host specificity, antigenic diversity, metabolism, developmental rate, reproductive biology, growth *in vitro*, infectivity, morphology and protein and isoenzyme analysis have been usefully employed for characterizing *Echinococcus* (Thompson and Lymbery, 1988). However, DNA sequence analysis offers the most direct approach to the characterization of distinct species, subspecies and strain groups. For the purpose of identification and assessing genetic variability, it is not necessary that genetic characters are directly related to functional or other phenotypic differences between forms. It is sufficient that they consistently act as markers of particular known epidemiological types.

Methods which examine the parasite genome directly are not confounded by variability induced by environment or host, or associated with life cycle stage or post-translational modification. Genetic variation can be investigated in either the mitochondrial or nuclear genome. Mitochondrial DNA (mtDNA) is useful for the discrimination of closely related organisms because of its relatively rapid rate of evolution. Furthermore, as mtDNA is haploid, allele haplotypes can be determined unambiguously. mtDNA has the additional advantage that it does not recombine, thus simplifying analysis. The nuclear ribosomal RNA (rRNA) gene repeat unit has different regions evolving at varying rates and so has also been used extensively to study variation and phylogeny at a number of different taxonomic levels.

RFLP and PCR-RFLP analysis

Early DNA studies of genetic variation in *Echinococcus* (McManus and Simpson, 1985; McManus *et al.*, 1987; McManus and Rishi, 1989) involved restriction fragment length polymorphism (RFLP) analysis using a Southern blotting approach whereby genomic DNA is restriction enzyme-digested and the resulting restriction fragments are electrophoretically separated (by virtue of size) on an agarose gel. The DNA fragments are transferred to a nitrocellulose or nylon filter and hybridized with a specific DNA probe that has been radioactively or otherwise labelled. Under the appropriate conditions of salt and temperature, the probe binds to complementary DNA sequences in fragments immobilized on the filter. After washing to remove non-bound probe and non-specific associations, the filter is exposed to X-ray film or a colorimetric detection system to reveal RFLPs. The banding pattern which results is determined by the absence or presence of restriction enzyme sites in DNA sequences that hybridize to the probe. Characteristic banding patterns are detected and the identity or level of similarity of isolates can be inferred by comparison of these patterns. The approach is able to distinguish unambiguously a number of distinct strains of *E. granulosus* and an extensive study (McManus and Rishi, 1989) of hydatid material collected in the United Kingdom has shown that the RFLP patterns are stable within a particular strain. The method, therefore, provides the basis of a relatively simple method which can distinguish between discrete strains of *E. granulosus*.

The advent of PCR has revolutionized genetic studies. It allows the rapid selection, isolation, amplification and sequencing of DNA from small amounts of tissue. In a modified version of the conventional RFLP procedure, PCR fragments are digested with restriction endonucleases, electrophoretically separated and their characteristic patterns visualized directly after ethidium bromide staining.

As described above, *E. granulosus* strains can be readily distinguished by conventional RFLP analysis. This RFLP procedure has now been greatly simplified, without loss of resolution or accuracy, by linking rDNA RFLP analysis with the polymerase chain reaction (PCR) (Bowles and McManus, 1993b). A great advantage of this method is its remarkable sensitivity.

Random amplified polymorphic DNA

The random amplified polymorphic DNA (RAPD)-PCR also called arbitrary priming-PCR (AP-PCR), is a DNA polymorphism assay based on the amplification of genomic DNA using a single oligonucleotide primer of arbitrary nucleotide sequence. Using carefully controlled conditions, RAPD-PCR has proved useful for distinguishing *Echinococcus* species and genetically distinct strains of *E. granulosus* (Scott and McManus, 1994). Although the method amplifies vertebrate DNA, the resulting patterns do not interfere with the identification of *Echinococcus* isolates. Consequently, the technique can still be used effectively to distinguish *Echinococcus* species and strains even if host contamination occurs during the course of parasite collection. Despite the advantages of using RAPD-PCR for characterizing isolates of *Echinococcus*, great care must be taken to use standardized reaction conditions, particularly the quantity of template DNA, to ensure

Table 19.1. *Echinococcus* genotypes as determined by mitochondrial DNA[1] sequencing (data from Bowles *et al.*, 1992a: Bowles and McManus, 1993c).

Genotype[2]	Host infected
G1	Sheep, human, kangaroo, cattle, camel, pig, goat
G2	Sheep (Tasmania)
G3	Buffalo (India)
G4	Horse (UK, Ireland, Switzerland), donkey (Ireland)
G5	Cattle (Switzerland, Holland), buffalo (India), human (Holland)
G6	Camel (Kenya, Somalia, Sudan), goat (Kenya)
G7	Pig (Poland)
G8	Moose (USA)
M1	Human, rodent (China, North America)
M2	Human, rodent (Europe)
V	Rodent (South America)
O	Rodent (Panama)

[1] A comparison of two genes, CO1 and NADH dehydrogenase 1, independently grouped the *Echinococcus* isolates into the genotypes shown with the exception that the G2 and G3 groups could not be distinguished by NADH dehydrogenase 1 sequencing.
[2] G1–G8, *E. granulosus*; M1, M2, *E. multilocularis*; V, *E. vogeli*; O, *E. oligarthrus*.

consistent results. The approach is best used in tandem with one of the other DNA-based methods recently developed, notably PCR-RFLP analysis (Bowles and McManus, 1993b), or mitochondrial DNA sequencing (see below) for identification of newly collected isolates.

Primary sequence comparison

Comparison, between organisms, of the nucleotide sequence of defined DNA segments offers the most direct and sensitive means of detecting genetic variation. The advent of PCR has made sequence comparison a feasible approach to the study of genetic variation and data for a portion (366 base pairs (bp)) of the mitochondrial protein-coding gene, cytochrome *c* oxidase subunit 1 (CO1), has been sequenced for 60 *Echinococcus* isolates (Bowles *et al.*, 1992a, 1994). Twelve different genotypes were detected, including eight within *E. granulosus* (Table 19.1). The four generally accepted *Echinococcus* species (see above) were clearly distinguishable using this approach, with levels of nucleotide variation of the order of 6–11.5%. Interestingly, the level of nucleotide variation between the sheep, horse, cattle and camel strains of *E. granulosus* is about 5–9%, which is not substantially lower than that observed between species. The consensus view of the strain pattern within *E. granulosus*, based on a variety of criteria, is broadly upheld. Furthermore, remarkable intrastrain homogeneity was found at the DNA sequence level. Two variants were found among *E. multilo-*

cularis isolates and only a single sequence type in each of *E. vogeli* and *E. oligarthrus*.

An examination of sequence variation (70 isolates) in a 471 bp region of the mitochondrial NADH dehydrogenase I (ND1) gene gave the same groupings of variants (Bowles, 1993; Bowles and McManus, 1993c; Bowles *et al.*, 1994), but with slightly less resolution. Two very similar variants (G2 and G3) detected in the CO1 study could not be separated from one another using the ND1 data.

The application of DNA approaches in epidemiological studies on *E. granulosus*

The value of using DNA methodologies in epidemiological surveys on *E. granulosus* has fully confirmed the distinctiveness between, but uniformity within, the sheep/dog and horse/dog forms of *E. granulosus* in the United Kingdom (McManus and Simpson, 1985; McManus and Rishi, 1989; Bowles *et al.*, 1992a; Bowles and McManus, 1993b), confirmed the presence and revealed the host preferences of sheep and camel strains in Kenya (Wachira *et al.*, 1993), confirmed that a single genotypic strain of *E. granulosus* cycles in domestic and sylvatic hosts on mainland Australia (Hope *et al.*, 1991, 1992), and shown that the cattle strain is infective to humans (Bowles *et al.*, 1992b). In addition, it was shown in a molecular genetic survey that all of 117 *E. granulosus* isolates examined from a range of intermediate hosts, including humans, in northwestern China were infected with the sheep (G1 genotype) of *E. granulosus* (McManus *et al.*, 1994). The findings of each of these studies have considerable implications for public health.

Characterization of the cervid strain of *E. granulosus*

Genetic methods have also been used to characterize the cervid strain or 'northern form' of *E. granulosus* from North America and Eurasia (Bowles *et al.*, 1994). The wolf is the principal definitive host and moose and reindeer (*F. cervidae*) serve as intermediate hosts; cycles involving dogs and domesticated reindeer also occur (Rausch, 1986). *E. granulosus* of cervid origin differs in a number of biological and clinical respects from domestic strains of the parasite (Rausch, 1986; Thompson and Lymbery, 1988).

Four isolates of the cervid strain were obtained from an infected moose and examined. PCR-RFLP analysis of the nuclear internal transcribed spacer (ITS1) region of the rDNA repeat was able to readily distinguish the cervid form from other strains of *E. granulosus* (Fig. 19.1). The complexity of the RFLP patterns obtained with the moose isolates suggested that a number of distinct ITS1 types are present. In fact, four cervid strain clones of PCR-amplified ITS1 were shown to be different

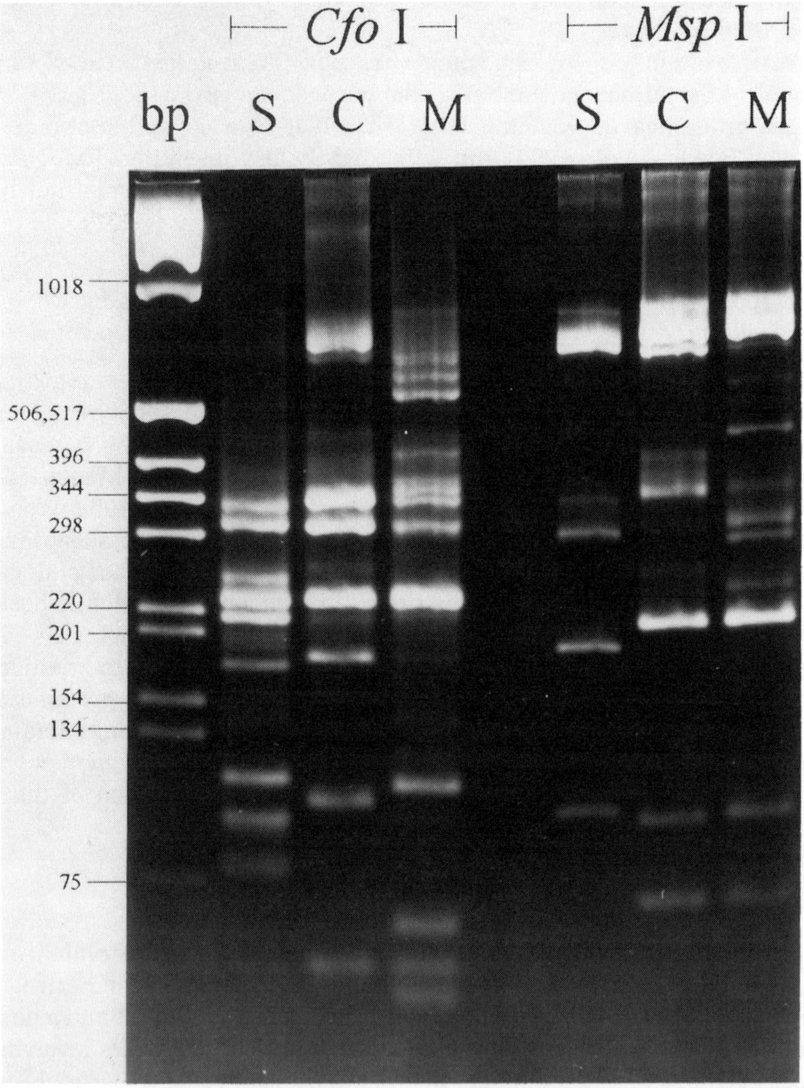

Fig. 19.1. A comparison of *E. granulosus* cervid (G8 genotype), sheep (G1 genotype) and camel (G6 genotype) strain PCR-RFLP patterns. Restriction endonucleases used to digest PCR amplified ITS1 fragments obtained with DNA samples isolated from each of the strains are indicated. Full details of the method are described in Bowles and McManus (1993b). Lane bp, molecular weight markers, lane S, sheep strain; lane C, camel strain; lane M, cervid strain.

by PCR-RFLP analysis. In subsequent analysis (Bowles *et al.*, 1995), a wide range of ITS1 sequence variation was shown among these clones. The cervid genotype is of further interest in terms of its mitochondrial ND1 and CO1 sequence. The four moose isolates gave identical sequences for both mitochondrial regions but each of the two regions implied different relationships for the strain. The CO1 sequence had ambiguities at a number of sites but was most similar to that of the sheep strain (G1 and G2) whereas the ND1 sequence most resembled camel and pig strains (G6, G7).

These data suggest that the cervid strain may represent an *E. granulosus* form of hybrid origin, but the situation is complex and difficult to explain at present. The persistence of at least four distinct types of ITS1 sequence, and the complexity of the ITS1 PCR-RFLP patterns, suggest that the cervid strain might be a relatively recently evolved variant. It appears that the process of concerted evolution has not yet homogenized different types of rDNA repeat within the genome. The moose isolates do appear to represent a distinct genetic form as evidenced by a unique ND1 sequence and ITS1 PCR-RFLP patterns. Rausch (1986) considered the northern form to be ancestral to the domestic strains of *E. granulosus* which he contended became adapted to synanthropic hosts with the development of animal husbandry. However, although more molecular data is required before firm conclusions can be made on the status of the cervid strain, phylogenetic analysis of available sequence information (Bowles *et al.*, 1995) does not support the hypothesis that this strain represents the ancestral form of *E. granulosus*.

Further comments on the phylogeny of the genus *Echinococcus*

Phylogenetic relationships among *Echinococcus* species and strains are far from clear based on morphological data (see Lymbery, 1992) and adequate biochemical data are not available. The very large number of characters made available by molecular analysis, and in particular DNA sequence analysis, can prove very useful in the refining of the available morphologically based phylogenies. Some analysis of sequence data, involving three nucleotide datasets – two mitochondrial (CO1 and ND1) and one nuclear (ribosomal ITS1) – has commenced in an attempt to further define the interrelationships amongst the species and strains of *Echinococcus* (Bowles *et al.*, 1995). The data has failed to support the hypothesis that *E. granulosus*, as it is currently viewed, is a single valid species. Rather, the strains of *E. granulosus* seem to comprise at least three evolutionarily diverse groups (sheep, bovine and horse strain groups) with molecular distances between them being comparable to, or greater than, molecular evolutionary distances observed between recognized species. The data also suggest that *E. multilocularis* may not be distinct from *E. granulosus* although *E. vogeli* and *E. oligarthrus* appear distinct and rather distant from the first two. Overall, the molecular data indicates that taxonomic revision of the *Echinococcus* genus is clearly warranted.

In conclusion, it should also be emphasized that direct DNA analysis has not confirmed the very high level of genetic polymorphism that has been reported within *E. granulosus* at the population level using protein electrophoresis techniques. Presumably, by looking at even more variable regions of the genome, it should be possible to detect variation between any two non-clonal individuals. Such study is, however, not warranted unless further refinements are made to our knowledge of the biological and epidemiological features of *Echinococcus* species and strains.

Genetic Variability in *Taenia*

Similar molecular approaches to those used to characterize *Echinococcus* species and strains have been used to determine the taxonomic status of a third form of human *Taenia* which occurs in Southeast Asia and which is distinguishable from *Taenia saginata* and *Taenia solium*. The precise taxonomic classification of this Asian *Taenia* has stimulated wide interest and, as two recent publications (Bowles and McManus, 1994; McManus and Bowles, 1994), have covered the area in great depth, only a brief summary of the recent findings is presented here.

Taenia saginata and *T. solium* have a cosmopolitan distribution, but in the Asian–Pacific region, *T. saginata* has long been considered the dominant species. This is despite the fact that in Taiwan, Korea, Indonesia and the Philippines, people frequently eat raw or undercooked pork, but rarely eat beef. Extensive studies on the epidemiological pattern of taeniasis in Southeast Asia have shown that it is another form of human *Taenia* which probably accounts for this paradox. This recently recognized form, originally described in Taiwanese aboriginals and hence termed 'Taiwan *Taenia*', has now been recorded in Korea, Indonesia, the Philippines and Thailand, and is now generally referred to as Asian *Taenia*.

A number of features distinguish Asian *Taenia* from the other human taeniids. It can infect cattle, goats, monkeys and wild boars, but pigs are the dominant intermediate hosts and the cysticerci are found generally in the liver although experimental infection of pigs has shown that extrahepatic organs (omentum, lungs and serosa of the colon) can account for at least 30% of the infection. These extrahepatic organs may, in fact, be the primary source of natural human transmission of

Asian *Taenia* given the eating habits of infected individuals. Earlier DNA studies with Asian *Taenia* suggested similarity with *T. saginata*, as does its gross adult morphology, a factor that has clearly confused diagnosis of the parasite in the past. However, a conspicuous rostellum occurs on the scolex of the adult worm and cysticercus of Asian *Taenia* and uterine buds (very short uterine branches without further branching) occur in the gravid proglottides. Furthermore, in contrast to *T. saginata* cysticerci, those of Asian *Taenia* are smaller, they are armed with two rows of (rudimentary) hooklets and have a shorter developmental period (4 weeks, compared with 8–18 weeks). In addition to the larval morphology, immunological evidence and the principal intermediate host – the pig – used by the Asian *Taenia* suggest similarity with *T. solium*.

Thus, until recently, the precise taxonomic status of Asian *Taenia* remained unresolved. The new species name of *Taenia asiatica* has recently been proposed, based on morphological criteria, although a new molecular genetic study (Bowles and McManus, 1994) does not support the reclassification of Asian *Taenia* as a distinct species. The new study examined sequence variation in the 5′ (large subunit) rRNA gene, including most of the rapidly evolving D1 region, and the CO1 gene for a number of taeniid species so that the genetic distances between *T. saginata*, *T. solium* and the Asian *Taenia* could be compared with the distances between accepted species in the family Taeniidae (Table 19.2). In general, distinct sequences were obtained with inter-species differences for recognized species within the genus *Taenia* ranging from 5 to 19 nucleotides for the former and 23 to 58 nucleotides for the latter gene. However, the 5′ 28S rRNA sequence for Asian *Taenia* is identical to that of *T. saginata* although the CO1 sequence differs in nine of the 366 nucleotide positions examined with one of these changes resulting in a coding change. The sequence information indicates *T. saginata* and Asian *Taenia* appear to be more closely related than the other species within the genus.

The molecular genetic study fully supports earlier conclusions that the Asian *Taenia* is a genetically distinct entity but is closely related to *T. saginata*. The evidence suggests that classification of the Asian *Taenia* as a subspecies or strain of *T. saginata* is more appropriate than its designation as a new species. Whereas *T. solium* cysticercosis is a serious and chronic disease, *T. saginata* cysticercosis, if it occurs at all, is an extremely rare phenomenon in humans. The molecular data supports the evidence that the Asian *Taenia* is much more closely related to *T. saginata* than to *T. solium*, the public health implication being that the Asian form is unlikely to be an important cause of human cysticercosis. This argument is strengthened by the fact that, despite the high prevalence and intensity of adult Asian *Taenia*

Table 19.2. Nucleotide differences between pairs of taeniid species for the mitochondrial CO1 gene (above the diagonal) and for the 5′ 28S rDNA region (below the diagonal) (data from McManus and Bowles, 1994).[1]

	Taenia saginata	Asian *Taenia*	*Taenia solium*	*Taenia multiceps*	*Taenia ovis*	*Taenia hydatigena*	*Taenia pisiformis*	*Taenia crassiceps*	*Taenia taeniaeformis*	*Echinococcus granulosus*
Taenia saginata	–	9	47	23	41	46	41	44	55	68
Asian *Taenia*	0	–	50	28	42	47	45	45	58	71
Taenia solium	9	8	–	39	39	52	45	55	57	68
Taenia multiceps	6	6	4	–	38	45	42	47	54	75
Taenia ovis	5	5	7	7	–	46	50	53	58	61
Taenia hydatigena	10	10	10	10	9	–	57	42	56	60
Taenia pisiformis	17	17	15	14	14	14	–	53	42	70
Taenia crassiceps	13	13	12	12	8	11	14	–	57	60
Taenia taeniaeformis	19	19	18	17	15	17	16	14	–	63
Echinococcus granulosus	21	21	18	18	17	18	17	16	18	–

[1] DNA was extracted from frozen, ethanol-preserved or freshly obtained isolates using standard methods. Partial sequences were obtained for the CO1 gene (~ 366 bp) and the 28S rDNA D1 region (~ 300 bp) of the various taeniids by PCR and direct sequencing (see Bowles and McManus, 1994 for experimental details). Sequences for the related taeniid, *E. granulosus* (sheep strain), were included for comparison. Sequence for the 28S 5′ region was obtained for three Asian *Taenia* isolates (invariant sequence obtained) and a single representative of each of the other taeniid species. Mitochondrial CO1 sequences were obtained for five Asian *Taenia* (invariant sequence obtained), three *T. saginata* (invariant sequence) and three *T. solium* (two identical sequences, minor sequence variation in the other) isolates together with a sequence for a single representative of each of the other taeniid species.

reported in Taiwanese aborigines, there appears to be no evidence of significant numbers of cysticercosis or neurocysticercosis cases in these populations.

Genetic Variability in *Schistosoma japonicum*

It is now recognized that intraspecific variation is widespread in the human schistosome species. A recent review (McManus and Hope, 1993) has described advances in the detection and characterization of variants within the human schistosomes and the implications of this variation in terms of infectivity and development in intermediate and definitive hosts, drug responsiveness, pathogenicity and immunogenicity. A brief discussion of intraspecific variation in the Asian schistosome, *S. japonicum*, is provided here along with some recent comparative data (see Bowles *et al.* 1993) obtained for geographical isolates of *S. japonicum* examined at the molecular level.

Geographic strains of *S. japonicum* have been described from China, the Philippines, Japan, Taiwan and Indonesia. The strains differ in their infectivity to the various subspecies of the snail intermediate host, *Oncomelania hupensis*, in adult worm morphology, in virulence and prepatent period, and in protein and isoenzyme profiles. The Philippine and mainland Chinese strains of *S. japonicum*, which are responsible for the majority of human schistosomiasis japonica cases, are considered to be the most distinct. Based on isoenzyme data, it has been estimated that these strains diverged one to two million years ago. Immunological differences have also been described. In particular, vaccination with irradiated Chinese *S. japonicum* cercariae protects against homologous but not heterologous challenge by Philippine *S. japonicum*, suggesting that the antigens involved in inducing protection have diverged significantly in the two strains. This lack of cross-protection suggests that a single vaccine may not be effective against both of the important *S. japonicum* strains. The reverse experiment has not been done, however, and it is possible that the Philippine *S. japonicum* is less susceptible to host immune defence mechanisms whether induced by homologous or heterologous cercariae.

Geographical isolates of *S. japonicum*, and particularly isolates from China and the Philippines, were examined at the molecular level for genetic divergence. Sequences from both nuclear and mitochondrial genomes were selected as markers of evolutionary divergence and *S. mekongi* and *S. mansoni* were included in the study for comparison purposes. The genomic regions chosen had proved useful as markers of species and strain identity in other schistosomes and, as described above, in *Echinococcus*. RFLP and PCR-RFLP analysis of the rDNA repeat unit and sequence analysis of the second ITS region (ITS2) within the rDNA repeat and the CO1 gene of the mitochondrial genome were performed. No intraspecific variation in *S. japonicum* was found in the rDNA repeat and only very slight variation was detected within the CO1 sequence (Fig. 19.2). A survey of the entire genome, using RAPD analysis, again showed that Chinese and Philippine *S. japonicum* are remarkably similar at the DNA sequence level.

It has been speculated that laboratory maintained strains of schistosomes cannot be considered representative of natural populations, probably because of interbreeding, selection and founder effects. It is possible that an analogous situation, on a larger time scale, has occurred in the evolution and divergence of *S. japonicum* strains. Parasites displaced from one geographic location to another would suddenly be subjected to a range of new conditions. In the case of *S. japonicum*, the major obstacle to survival in a new environment is likely to be the absence of the normal intermediate host. Rapid adaptive changes in the parasite phenotype, made possible by the presence of minor interindividual variation, could occur. If such changes are advantageous in the new environment, they would tend to become fixed in the population. As has been postulated for laboratory strains, inbreeding and founder effects might also play a role in the rapid alteration of the parasite phenotype. Because these changes would have occurred over a short period of time, the genome as a whole would not have evolved with them. Only minor differences would exist between individuals of the original and the new population.

The molecular analysis, involving direct methods and referred to above, suggests that, in general, the genomes of the Chinese and Philippine strains of *S. japonicum* are very similar. The results were unexpected but the data do not alter the fact that several other methods of characterization, particularly those based on isoenzyme analysis, have demonstrated that these organisms are morphologically and biologically distinct. It does suggest that, in some cases, sequence evolution may not be relevant to organism evolution. On a practical level, the conservation of much of the *S. japonicum* genome suggests that many potential vaccine candidiate genes, will similarly be conserved between strains.

Conclusions

Species and strains within the genera *Echinococcus* and *Taenia* have been defined previously by various identification and characterization methods. Here, the usefulness of some direct DNA approaches in the further characterization of strains and in solving particular systematic

```
S.mansoni                     GGG TTT GGT ATA GTT AGG CAT ATC TGT ATN AGT CTA
S.japonicum(China)            ..A ... ... ... ... ..T ... ..A ... ..G TC. T..
S.japonicum(Taiwan)           ..A ... ... ... ... ..T ... ..A ... ..G TC. T..
S.japonicum(Indonesia)        ..A ... ... ... ... ..T ... ..A ... ..G TC. T..
S.japonicum(Japan)            ..A ... ... ..  ... ..T ... ..A ... ..G TC. T..
S.japonicum(Philippines)      ..A ... ... ... ... ..T ... ..A ... ..G TC. T..
S.mekongi                     ..T ... ..A ... ..G ..T ... ..A ... ..G .C. T.G

AGG AAT AAA GAT TCG TNG TTT GGT TAT TAT GGA TTG ATT TGC GCT ATG GCT TCC ATA
..T ... ..T A.. ..T .C. ... ..A ... ... ..G ..A G.. ..T ... ... .G. ..T ..T
..T ... ..T A.. ..T .CA ... ..A ... ... ..G ..A G.. ..T ... ... .G. ..T ...
..T ... ..T A.. ..T .C. ... ..A ... ... ..G ..A G.. ..T ... ... .G. ..T ..T
..T ... ..T A.. ..T .C. ... ..A ... ... ..G ..A G.. ..T ... ... .G. ..T ..T
..T ... ..T A.. ..T .C. ... ..A ... ... ..G ..A G.. ..T ... ... .G. ..T ..T
..N ..A ..T A.. ..T .T. ... ... ... ... ..T ..A G.G ..T ... ... .G. ..T ...

GTA TGC TTA GGT AGA GTA GTA TGG GGT CAT CAT ATG TTT ATG GTT GGC TTT GAT TCG
..G ..T ..G ..G ... ..T ..T ... .C. ..C ... ... ... ... ... ..T A.G ... GTA
..G ..T ..G ..G ... ..T ..T ... .C. ... ... ... ... ... ... ..T A.G ... GTA
..G ..T ..G ..G ... ..T ..T ... .C. ... ... ... ... ... ... ..T A.G ... GTA
..G ..T ..G ..G ... ..T ..T ... .C. ... ... ... ... ... ... ..T A.G ... GTA
..G ..T ..G ..G ... ..T ..T ... .C. ... ... ... ... ... ... ..T A.G ... GTA
..G ..T ..G ... ... ... ..T ..A .C. ... ... ... ... ... ... ..T A.G ... GT.

TTA ACT GGA GTG TTT TTT AGT TCT ATT ACT ATG ATA ATA GGT GTT CCT ACT GGT ATT
AAG ... TCT ..T ... ... ... ... G.A ..A ... ..T ... ... A.A ..A ..A ... ...
AAG ... TCT ..T ... ... ... ... G.A ..A ... ..T ... ... A.A ..A ..A ... ...
AAG ... TCT ..T ... ... ... ... G.A ..A ... ..T ... ... A.A ... ..A ... ...
AAG .T. TTT ..T ..C ... ... ... G.A ..A ... ..T ... ... A.A ... ..A ... ...
AAG ... TCT ..T ... ... ... ... G.A ..A ... ..T ... ... A.A ..G ..A ... ...
AAG ... .C. A.T ... ... ... ... G.. ... ... ... ... ... A.A ... ... ... ...

AAG GTG TTT TCA TGA CTT TAT ATG TTG AAT AGT TGT GGT ATG CGG GTT TTA GAT CCC
... ... ... ..T ... T.A ... ... ..A GGG ... A.. ..G T.. ..T .CG GCT ... ..A
... ... ... ..T ... T.A ... ... ..A GGG ... A.. ..G T.. ..T .CG GCT ... ..A
... ... ... ..T ... T.A ... ... ..A GGG ... A.. ..G T.. ..T .CA GCT ... ..A
... ... ... ..T ... T.A ... ... ..A GGG ... A.. ..G T.. ..T .CA GCT ... ..A
... ... ... ..T ... T.A ... ... ..A GGG ... A.. ..G T.. ..T .CG GCT ... ..A
... ... ... ..T ... T.A ... ... ... .G. G.. ... ..G T.A ... ... A.T ... ..T

ATA GTA TGG TGA TTA GTC GGT TTT ATA TTT TTA TTT ACG GTT GGT GGT GTC ACA GGG
... C.T ... ... A.T ..T ... ... ... ... ..G ... ..A ... ... ... ..T ..T ...
... C.T ... ... A.T ..T ... ... ... ... ..G ... ..A ... ... ... ..T ..T ...
... C.T ... ... A.T ..T ... ... ... ... ..G ... ..A ... ... ... ..T ..T ...
... C.T ... ... A.T ..T ... ... ... ... ..G ... ..A ... ... ... ..T ..T ...
... C.T ... ... A.T ..T ... ... ... ... ..G ... ..A ... ... ... ..T ..T ...
G.. ..T ..A ... A.T ..T ... ... ... ... ... ... ... ..G ... ..A ..T ..T ..T

GTG GCT TTA TCT GCA TCT GCT TTA GAT ATA CTG TTT CAT GAT ACT TGG TTT
A.A .T. ... ... ..T ... ... ..G ... .G. T.A ... ..C ... ... ..A ...
A.A .T. ... ... ..T ... ... ..G ... .G. T.A ... ..C ... ... ..A ...
A.A .T. ... ... ..T ... ... ..G ... .G. T.. ... ..C ... ... ..A ...
A.A .T. ... ... ..T ... ... ..G ... .G. T.. ... ..C ... ... ..A ...
A.A .T. ... ... ..T ... ... ..G ... .G. T.A ... ..C ... ... ..A ...
A.A .T. ... ... ..T ..G T.. ..G ... .GG T.A ... ... ... ... ..A ...
```

Fig. 19.2. The sequences of a 372 bp fragment of the mitochondrial CO1 gene from *S. mansoni*, a number of isolates of *S. japonicum* and *S. mekongi* (after Bowles *et al.*, 1993). A dot represents homology with the sequence given for *S. mansoni* whilst 'N' denotes an unidentified nucleotide. Note that the *S. japonicum* sequences differ from those of *S. mekongi* and *S. mansoni* in approximately 13% and 24% of nucleotide positions, respectively, whilst *S. mansoni* and *S. mekongi* differ in 21% of the CO1 gene sequence.

problems within these two groups and in *S. japonicum* have been assessed.

In some of the studies described, conclusions drawn on the basis of DNA sequence data were in accord with results obtained using more traditional methods. In others, some apparent discrepancies have been found. For example, all strains of *E. granulosus* have been defined on the basis of biological and epidemiological distinctiveness. Some of these variants were found to be very distinct when markers of overall genomic evolution were compared. Others, such as the Tasmanian sheep (G2 genotype) and common sheep (G1 genotype) strains (Table 19.1) have been found to vary only slightly. Similarly, geographic strains of *S. japonicum* could not be discriminated using these techniques although, on the basis of biological criteria, it was expected that they could be.

Isoenzyme approaches have proved particularly useful in studies of population genetics. The studies described here suggest, however, that, for determining the actual genetic identity of a particular organism, DNA-based methods are more suitable. The genetic characterization of the Asian *Taenia* testifies to the power of direct DNA analysis. One major advantage over alternative methodologies is that DNA sequence comparison allows a direct examination of the genome independently of environmental and ontogenic influences. Another strength of sequence data is that it is absolute rather than relative. Once sequence information is obtained, it is always available for use in comparative studies. Other advantages are that levels of divergence are relatively easy to quantitate and that the range of nucleotide characters for phylogenetic analysis is practically unlimited. Largely because of the minimal requirements for material, PCR-based methods, in particular, should prove of especial value in future studies of molecular variation in parasitic helminths.

Acknowledgements

I am indebted to Dr Josephine Bowles who was instrumental in producing much of the molecular data described in this paper. I am also grateful to the Australian National Health and Medical Research Council of Australia and the WHO(TDR)-Rockefeller Foundation Health Sciences for the Tropics Partnership in Research Program for support of some of the work described in this paper. I would like also to thank the organising committee for kindly inviting me to present this paper at the Eighth International Congress of Parasitology in Izmir, Turkey.

References

Bowles, J. (1993) Genetic approaches to the identification and evolutionary study of helminth parasites. Unpublished PhD Thesis, University of Queensland, Australia.

Bowles, J. and McManus, D.P. (1993a) Molecular variation in *Echinococcus*. *Acta Tropica* 53, 291–305.

Bowles, J. and McManus, D.P. (1993b) Rapid discrimination of *Echinococcus* species and strains using a PCR-based RFLP method. *Molecular and Biochemical Parasitology* 57, 231–239.

Bowles, J. and McManus, D.P. (1993c) NADH dehydrogenase 1 gene sequences compared for species and strains of the genus *Echinococcus*. *International Journal for Parasitology* 23, 969–972.

Bowles, J. and McManus, D.P. (1994) Genetic characterization of the Asian *Taenia*, a newly described taeniid cestode of humans. *American Journal of Tropical Medicine and Hygiene* 50, 33–44.

Bowles, J., Blair, D. and McManus, D.P. (1992a) Genetic variants within the genus *Echinococcus* identified by mitochondrial sequencing. *Molecular and Biochemical Parasitology* 54, 165–174.

Bowles, J., Van Knapen, F. and McManus, D.P. (1992b) Cattle strain of *Echinococcus granulosus* and human infection. *Lancet* 339, 1358.

Bowles, J., Hope, M., Tiu, W.U., Liu, X.S. and McManus, D.P. (1993) Nuclear and mitochondrial genetic markers highly conserved between Chinese and Philippine *Schistosoma japonicum*. *Acta Tropica* 55, 217–229.

Bowles, J., Blair, D. and McManus, D.P. (1994) Molecular genetic characterisation of the cervid strain ("northern form") of *Echinococcus granulosus*. *Parasitology* 10, 215–221.

Bowles, J., Blair, D. and McManus, D.P. (1995) A molecular phylogeny of the genus *Echinococcus*. *Parasitology* 110, 317–328.

Hope, M., Bowles, J. and McManus, D.P. (1991a) A reconsideration of the *Echinococcus granulosus* strain situation in Australia following RFLP analysis of cystic material. *International Journal for Parasitology* 21, 471–475.

Hope, M., Bowles, J., Prociv, P. and McManus, D.P. (1991b) A genetic comparison of human and wildlife isolates of *Echinococcus granulosus* in Queensland and the public health implications. *Medical Journal of Australia* 156, 27–30.

Lymbery, A.J. (1992) Interbreeding, monophyly and the genetic yardstick: species concepts in parasites. *Parasitology Today* 8, 208–211.

McManus, D.P. and Bowles, J. (1994) Asian (Taiwan) *Taenia*: species or strain. *Parasitoloey Today* 10, 273–275.

McManus, D.P. and Hope, M. (1993) Molecular variation in the human schistosomes. *Acta Tropica* 53, 255–276.

McManus, D.P. and Rishi, A.K. (1989) Genetic heterogeneity within *Echinococcus granulosus*: isolates from different hosts and geographical areas characterised with DNA probes. *Parasitology* 99, 17–29.

McManus, D.P. and Simpson, A.J.G. (1985) Identification of the *Echinococcus* (hydatid disease) organisms using cloned DNA markers. *Molecular and Biochemical Parasitology* 17, 171–178.

McManus, D.P., Simpson, A.J.G. and Rishi, A.K. (1987) Characterisation of the hydatid disease organism *Echinococcus granulosus*, from Kenya using cloned DNA markers. In: Geerts, S., Kumar, V. and Brandt, J. (eds) *Helminth Zoonoses*. Martinus Nijhoff, Dordrecht, pp. 29–36.

McManus, D.P., Ding, Z. and Bowles, J. (1994) A molecular genetic survey indicates the presence of a single, homogeneous strain of *Echinococcus granulosus* in north-western China. *Acta Tropica* 56, 7–14.

Rausch, R.L. (1986) Life-cycle patterns and geographic distribution of *Echinococcus* species. In: Thompson, R.C.A. (ed.) *The Biology of Echinococcus and Hydatid Disease*. George Allen and Unwin, London, pp. 44–80.

Scott, J.C. and McManus, D.P. (1994) The random amplification of polymorphic DNA can discriminate species and strains of *Echinococcus*. *Tropical Medicine and Parasitology* 45, 1–4.

Thompson, R.C.A. and Lymbery, A.J. (1988) The nature, extent and significance of variation within the genus *Echinococcus*. *Advances in Parasitology* 27, 209–258.

Thompson, R.C.A. and Lymbery, A.J. (eds) (1995) *Echinococcus and Hydatid Disease*. CAB International, Wallingford, Oxon, UK.

Wachira, T.M., Bowles, J., Zeyhle, E. and McManus, D.P. (1993) Molecular examination of the sympatry and distribution of sheep and camel strains of *Echinococcus granulosus* in Kenya. *American Journal of Tropical Medicine and Hygiene* 48, 473–479.

20

Allergic Reactivity and Helminthic Infection

N.R. Lynch

Instituto de Biomedicina, Aptdo 4043, Carmelitas, 1010A Caracas, Venezuela

Numerous studies have indicated that the prevalence of allergic diseases in rural populations in tropical or subtropical zones can be far lower than in industrialized countries with more temperate climates, and that this could be related to the endemic helminthic infections in these regions (Editorial, 1976; Turner, 1980; Lynch, 1987). The best substantiated explanation for this is that the highly elevated IgE levels that can be produced by helminthic parasites result in a saturation of the mast cell Fcϵ receptors, thus producing mast cell blockade, which is reflected by a reduced allergic reactivity (Godfrey and Gradidge, 1976).

These proposals have, however, stimulated considerable controversy (Turner, 1980; Grove, 1982; Moqbel and Pritchard, 1990; Lynch, 1992), and in an attempt to obtain a more comprehensive view of the prevalence of allergic disease in the tropical situation, and to determine the possible influence of helminthic infection, we have, over the past decade, undertaken studies in the different socioeconomic sectors of the population of tropical Venezuela (latitude 2 to 12°N).

In our earliest studies (Lynch and DiPrisco-Fuenmayor, 1984), that we have confirmed more recently (Lynch *et al.*, 1987a; Hagel *et al.*, 1993a), we evaluated Venezuelans of medium-high socioeconomic level (HSEL) in the urban environment of Caracas, and compared these to a group of persons of equivalent age and social status, but who were native to more temperate and industrialized countries, and who were examined within 6 weeks of their arrival in Venezuela. A surprisingly high 43% of the Venezuelan subjects were found to have suffered, within the previous 5 years, symptoms that were consistent with allergic disease. This prevalence was almost identical to that independently obtained by other groups in Venezuela (Anderson *et al.*, 1980; Ponce *et al.*, 1984; Massicot and Cohen 1986). Applying the same clinical criteria as for the Venezuelan study group, we found that a significantly lower proportion (29%) of the non-tropical subjects evaluated in parallel reported a history of allergic disease. This difference was mirrored by a considerably higher degree of positivity in immediate hypersensitivity skin tests to environmental allergens in the Venezuelans (63%), compared to the non-tropical group (37%). Consistent with the higher allergic reactivity of the Venezuelan group was the fact that their total serum IgE levels (geometric mean 369 international units (IU) ml^{-1}) were significantly elevated, compared with the non-tropical subjects (68 IU ml^{-1}). The Venezuelans were also found to suffer light and sporadic helminthic infections at a frequency (approximately 10%) that was notably higher than in the non-tropical subjects (3%). This may explain the increased allergic reactivity of the former group, because helminthic infection can non-specifically and polyclonally potentiate the synthesis of IgE (Jarrett and Bazin, 1974; Turner *et al.*, 1979), via the production of interleukin (IL)-4 by the Th2 subset of CD4 T-lymphocytes (King *et al.*, 1990; Romagnani, 1990; Finkelman *et al.*, 1991). This raises the serum IgE levels in infected individuals (Johansson *et al.*, 1968), and part of this IgE can be directed against environmental allergens (Jarrett and Stewart, 1972; Joubert *et al.*, 1980). As helminthic infection can also stimulate local mast cell proliferation (Jarrett and Miller, 1982), and a number of the most common geohelminths have pulmonary phases, it has been postulated that such parasites can increase the allergic reactivity of infected individuals (Huntley, 1976; Joubert *et al.*, 1980).

As a continuation of these studies (Lynch *et al.*, 1984), we also evaluated the allergic reactivity of persons of low socioeconomic level (LSEL) living in slum areas of Caracas, who have racial and cultural characteristics in common with the HSEL group, but because of their substandard living conditions are exposed to a much higher degree of parasitic infection. The prevalence of intestinal helminthiasis in this group was 80%, *Ascaris lumbricoides* being the most common infection (71%). The total serum IgE levels were correspondingly high (745 IU ml^{-1}), but both the prevalence of allergic disease (24%) and the positivity in skin tests to common environmental allergens (27%), were significantly lower than in the HSEL group.

In recent studies (Lynch *et al.*, 1992b, 1993b; Hagel *et al.*, 1993b) we have noted a dramatic increase in the levels of serum IgE in children living in urban slum areas of Caracas, and have related this directly with their degree of poverty, deficient sanitary conditions and prevalence of helminthic infection. We also demonstrated a direct correlation ($r = 0.78$) between the serum levels of IgE and the IL-4 concentrations in such children.

When we evaluated rural subjects in a subsistence farming situation (Lynch *et al.*, 1984), we found that the prevalence of intestinal helminthic infection (78%) and the total serum IgE levels (975 IU ml^{-1}) were broadly comparable to the urban slum-dwellers, but both the prevalence of allergic disease (12%) and the positivity in immediate hypersensitivity skin tests (10%) were significantly lower in the rural environment. In later studies (Hagel *et al.*, 1993b) we showed that the total IgE levels in the rural subjects have risen even more significantly than in the urban situation over recent years, and that their allergic reactivity has decreased even futher. This situation was even more evident in the hunter–gatherer Amerindian communities that are still to be found in the Venezuelan sector of the Amazon basin (Lynch *et al.*, 1993). These individuals showed extremely high prevalences (88%) and intensities of helminthic infection, and total IgE levels that, together with those of the Waroni indians of Ecuador (Kaplan *et al.*, 1980), are the highest yet reported for any population group (geometric mean 13,088 IU ml^{-1}). While the positivity in skin tests with common environmental allergens was very low in these indians (7%), the proportion of the subjects with significant levels of specific IgE antibody against these was almost tenfold higher (Lynch *et al.*, 1983). As the cutaneous sensitivity of these indians to histamine was normal, these results suggest that the low allergic reactivity of the Amazon indians is not due to an inherent lack of response to envirormental allergens, but rather to a deficiency in mast cell function.

A number of experimental studies have revealed that high levels of IgE can cause saturation of the Fcϵ receptors of mast cells, thus reducing their sensitivity to degranulation by allergens (Godfrey and Gradidge, 1976). One manner of determining the existence of mast cell blockade *in vivo* is by Prausnitz–Küstner (P–K) testing; the saturation of Feϵ receptors being revealed by the inhibition of the capacity of passively transferred IgE to sensitize skin mast cells (Bazaral *et al.*, 1973). We have applied these tests to persons from different socioeconomic levels in Caracas (Lynch *et al.*, 1987a), and found that while essentially all of the subjects showed normal cutaneous reactivity to histamine, codeine (a non-specific pharmacological stimulator of mast cell degranulation) and anti-IgE (a non-specific immunological mast cell degranulator), the P–K positivity was considerably lower in the LSEL slum children, compared to HSEL group. In recent studies (Hagel *et al.*, 1993a,b) of the slum children we have demonstrated an inverse correlation between total serum IgE levels and wheal diameters both in skin tests to environmental allergens ($r = -0.5$) and P–K tests ($r = -0.75$).

It is important to note that suppression of P–K positivity did not occur in all of the subjects of LSEL that we evaluated, thus indicating that some other factor might also be operating to lower the allergic reactivity of this group. A closer analysis of the specific IgE levels, measured by the radio-allergosorbant test (RAST) in our study groups revealed that in the LSEL there was a disproportion between positivity at the two cut-off levels in the RAST, when compared to the HSEL. For example, in allergic subjects of HSEL, the positivity at 0.35 PRU ml^{-1} (RAST level 1) was 55.6%, and at 0.70 PRU ml^{-1} (RAST level 2) was 50.8%. This signifies that in this group the vast majority of persons with detectable specific IgE also had clinically significant levels. In contrast, in people of LSEL, although 75.6% had detectable IgE ($\geq$ level 1), only 24.4% had levels that are generally considered to be clinically relevant ($\geq$ level 2). Thus, in the LSEL group it appears that whilst a general stimulation of IgE synthesis occurs, the actual magnitude of the response against any given environmental allergen is relatively low. In support of this hypothesis, we have demonstrated (Hagel *et al.*, 1993b) the existence of an inverse correlation ($r = -0.5$) between total serum IgE levels and the concentrations of specific IgE against environmental allergens.

We have proposed (Lynch, 1992) a hypothesis to explain this effect; all individuals, and especially atopics, who are not exposed to helminthic parasites probably have a certain degree of differentiation of $B\epsilon$ precursors, particularly of specificities against common environmental allergens. Persons with mild helminthic infection undergo a non-specific stimulation of IgE synthesis, that probably affects specificities that are already being expressed (Orr and Blair, 1969; Jarrett and Stewart, 1972), and thus have high levels of specific IgE antibody against common environmental allergens. Under conditions of more intense helminthic infection, it is possible that the stimulation of IgE synthesis is much more widely polyclonal, and that the elevated total IgE levels observed are due to the summation of the limited expression of a large number of different specificities. This could, therefore, explain why a high proportion of heavily parasitized persons have detectable IgE against environmental allergens, but relatively few demonstrate clinically relevant levels. This highly polyclonal IgE can also cause mast cell saturation.

In subsequent studies we have substantiated various aspects of this model in a group of parasitized children who were evaluated, then treated for 22 months by monthly anthelmintic (oxantel/pyrantel) administration (Lynch *et al.*, 1993b). Their total serum IgE levels fell from intial values of 2543 IU ml^{-1} to 1124 IU ml^{-1} after treatment, and in parallel significant increases occurred in the skin test reactivity to environmental allergens (from 17% to 68% positive), in positivity in P–K tests (from 52% to 79%), and in the serum concentrations of specific IgE antibody against environmental allergens (from 32% to 63% positive at RAST level 2). In fact, the initial allergic reactivity that was characteristic of the LSEL group, was converted to a situation comparable to that of the HSEL group.

Despite all of the above considerations, there is also considerable evidence that helminthic infections can have a more direct role in the expression of allergic reactions (Zvaifler, 1976; Lynch, 1987). The existence of allergic reactivity against helminthic parasites is well known (Ball *et al.*, 1971; Zvaifler, 1976; Jarrett and Miller, 1982) and although a report of an almost constant cause–effect relationship between these (Tullis, 1970) has been dismissed (Van Dellen and Thompson, 1971), the association of helminthic infection with clinically significant allergic disease has been demonstrated (Phills *et al.*, 1972; Dao, 1979; Desowitz *et al.*, 1981; Pawlowski and Arfaa, 1984).

We have noted that in the tropical situation, patients with clear symptoms of allergic disease can present negative skin test reactions toward a wide variety of environmental allergens, but are strongly positive with extracts of *Ascaris*, and thus pose the question as to the clinical significance of this reactivity against the parasite (Lynch *et al.*, 1987b). We found that such allergic patients have significantly higher serum levels of anti-*Ascaris* IgE antibody than subjects who demonstrate comparable cutaneous reactivity but do not suffer allergic symptoms (Lynch *et al.*, 1987b).

In order to further examine this situation, we have evaluated the pulmonary response to the inhalation of a bronchodilator (salbutamol) before and after long-term (22 months) anthelmintic treatment (monthly oxantel/pyrantel) in apparently normal, urban slum children, endemically exposed to helminthic infection (Lynch *et al.*, 1992c). Before anthelmintic treatment we found a direct association between the degree of helminthic infection and the increase in peak expiratory flow rates caused by the bronchodilator. The long-term elimination of the helminthic infections was accompanied by a significant decrease in the response to the bronchodilator. These results indicate a state of latent bronchoconstriction in the infected children.

We have also shown (Lynch *et al.*, 1992a) that tropical asthmatics from areas endemic for intestinal helminths can undergo significant changes in pulmonary function after the inhalation of minute quantities of *Ascaris* allergens. Significant reductions in pulmonary function occurred within 30 min of the *Ascaris* challenge in 25% of the patients, between 30 min and 6 h in 45%, and 60% showed late responses (6–16 h post-challenge). Similar challenges did not produce significant alterations in non-asthmatic subjects from the same endemic area, or in asthmatic patients who had not experienced prior contact with the parasite. As *Ascaris* larvae migrate through the lung, and soluble parasite antigen is found in the blood of infected individuals (Tanaka *et al.*, 1983), this hypersensitivity to the helminth could contribute to the allergic reactivity of tropical populations.

The fact that the allergic reactivity of tropical populations could depend upon the intensity, and possibly type, of helminthic infection that they experience might explain much of the controversy surrounding the prevalence of allergic diseases in such groups, and suggests that this factor should be taken into account in evaluations.

Acknowledgements

The studies described here are the result of a collaborative programme in which the contribution of the following people has been essential: M.C. DiPrisco-Fuenmayor, R.I. Lopez, I. Hagel, N. Garcia, G. Isturiz, M. Perez and L. Medouze. Financial assistance was received from the CDCH/UCV and CONICIT project RPV 170041.

References

Anderson, O., Orozco, A. and Ponce, D. (1980) Incidencia de enfermedades atópicas en la población escolar venezolana. *Acta Científica Venezolana* 31, 145.

Ball, P.A., Voller, A. and Taffs, L.F. (1971) Hypersensitivity to some nematode antigens. *British Medical Journal* 1, 210.

Bazaral, M., Orgel, H.A. and Hamburger, R.N. (1973) The influence of serum IgE levels of selected recipients, including patients with allergy, helmintiasis and tuberculosis, on the apparent P–K titre of a reaginic serum. *Clinical and Experimental Immunology* 14, 117.

Dao, L. (1979) Parasitosis intestinal: Su influencia en el asma bronquial. *Tribune Medica* 603, 41.

Desowitz, R.S., Rudoy, R.and Barnwell, J.W. (1981) Antibodies to canine helminthic parasites in asthmatic and non-asthmatic children. *International Archives of Allergy and Applied Immunology* 65, 361.

Editorial (1976) IgE, parasites, and allergy. *Lancet* i, 894.

Finkelman, F.D., Pearce, F., Urban, J. and Sher, A. (1991) Regulation and biological function of helminth-induced cytokine responses. *Immunology–Parasitology Today* 12, A62.

Godfrey, R.C. and Gradidge, C.F. (1976) Allergic sensitisation of human lung fragments prevented by saturation of IgE binding sites. *Nature* 259, 484.

Grove, D.I. (1982) What is the relationship between asthma and worms? *Allergy* 37, 139.

Hagel, I., Lynch, N.R., Perez, M., DiPrisco, M.C., Lopez, R. and Rojas, E. (1993a) Modulation of the allergic reactivity of slum children by helminthic infection. *Parasite Immunology* 15, 311.

Hagel, I., Lynch, N.R., DiPrisco, M.C., Lopez, R.I. and Garcia, N. (1993b) Allergic reactivity of children of different socio-economic level in tropical populations. *International Archives of Allergy and Immunology* 101, 209.

Huntley, C.C. (1976) Of worms and asthma or Tullis revisited. *New England Journal of Medicine* 249, 1295.

Jarrett, E.E. and Bazin, H. (1974) Elevation of total serum IgE in rats following helminth parasite infection. *Nature* 251, 613.

Jarrett, E.E. and Miller, B.R. (1982) Production and activities of IgE in helminth infection. *Progress in Allergy* 31, 178.

Jarrett, E.E. and Stewart, D.C. (1972) Potentation of rat reaginic (IgE) antibody by helminth infection. Simultaneous potentation of separate reagins. *Immunology* 23, 749.

Johansson, S.G., Melbin, T. and Vahlquist, B. (1968) Immunoglobulin levels in Ethiopian preschool children with special reference to high concentrations of immunoglobulin E (IgND). *Lancet* i, 1118.

Joubert, J.B., Van Schalkwyk, D.J. and Turner, K.J. (1980) *Ascaris lumbricoides* and the human immunogenic response: enhanced IgE-mediated reactivity to common inhaled allergens. *South African Medical Journal* 57, 409.

Kaplan, J.E., Larrick, J.W. and Yost, J.A. (1980) Hyperimmunoglobulinema E in the Waorani, an isolated Amerindian population. *American Journal of Topical Medicine and Hygiene* 29, 1012.

King, C.L., Ottesen, E.A. and Nutman, T.B. (1990) Cytokine regulation of antigen-driven immunoglobulin production in filarial parasite infections in humans. *Journal of Clinical Investigation* 85, 1810.

Lynch, N.R. (1987) Immediate hypersensitivity (allergic) reactions to intestinal helminthic infections. *Baillère's Clinical Tropical Medicine and Communicable Diseases* 2, 573.

Lynch, N.R. (1992) Influence of socio-economic level on helminthic infection and allergic reactivity in tropical countries. In: Moqbel, R. (ed.) *Allergy and Immunity to Helminthic Infection: Common Mechanisms or Divergent Pathways?* Taylor & Francis, London.

Lynch, N.R. and DiPrisco-Fuenmayor, M.C. (1984) High allergic reactivity in a tropical environment. *Clinical Allergy* 14, 233.

Lynch, N.R., Lopez, R.L., Isturiz, G. and Tenias-Salazar, E. (1983) Allergic ractivity and helminthic infection in Amerindians of the Amazon basin. *International Archives of Allergy and Applied Immunology* 72, 369.

Lynch N.R., Medouze, L., DiPrisco-Fuenmayor, M.C., Verde, O., López, R.I. and Malavé, C. (1984) Incidence of atopic disease in a tropical environment: partial independence from intestinal helminthiasis. *Journal of Allergy and Clinical Immunology* 73, 229.

Lynch, N.R., Lopez, R.I., DiPrisco-Fuenmayor, M.C., Hagel, I., Medouze, L., Viana, G., Ortega, C. and Prato, G. (1987a) Allergic reactivity and socio-economic level in a tropical environment. *Clinical Allergy* 17, 199.

Lynch, N.R., Perez, M., Lopez, R.I. and Turner, K.J. (1987b) Measurement of anti-*Ascaris* antibody levels in tropical allergic patients, using a modified ELISA. *Allergology et Immunopathology* 15, 19.

Lynch, N.R., Isturiz, G., Sanchez, Y., Perez, M., Martinez, A. and Castes, M. (1992a) Bronchial challenge of tropical asthmatics with *Ascaris lumbricoides. Journal of Investigational Allergology and Clinical Immunology* 2, 97.

Lynch, N.R., Hagel, I., DiPrisco, M.C., Lopez, R.I., Garcia, N. and Perez, M. (1992b) Serum IgE levels, helminthic infection and socio-economic change. *Parasitology Today* 8, 166.

Lynch, N.R., Hagel, I., Perez, M., DiPrisco, M.C., Alvarez, N. and Rojas, E.

(1992c) Bronchoconstriction in helminthic infection. *International Archives of Allergy and Immunology* 98, 77.

Lynch, N.R., Hagel, L., Vargas, M., Perez, M., Lopez, R.I., Garcia, N., DiPrisco, M.C. and Arthur, I.H. (1993a) Effect of age and helminthic infection on the IgE levels of slum children. *Journal of Investigational Allergology and Clinical Immunology* 3, 96.

Lynch, N.R., Hagel, I., Perez, M., DiPrisco, M.C., Lopez, R.I. and Alvarez, N. (1993b) Effect of anthelmintic treatment on the allergic reactivity of children in a tropical slum. *Journal of Allergy and Clinical Immunology* 92, 404.

Massicot, J.G. and Cohen, S.G. (1986) Epidemiologic and socioeconomic aspects of allergic diseases. *Journal of Allergy and Clinical Immunology* 78, 954.

Moqbel, R. and Pritchard, D.I. (1990) Parasites and allergy: evidence for a cause and effect relationship. *Clinical and Experimental Allergy* 20, 611.

Orr, T.S. and Blair, M. (1969) Potentiated reagin response to egg albumin and conalbumin in *Nippostrongylus brasiliensis* infected rats. *Life Sciences* 8, 1073.

Pawlowski, Z.S. and Arfaa, F. (1984) Ascariasis. In: Warren, K.S. and Mahmound, A.A. (eds) *Tropical and Geographic Medicine*. McGraw-Hill, New York, 347 pp.

Phills, J.A., Harrold, A.J., Whiteman, G.V. and Perelmutter, L. (1972) Pulmonary infiltrates, asthma and eosinophilia due to *Ascaris suum* infestation in man. *New England Journal of Medicine* 286, 965.

Ponce, D.P., Muñoz, J.F., Quijada, L., Fernandez, R., Ossot, E.B. and Bianco, N.B. (1984) Prevalencia y caracteristicas immnuoclínicas de las enfermedades alergicas en Venezuela. Reporte preliminar. *Acta Científica Venezolana* 34, 302.

Romagnani, S. (1990) Regulation and deregulation of human IgE synthesis. *Immunology Today* 11, 316.

Tanaka, K., Kawamura, H., Tohgi, N., Tsumi, M., Miyachi, Y. and Miyoshi, A. (1983) The measurement of *Ascaris suum* protein by radioimmunoassay in sera from patients with helminthiasis and with gastrointestinal disease. *Parasitology* 86, 291.

Tullis, D.C. (1970) Bronchial asthma associated with intestinal parasites. *New England Journal of Medicine* 282, 370.

Turner, K.J. (1980) Is the prevalence of allergy related to parasitic disease? In: Oehling, A. (ed.) *Advances in Allergology and Immunology*. Oxford, Pergamon Press, 279 pp.

Turner, K.J., Feddma, L., and Quinn, E.H. (1979) Non-specific potentation of IgE by parasitic infections in man. *International Archives of Allergy and Applied Immunology* 58, 232.

Van Dellen, R.G. and Thompson, J.H. (1971) Absence of intestinal parasites in asthma. *New England Journal of Medicine* 285, 146.

Zvaifler, N.J. (1976) Immediate hypersensitivity (type I) reactions. In: Cohen, S. and Sadun, E.H. (eds) *Immunology of Parasitic Infections*. Blackwell Scientific Publications, Oxford, 410 pp.

21

Protective Immune Mechanisms against *Echinococcus multilocularis*

B. GOTTSTEIN

Institute of Parasitology, University of Bern, P.O. Box 8466, Bern, Switzerland

Alveolar echinococcosis (AE), caused by the metacestode stage of *Echinococcus multilocularis*, is one of the most lethal helminthic infections of humans. The natural life cycle involves predominantly rodents as intermediate hosts, where the metacestode parasite develops mainly in the liver as a solid tumour-like and vascularized tissue. Metastasis to various other sites, such as the peritoneal cavity, the diaphragm, the lungs and the kidney, can also occur. In human patients and in experimentally infected mice, the growth of the metacestode tissue is always associated with an intense humoral and cell-mediated immune response (Gottstein *et al.*, 1984, 1989, 1992; Bresson-Hadni *et al.*, 1989; Liance *et al.*, 1990; Rakha *et al.*, 1991). Generally, humoral and cellular immune reactions, rather than helping the host in controlling parasite proliferation, appeared to be involved in immunopathological mechanisms responsible for the occasional chronic granulomatous course of the disease. In some cases, however, the immune system might be able to prevent disease. This is suggested by the finding of patients (upon serological screening by Em2-enzyme-linked immunosorbent assay (ELISA)) with lesions that have 'aborted' or 'died-out' (lesions are considered to be abortive when no viability can be shown after surgical resection of the parasite lesion and subsequent transplantation of the parasite to susceptible laboratory rodents) (Rausch *et al.*, 1987). Alaskan patients with a 'died-out' metacestode exhibited high *in vitro* lymphoproliferative responses to *E. multilocularis* crude antigen, in contrast to patients with a metacestode which is still active (Gottstein *et al.*, 1992a). Vuitton *et al.* (1989) showed that the periparasitic granuloma, which is mainly composed of macrophages, myofibroblasts and T-cells, contained a large number of $CD4^+$ lymphocytes in (French) patients with lesions that had 'died-out'. In patients with active parasite tissue, on the other hand, the number of $CD8^+$ cells was increased.

From the epidemiological standpoint, the wide distribution of foxes infected with egg-producing *E. multilocularis* in Central Europe (prevalence 20–40%) stands in marked contrast to the respective disease in humans which is relatively rare (annual morbidity rate in Switzerland: 0.18 AE cases per 100,000 inhabitants). Therefore, resistance to disease following infection may be an immunological feature to be investigated. The finding of early cases of active but asymptomatic AE and of self-healed inactive cases of AE in humans indicated that the incidence of disease may be restricted genetically or by other means. We have addressed this question by determining the gene frequency (typed by polymerase chain reaction-sequence-specific oligotyping (PCR-SSO)) of certain lymphoid cell-surface proteins encoded in the HLA-DR region (Gottstein and Bettens, 1995). Blood samples from 32 patients with hepatic AE were collected over a 3-year period. In our patient group the frequency of HLA-DR13 antigen and susceptibility to disease following infection with *E. multilocularis* was significantly increased (patients, 34.4%, local random panel, 13.6%; $\chi^2 = 7.0$, $P = 0.008$). The association found between an HLA antigen and susceptibility to AE is highly significant. Interestingly, subgrouping of patients with respect to their individual clinical status/course (see above) provided more insight into the distribution of the respective DR13. From the group of inoperable patients, 2 out of 14 (=12%) were expressing DR13, from the group of patients with partial surgery 4 out of 10 (=40%) were expressing DR13, and from the group of patients with a radical resection, 7 out of 8 (=87%) were expressing DR13. Consequently, the results suggest that a person expressing DR13 will develop AE with a much higher probability than persons not expressing DR13, but that radical surgical resection will become more likely as evidenced by our patients with DR13. Converse to the findings related to susceptibility, healthy control persons expressed DR4 at a significantly (patients, 15.6%, local random panel, 41.3%; $\chi^2 = 6.5$, $P = 0.01$) higher rate, thus providing an indication for the existence of markers with potential functional implications regarding resistance to disease.

Liance *et al.* (1984, 1990) and Bresson-Hadni *et al.* (1990) developed murine models which reflect susceptibility (AKR and C57BL/6) or relative resistance (C57BL/10) to secondary AE. Using these models, they found that a delayed-type hypersensitivity (DTH) *in vivo* response upon antigen challenge was significantly higher in infected resistant mice than in susceptible mice. A subsequent analysis of the collagen types and of the phenotype of cells within the periparasitic granuloma showed that susceptibility was associated with a persistence of numerous $CD4^+$ lymphocytes and low macrophage number, whereas the periparasitic granuloma of resistant animals showed elevated numbers of $CD8^+$ T-cells. Kroeze and Tanner (1987) suggested that in mice susceptibility or resistance is probably controlled by non-*H-2* gene(s). Recent reports showed that $CD8^+$ T-suppressor (Ts) cells with a low density of the CD8-antigen ($CD8^{dull}$ cells) were detectable in spleens from mice infected with *E. multilocularis* protoscoleces (Kizaki *et al.*, 1991) and in polyclonally activated lymphocytes from normal mice. Further analysis showed that protoscoleces induced $CD8^+$ Ts cells *in vitro* and were assumed to be responsible for immune suppression in mice with alveolar echinococcosis (Kizaki *et al.*, 1993). *Scid*-mice reconstituted with purified lymphocytes from infected donor mice were able to suppress metacestode proliferation and protoscolex formation. In contrast, reconstitution with non-immune lymphocytes resulted only in the suppression of metacestode growth (Playford *et al.*, 1993). In natural infections, protoscolex formation occurs only in some rodent species and very rarely in human patients. It is widely believed that the laminated layer of the *E. multilocularis* metacestode protects the parasite from host immune reactions (Gottstein *et al.*, 1992b). In this respect we have identified an immunodominant antigen of *E. multilocularis* (Em2-antigen) against which we have subsequently generated a monoclonal antibody (mAb) G11. Immunobiological and immunochemical studies revealed that mAb G11 recognizes a dominant epitope in the laminated layer of the metacestode tissue, which could not be detected in protoscoleces. We obtained experimental evidence that only those *E. multilocularis* metacestode structures which were able to synthesize a laminated layer containing Em2 (as recognized by mAb G11) exhibited the potential to induce secondary alveolar echinococcosis in rodents (Gottstein *et al.*, 1992b). With respect to immunity in AE, parasite-specific humoral and cell-mediated immune responses were investigated in susceptible (AKR and C57BL/6J) and relatively resistant (C57BL/10) mice undergoing secondary AE (Gottstein *et al.*, 1994). The parasite-specific proliferative immune response of lymph node cells upon *in vitro* antigen stimulation remained weak in all three mouse strains. By day 30 post-infection (p.i.), $CD4^+$ lymphoblast cells dominated the total population of blast cells in all three mouse strains. There was, however, an unexpectedly high proportion of $CD8^+$ blast cells; by day 90 p.i., a marked proportional increase in $CD8^+$ cells was seen in susceptible (AKR and C57BL/6J), but not in resistant (C57BL/10) mice. Susceptible, but not resistant, mice exhibited a significantly decreased responsiveness of lymph node cells to Concavalin A stimulation on day 90 p.i. Analysis of the humoral immune response by ELISA showed that resistance in C57BL/10 mice was associated with the ability of the host to synthesize antibodies to Em2 of the IgG_3 and IgG_1 isotype. Em2 is a lectin-binding carbohydrate antigen of the laminated layer. In susceptible AKR and

C57BL/6J mice, low levels of anti-Em2 antibodies of the IgG_{2a} isotype were detected. Anti-Em2 antibodies of the IgG_3/IgG_1 isotype, however, were absent. Differences in subclass-specific IgG responses were confirmed by immunoblot analyses. Our findings suggest that differences in antigen recognition (with respect to subsets of humoral and cellular immune components) probably controlled by non-*H-2* gene(s), coupled to immune suppression modulated by $CD8^+$ cells and/or respective cytokines, may determine susceptibility or resistance in experimental infection with *E. multilocularis*.

References

Bresson-Hadni, S., Vuitton, D.A., Lenys, D., Liance, M., Racadot, E. and Miguet, J.P. (1989) Cellular immune response in *Echinococcus multilocularis* infection in humans. I. Lymphocyte reactivity to *Echinococcus* antigens in patients with alveolar echinococcosis. *Clinical and Experimental Immunology* 66, 78–61.

Bresson-Hadni, S., Liance, M., Meyer, J.P., Houin, R., Bresson, J.L. and Vuitton, D.A. (1990) Cellular immunity in experimental *Echinococcus multilocularis* infection. II. Sequential and comparative phenotypic study of the periparasitic mononuclear cells in resistant and sensitive mice. *Clinical and Experimental Immunology* 82, 378–383.

Gottstein, B. and Bettens, F. (1995) Association between HLA-DR13 and susceptibility to alveolar echinococcosis. *Journal of Infectious Diseases* 169, 1416–1417.

Gottstein, B., Eckert, J. and Woodtli, W. (1984) Determination of parasite-specific immunoglobulins using the ELISA in patients with echinococcosis treated with mebendazole. *Parasitology Research* 70, 385–389.

Gottstein, B., Tschudi, K., Eckert, J. and Ammann, R. (1989) Em2-ELISA for the follow-up of alveolar echinococcosis after complete surgical resection of liver lesions. *Transactions of the Royal Society of Tropical Medicine and Hygiene* 83, 389–393.

Gottstein, B., Mesarina, B., Tanner, I., Ammann, R.W., Eckert, J., Wilson, J.F., Lanier, A. and Parkinson, A. (1992a) Specific cellular and humoral immune responses in patents with different long-term courses of alveolar echinococcosis (infection with *Echinococcus multilocularis*). *American Journal of Tropical Medicine and Hygiene* 45, 734–742.

Gottstein, B., Deplazes, P. and Aubert, M. (1992b) *Echinococcus multilocularis*: Immunological study on the 'Em2-positive' laminated layer during *in vitro* and *in vivo* post-oncospheral and larval development. *Parasitology Research* 78, 291–297.

Gottstein, B., Wunderlin, E. and Tanner, I. (1994) *Echinococcus multilocularis*: Parasite-specific humoral and cellular immune response subsets in mouse strains susceptible (AKR, C57BL/6J) or 'resistant' (C57BL/10) to secondary alveolar echinococcosis. *Clinical and Experimental Immunology* 96, 245–252.

Kizaki, T., Kobayashi, S., Ogasawara, K., Day, N.K., Good, R.A. and Onoe, K. (1991) Immune suppression induced by protoscoleces of *E. multilocularis* in mice. Evidence for the presence of CD8dull suppressor cells in spleens of mice intraperitoneally infected with *E. multilocularis*. *Journal of Immunology* 147, 1659–1666.

Kizaki, T., Ishige, M., Bingyan, W., Kumagai, M., Day, N.K., Good, R.A. and Onoe, K. (1993) Interleukin-1-dependent mitogenic responses induced by protoscoleces of *Echinococcus* in murine lymphocytes. *Journal of Leukocyte Biology* 53, 233–239.

Kroeze, W.K. and Tanner, C.E. (1987) *Echinococcus multilocularis*: susceptibility and responses to infection in inbred mice. *International Journal of Parasitology* 17, 873–883.

Liance, M., Vuitton, D., Guerret-Stocker, S., Carbillet, J.P., Grimaud, J.A. and Houin, R. (1984) Experimental alveolar echinococcosis. Suitability of a murine model of intrahepatic infection by *Echinococcus multilocularis* for immunological studies. *Experientia* 40, 1436–1439.

Liance, M., Bresson-Hadni, S., Meyer, J.P., Houin, R. and Vuitton, D.A. (1990) Cellular immunity in experimental *Echinococcus multilocularis* infection. I. Sequential and comparative study of specific *in vivo* delayed-type hypersensitivity against *E. multilocularis* antigens in resistant and sensitive mice. *Clinical and Experimental Immunology* 82, 373–377.

Playford, M.C., Ooi, H.K., Ito, M. and Kamiya, M. (1993) Lymphocyte engraftment conveys immunity and alters parasite development in *scid* mice infected with *Echinococcus multilocularis*. *Parasitology Research* 79, 261–268.

Rakha, N.K., Dixon, J.B., Carter, S.D., Craig, P.S., Jenkins, P. and Folkard, S. (1991) *Echinococcus multilocularis* antigens modify accessory cell function of macrophages. *Immunology* 74, 652–656.

Rausch, R.L., Wilson, J.F., Schantz, P.M. and McMahon, B.J. (1987) Spontaneous death of *Echinococcus multilocularis*: cases diagnosed serologically by Em2-ELISA and clinical significance. *American Journal of Tropical Medicine and Hygiene* 36, 576–585.

Vuitton, D.A., Bresson-Hadni, S., Laroche, L., Kaiserlian, D., Guerret-Stocker, S., Bresson, J.L. and Gillet, M. (1989) Cellular immune response in *Echinococcus multilocularis* infection in humans. II. Natural killer cell activity and cell subpopulations in the blood and in the periparasitic granuloma of patients with alveolar echinococcosis. *Clinical and Experimental Immunology*, 78, 67–74.

22

Host–Metacestode Interplay: The Role of Macrophages

J.B. Dixon

Department of Veterinary Pathology, University of Liverpool, P.O. Box 147, Liverpool L69 3BX, UK

Recent work presented at ICOPA VIII shows that vaccination against infection with hydatid organisms is now a real possibility. The existence of recombinant helminth antigens also means that potentially the immunologist is in a position to provide adequate supplies of vaccine. The arrival of immunoparasitology at this undoubted landmark may prompt us to reappraise certain fundamental aspects of hydatid immunology.

The problem is the established metacestode. In the presence of natural immunity to reinfection, hydatids can survive, grow and metastasize. Through natural acquired immunity to infection, the parasite secures its dwelling against promiscuous invasion from outside, but it then sets about modifying its new accommodation to suit its own purposes. In controlling the growth of hydatids, we do not know whether artificial immunity can succeed where natural immunity fails. As matters stand, it might appear that a single oncosphere which survives against a vaccine-induced acquired immunity could give rise to full-blown hydatidosis. The task remains, as much for the applied as for the fundamental immunologist, to understand how the hydatid establishes its enduring relationship with the host, and how it survives in the face of potentially lethal effector mechanisms.

This chapter sets out to show that modified immunity is an important factor in the survival of hydatids. The modification involves macrophages and their interaction with T-cells, and similar cellular alterations can be demonstrated in infections with cysticerci, coenuri and tetrathyridia as well as with hydatids. While evidence for this view is growing fast, it does not exclude a number other explanations which have powerfully influenced immunoparasitology in the past. These are considered here in the context of immunity to hydatids and other species of metacestode.

Limited antigenicity

Antigens evidently make up less than 3% of the total protein in hydatids (D'Amelio *et al.*, 1985), and protein antigens (apart from 5 and B) cannot be identified consistently in all samples of hydatid fluid (Shepherd and McManus, 1987). Metacestode antigens also show surprising quantitative and qualitative variability. Antigenic variation is also likely to occur during the differentiation of metacestodes (Craig and Rickard, 1982; Craig, 1988), and parasites may even adapt their secretory metabolism to present the minimum antigenic stimulus (Pezzella *et al.*, 1984). Accordingly, Gottstein (1991) and Kroeze and Tanner (1987) have reported antigenic variance between different isolates of *Echinococcus multilocularis* which cannot be explained by their genetic strain, host origin, organ origin, fertility or protein content. Differences have even been found between protoscoleces from different cysts within the same individual host (Thompson, 1991). Furthermore, amino acid sequencing of a 12 kDa subunit of *Echinococcus granulosus* antigen B has indicated sequence diversity in a limited number of clonal populations of protoscoleces (Shepherd *et al.*, 1991). The implications of this discovery are considerable. Host defence may exert pressure for a high rate of mutation in the 12 kDa antigen gene and a consequent 'fluidity' in parasite escape mechanisms. The idea that parasitic differentiation may involve actual genomic modulation of the antigen profile is not totally unprecedented. Jones and Kusel (1989) have found variation in expression of surface epitopes among cercariae and schistosomula of *Schistosoma mansoni*, even among members of a clone. They point out that the variability could be a response to 'unknown host or environmental triggers'.

In an extensive study of antigenic variation, Yakoleff-Greenhouse *et al.* (1982) randomly collected *Taenia solium* cysticercal extracts and precipitated them with different individual mouse antisera. They showed by variance analysis that both the extracts and the antisera had widely variable precipitating properties and that both sources of variation were independent of each other, and of the protein concentration of the extracts. This emphasizes the host's contribution to antigenic variation. It thus appears that metacestode antigens are few, variable and are not always recognized.

Sequestration

It is widely assumed that the surrounding membranes of larvae shield the host from the parasite antigens (Mackenzie, 1984). Since host macromolecules can enter the larval cyst and parasite antigens are found in the circulation, the laminar and germinal membranes do not seem to form absolute barriers to macromolecular exchange. Also the protective function of the membranes against host aggression is evidently a relative one at most: thus it is common knowledge that metastasis of hydatid contents takes place without the intervention of protective membranes. Nevertheless the membranes may exert some control over the flow of molecules to and from the cyst cavity (Coltorti and Varela-Diaz, 1974) and provide a further source of variation in the availablility of antigen to the host.

Disguise

Another explanation for the equivocal nature of metacestode antigenicity is that these parasites, like schistosomes, can disguise themselves by accretion of host protein (Coltorti and Varela-Diaz, 1975; Vidor *et al.*, 1987). The adhered protein on schistosomes can induce a protective immune response when these parasites are transplanted to allogeneic hosts (Smithers and Terry, 1976). The only experiment in which this has been attempted with hydatids failed to induce a clear protection in the recipient (Varela-Diaz *et al.*, 1974). The ultimate importance of disguise to the hydatid is therefore uncertain.

Mimicry

Other authors have referred to mimicry in metacestodes. Frosch *et al.* (1991) report that the recombinant antigen EM10 from *E. multilocularis* shares 42.6% identity with human cytovillin, a plasma membrane protein that may be associated with invasive growth in carcinomata. As reported for several other metazoon parasites, substances with blood-group specificity are found in *Echinococcus* spp. (Cameron and Staveley, 1957; Smyth, 1969; Schantz and Kagan, 1980). *Taenia* antigen B has fibronectin-like properties and binds to collagen (Plancarte *et al.*, 1982, 1983). This property has also been remarked on by Flisser *et al.* (1982) as a potential means by which the parasite might distract attention from itself. Rickard and Williams (1982) citing Willms *et al.* (1980) referred to an RNA which encoded mammalian-Ig-like molecules as an instance of mimicry.

The Immune Response Engendered by Hydatids

Given that metacestodes are organisms of equivocal antigenicity, it is interesting to summarize descriptions of immunity in hydatid infections which have been induced artificially by protoscolex injection. Such infections can induce relative resistance to subsequent infection both at the same and at distant sites (De Rycke and Pennoit-De Cooman, 1973; Rau and Tanner, 1973; Hinz, 1979). However, in some cases primary infection with protoscoleces enhanced secondary growth in another site (Hinz, 1985). The nature of the host response seems to be determined in large measure by the biomass of hydatid. Resection of large, established subcutaneous *E. multilocularis* cysts produced a 20-fold increase in weight of intrathoracic metastases (Rau and Tanner, 1976).

The Cellular Basis of the Host–Parasite Relationship

Histopathological observations of hydatidosis suggest the involvement of the lymphoid–macrophage system as a regulator. The dominant immunopathological event in hydatidosis is lymphoproliferation followed by cellular depletion of local lymph nodes (Ali-Khan, 1974, 1978a,b, 1980), the depletion of lymph node tissue being accentuated in secondary as compared to primary infection (Riley *et al.*, 1985, 1986, 1987). This paradoxical effect is associated with apparent diminution of response to antigens in the node (Allan *et al.*, 1981) and possibly with loss of resistance to local intercurrent infection (Ellis *et al.*, 1991). T-cell depletion is also seen in the circulation (Vuitton *et al.*, 1989). *E. multilocularis* hydatids have a biphasic pattern of growth. Early growth is restricted but later becomes progressive. The progressive phase is associated with proliferation of macrophage-like cell types in the locality, together with loss of ability to contain the parasite (Ali-Khan and Siboo, 1980), implying that macrophage effector function is a major regulator of growth.* On these grounds the macrophages and T-cells appear to be involved in a subversive exercise. At the same time macrophages are effector cells against *Echinococcus* protoscoleces and other cestode larvae, both *in vivo* and *in vitro*, and are enhanced in this action by BCG vaccination (Reuben and Tanner, 1983). Their effector activity can also be regulated by cytokines including interferon-gamma (INF-γ) and tumour necrosis factor-alpha (TNF-α) (Jenkins *et al.*, 1990, 1991, 1992). An intact T-cell system is also essential to prevent ungoverned parasitic growth and invasion (Kamiya *et al.*, 1982; Davis and Hammerberg, 1990; Liance *et al.*, 1992).

* Against this interpretation, Playford *et al.* (1992) found biphasic *E. multilocularis* growth in mice with severe combined immunodeficiency. This implies that such growth is an inherent parasite characteristic. Confirmation of this appears vital to our understanding of the host–hydatid relationship.

Lymphocyte Proliferative Responses

Several studies in cellular immunology indicate that cestode larvae activate and also modify lymphoid cell functions. Metacestode antigens induce specific lymphocyte proliferative responses, which have been studied for their immunodiagnostic potential (Gottstein *et al.*, 1991), but little is known about the function or phenotype of the specifically reactive populations. On the other hand there have been reports of the direct inhibition of lymphoid proliferation (Annen *et al.*, 1981; Osuna *et al.*, 1987).

In apparent contrast, Dixon and co-workers at Liverpool (1978, 1982) reported that protoscoleces of *E. granulosus* were directly and non-specifically mitogenic for murine lymph node cells *in vitro*. The proliferative reaction occurred in T- and B-cells. It was macrophage-dependent in T-cells and was restricted to Class II (Cox *et al.*, 1986). Similar activities were found in T-, but not B-, cells in response to *Taenia multiceps* (Rakha *et al.*, 1991b). Mitogenic activity was also detectable either in the fluids, culture fluids or living metacestodes of *Taenia hydatigena* and *Taenia pisiformis*, in the scolex of *Hymenolepis diminuta*, and in extracts of adult *Moniezia expanse*, *Anoplocephala perfoliata* (Judson *et al.*, 1987) and *T. solium* (Willms, 1975; Sealey *et al.*, 1981).

The action of the mitogenic principles appears independent of presensitization, since *T. multiceps* coenurus fluid activates blood lymphocytes from day-old calves. Activation by *E. granulosus* is also polyclonal in B-cells (Cox *et al.*, 1989). In T-cells the class II restriction of the response presumably implies the involvement of the antigen receptor. Metacestode mitogens have been partially characterized (Judson *et al.*, 1987; Rakha *et al.*, 1991b). Non-specific mitogenesis may be at least a contributary cause of the early expansion of $CD4^+$ or $Lyt1^+$ cells reported in early histopathological response to hydatid organisms (Riley *et al.*, 1987).

Non-specific mitogenic effects have also been studied by Kizaki *et al.* (1993a,b,c), who have confirmed most of the above *in vitro* findings including the role of macrophages in lymphocyte mitosis, which was at least partly interleukin (IL)-1-mediated. Kizaki *et al.*, unlike the Liverpool group, report mitosis in B-cells in response to *E. multilocularis* protoscoleces. The reason for this additional activity may be the higher levels of fetal calf serum used by the former workers in lymphocyte cultures. Kizaki *et al.* have extended their observations to show that protoscoleces are instrumental in inducing $CD8^+$ T-cells which suppress 3-day Concavalin A (Con A) or anti-CD3 responses of splenic lymphocytes by inhibiting IL-2 production. The expression of IL-2 receptor was also diminished in cultures co-stimulated with Con A and protoscoleces. The expansion of T-suppressor populations by protoscoleces *in vitro* is highly consistent with observations by Riley *et al.* (1987) on lymph nodes during *E. granulosus* infection, where initial expansion of the $Lytl^{+}$ population was followed by depression of the $Lytl^{+}$: $Lyt2^{+}$ ratio after 12 weeks of infection. Other apparent discrepancies between Kizaki's results and those of the Liverpool group are of interest: whereas Kizaki *et al.* (1993b) mainly report depression of Con A response in the presence of protoscoleces, Judson and Dixon (unpublished) largely recorded enhancement. This disagreement may disappear, however, when the time dimension is considered: Kizaki *et al.* themselves (1993b) record enhancement when 2-day, as distinct from 3-day, Con A cultures are under study. Rakha and Dixon (unpublished) found that unprimed responders to *E. granulosus* and *T. multiceps* are $CD4^{+}$, whereas Kizaki *et al.* emphasize the expansion of $CD8^{+}$ populations. These findings might also prove consistent if a $CD4^{+}$ population was responsible for the induction of the $CD8^{+}$ suppressor cells, and/or if the $CD4^{+}$ proliferation was self-destructive, a possibility raised by Allan *et al.* (1981). Bresson-Hadni *et al.* (1990) have inferred from histological observation on resistant and susceptible mouse strains that the combination of macrophages and $CD4^{+}$ cells could be deleterious to the hydatid-bearing host.

Other evidence can be cited for the participation of lymphocyte mitosis in immunosuppression. Judson *et al.* (1985), reanalysing previous work by Yusuf *et al.* (1975), found that a significant negative correlation was exhibited between lymphocyte transformation and indirect haemagglutination titre to hydatid fluid in ten human patients. Judson *et al.* (1985) also found negative correlation in sheep between transformation to hydatid fluid on one hand and antibody titre, dermal sensitivity and leucocyte migration on the other. They suggested that a reciprocal relationship exists between transformation and other responses to *Echinococcus*. Investigation by Siracusano *et al.* (1988), using purified antigen rather than hydatid fluid, failed to discover any correlation between transformation and antibody response in a group of hydatid patients. The tendency for antibody and proliferative responses to be reciprocal was, however, born out in a later study of human alveolar hydatid disease by Gottstein *et al.* (1991) who found very high lymphoproliferative responses to *E. multilocularis* antigen (Em2) in patients cured by radical surgery or in patients who appeared, on the grounds that their lesions contained dead parasites, to have undergone spontaneous cure. Antibody levels to Em2 antigen were high in patients who had had incomplete or no surgery, and low or absent in cured patients who had undergone radical surgery or were apparently self-cured. Gottstein *et al.* point out that the results could be consistent with cellular mechanisms killing the parasite, e.g. activated T-cells and their lymphokines acting on effector cells, or alternatively with suppression of cell-mediated immunity by the parasite.

Subsequent to the discovery of cestode mitogens, interest has grown in mitogenic activities from other organisms. Superantigens of bacterial and viral origin, which resemble cestode mitogens in their capacity for non-specific T-cell activation, are currently being considered as subversive mechanisms. Superantigens are credited with mediating autocytotoxicity against activated T-cells (reviewed by Dohlsten *et al.*, 1991), an effect previously proposed by Allan *et al.* (1981) to explain the suppression of rosette-forming cell response by *E. granulosus*. Causal connections between non-specific T-cell stimulation and T-cell depletion are also being urgently reviewed in the context of acquired immune deficiency syndrome (AIDS) research (Ameisen, 1992). $CD4^+$ T-cells from human immunodeficiency virus (HIV)-infected individuals undergo programmed cell death on stimulation with superantigens, and proteins of the HIV envelope will prime even normal T-cells for cell death on subsequent stimulation. Thus inappropriate T-cell activation could be the grounds for T-cell proliferation followed by T-cell depletion. A further likely possibility is that mitogen-activated proliferation in $CD4^+$ cells leads to significant alteration of the ratio of Th1 to Th2 populations (Coffman *et al.*, 1991). A predominance of Th2 cellular reaction is thought to exist in murine *Mesocestoides corti* infection (Zakroff *et al.*, 1989; Desai *et al.*, 1990).

Modification of Macrophage Accessory Activity

Since the mitogenic response of T-cells to *E. granulosus* was macrophage-dependent (Cox *et al.*, 1986), Rakha *et al.* (1991a) hypothesized that the interaction of the two cell types might also be subject to manipulation by metacestodes, and designed an assay for accessory cell activity of macrophages based on the normal ability of these cells to enhance mitogen-induced T-cell response. It was found that intraperitoneal injection of the supernatant of *T. multiceps* protoscolex cultures induced a population of abnormal peritoneal macrophages. The abnormal cells enhanced transformation up to a certain macrophage concentration and then progressively inhibited it (Rakha *et al.*, 1991a). The modification was apparent with T-cells stimulated by Con A, phytohaemagglutinin (PHA) or the native mitogen of *T. multiceps*. The modified cell was additionally found in mice infected with *E. multilocularis* (Rakha *et al.*, 1991c) and *M. corti* (Kadian *et al.*, 1994).

Fast protein liquid chromatography (FPLC) analysis of coenurus fluid clearly separated its immunological activities, with the mitogen in fraction 7 (F7), and the macrophage modifying activity in a cluster of fractions around 24 (F24). In addition, the macrophage–T-cell interaction was further impaired if the T-cells were activated with F7 and the macrophages with F24, suggesting the presence of an integrated mechanism for inhibiting the activation of T-cells. In accordance with this, F24 was shown to inhibit the rosette-forming-cell response to sheep red cells (Rakha *et al.*, 1991b). The altered macrophage activity could also be induced by 48 h incubation with F24 *in vitro* in the absence of lymphocytes, indicating that the action on the macrophages was direct (Rakha *et al.*, 1992).

Conclusion

The alterations in cell behaviour resulting from exposure to metacestodes are apparently induced by at least two separate components of the metacestode secretions. One of these acts on the T-cell as mitogen inducing a number of changes which are not completely understood, but which appear to be associated with the generation of suppressor cells and possibly with subsequent T-cell depletion. The other factor is active directly on macrophages with evident loss of their accessory interaction with T-cells. The two factors are also found to be synergistic in that their combined action brings about the maximum disturbance to the macrophage–T-cell interaction. In infection, the earliest change is T-cell proliferation; T-suppressor expansion and loss of macrophage interaction come later. The behaviour of the system suggests that the host's policy is one of collaboration rather than of confrontation. Such a reaction could have evolved if there were serious survival disadvantages to the host in ungoverned aggression. One such disadvantage could be the risk of autoimmune reaction. There is some evidence that this a real issue. Ameglio *et al.* (1987) and Quintieri *et al.* (1992) have demonstrated autoreactivity in humoral and cellular immune systems of hydatid patients.

The host–hydatid relationship, whatever its evolutionary origin, now seems clearly to involve immunoregulation by *Echinococcus*. Rogers (1962) regards a parasite as an organism requiring differentiational signals from a host. It should not be entirely surprising to find that the host–hydatid relationship requires the host to respond reciprocally to organizational signals from the parasite. These signals could provide the key to immunological intervention in established hydatidosis.

References

Ali-Khan, Z. (1974) Host–parasite relationship in echinococcosis. II. Cyst weight, haematologic alterations and gross changes in the spleen and lymph nodes of C57L mice against graded doses of *Echinococcus multilocularis* cysts. *Journal of Parasitology* 60, 236–242.

Ali-Khan, Z. (1978a) Pathological changes in the lymphoreticular tissues of Swiss mice infected with *Echinococcus granulosus* cysts. *Zeitschrift für Parasitenkunde* 58, 47–54.

Ali-Khan, Z. (1978b) Cellular changes in the lymphoreticular tissues of C57L/J mice infected with *Echinococcus multilocularis* cysts. *Immunology* 34, 831–839.

Ali-Khan, Z. and Siboo, R. (1980) Pathogenesis and host response in subcutaneous alveolar hydatidosis. I. Histogenesis of alveolar cyst and a qualitative analysis of the inflammatory infiltrates. *Zeitschrift für Parasitenkunde* 62, 241–254.

Allan, D., Jenkins, P., Connor, R.J. and Dixon, J.B. (1981) A study of immunoregulation of BALB/c mice by *Echinococcus granulosus equinus* during prolonged infection. *Parasite Immunology* 3, 137–142.

Ameglio, F., Saba, F., Bitti, A., Aceti, A., Tanigaki, N., Sorrentino, R., Dolei, A. and Tosi, R. (1987) Antibody reactivity to HLA classes I and II in sera from patients with hydatidosis. *Journal of Infectious Diseases* 156, 673–676.

Ameisen, J.C. (1992) Programmed cell death and AIDS: from hypothesis to experiment. *Immunology Today* 13, 388–391.

Annen, J.M., Kohler, P. and Eckert, J. (1981) Cytotoxicity of *Echinococcus granulosus* cyst fluid *in vitro*. *Zeitschrift für Parasitenkunde* 65, 79–88.

Bresson-Hadni, S., Liance, M., Meyer, J.P., Houin, R., Dresson, J.L. and Vuitton, D.A. (1990) Cellular immunity in experimental *Echinococcus multilocularis* infection. II. Sequential and comparative phenotypic study of the periparasitic mononuclear cells in resistant and sensitive mice. *Clinical and Experimental Immunology* 82, 378–383.

Cameron, G.L. and Staveley, M.J. (1957) Blood group P substance in hydatid cyst fluid. *Nature* 179, 147–148.

Coffman, R.L., Varkila, K., Scott, P. and Chatelain, R. (1991) Role of cytokines in the differentiation of CD4+ T-cell subsets *in vivo*. *Immunological Reviews* 123, 189–207.

Coltorti, E.A. and Varela-Diaz, V.M. (1974) *Echinococcus granulosus*. Penetration of macromolecules and their localisation on the parasite membranes of cysts. *Experimental Parasitology* 35, 225–231.

Coltorti, E.A. and Varela-Diaz, V.M. (1975) Penetration of host IgG molecules into hydatid cysts. *Zeitschrift für Parasitenkunde* 48, 47–51.

Cox, D.A., Dixon, J.B. and Marshall-Clarke, S. (1986) Transformation induced by *Echinococcus granulosus* protoscoleces in unprimed murine spleen cells: identity and MHC restriction of participating cell types. *Immunology* 57, 461–466.

Cox, D.A., Marshall-Clarke, S. and Dixon, J.B. (1989) Activation of normal murine B cells by *Echinococcus granulosus*. *Immunology* 67, 16–20.

Craig, P.S. and Rickard, M.D. (1982) Antibody responses of experimentally infected lambs to antigens collected during *in vitro* maintenance of the adult, metacestode or oncosphere stage of *Taenia hydatigena* and *Taenia ovis* with further observations on antioncospheral antibodies. *Zeitschrift für Parasitenkunde* 67, 197–209.

Craig, P.S. (1988) Surface-associated proteins and host IgG on early and late metacestode stages of *Taenia pisiformis*. *Parasite Immunology* 10, 243–254.

D'Amelio, R., Pontesilli, O., Dayal, R., de Rosa, F., Barnet, M., Teggi, A., Brighouse, G. and Lambert, P.H. (1985) Characterization of parasite antigens from human hydatid cyst fluid by SDS-PAGE and IEF. *Medical Microbiology and Immunology* 74, 43–50.

Davis, S.W. and Hammerberg, B. (1990) Effect of depletion of L3T4$^+$ cells on resistance to early primary infection of mice with *Taenia taeniaeformis* *International Journal for Parasitology* 20, 595–601.

Desai, B.B., Abraham, K.M. and Teale, J.M. (1990) The isotype potential of B cells present in BALB/c mice chronically infected with *Mesocestoides corti*. *Cellular Immunology* 130, 139–149.

De Rycke, P.H. and Pennoit-De Cooman, E. (1973) Experimental secondary echinococcosis of *Echinococcus granulosus*. IV. Vaccination of host mice. *Zeitschrift für Parasitenkunde* 42, 49–59.

Dixon, J.B., Jenkins, P. and Allan, D. (1978) Blastic stimulation of unprimed mouse lymphocytes by living protoscolices of *Echinococcus granulosus*: a possible connection with transplant immunity. *Journal of Parasitology* 64, 949–950.

Dixon, J.B., Jenkins, P. and Allan, D. (1982) Immune recognition of *Echinococcus granulosus*. 1. Parasite-activated, primary transformation by normal murine lymph node cells. *Parasite Immunology* 4, 33–45.

Dohlsten, M., Hedlund, G. and Kalland, T. (1991) Staphylococcal-enterotoxin-dependent cell-mediated cytotoxicity. *Immunology Today* 12, 147–149.

Ellis, M.E., Sinner, W., Ali, M.A. and Qadri, S.M.H. (1991) Echinococcal diseases and mycobacterial infection. *Annals of Tropical Medicine and Parasitology* 85, 243–251.

Flisser, A., Plancarte, A., Espinoza, B., Woodhouse, E., Rivera, L., Trueba, J., Yokoleff-Greenhouse, V., Sierra, A., Laclette, J.P., Guerra, C., Canedo, L. and Larralde, C. (1982) The immune status of patients with cysticercosis. *African Journal of Clinical and Experimental Immunology* 3, 183–190.

Frosch, P.M., Frosch, M., Pfister, T., Schaad, V. and Bitter-Suermann, D. (1991) Cloning and characterisation of an immunodominant major surface antigen of *Echinococcus multilocularis*. *Molecular and Biochemical Parasitology* 48, 121–130.

Gottstein, B. (1991) *Echinococcus multilocularis*: antigenic variance between different parasite isolates. *Parasitology Research* 77, 359–361.

Gottstein, B., Mesarina, B., Tanner, I., Ammann, R.W., Wilson, J.F., Eckert, J. and Lanier, A. (1991) Specific cellular and humoral immune responses in patients with different long-term courses of alveolar echinococcosis (infection with *Echinococcus multilocularis*). *American Journal of Tropical Medicine and Hygiene* 45, 734–742.

Hinz, E. (1979) *Echinococcus multilocularis*: superinfections in experimentally infected mice. *Tropenmedizin und Parasitologie* 30, 387–390.

Hinz, E. (1985) *Echinococcus multilocularis*: further investigations of the significance of superinfections in experimentally infected mice. *Mitteilungen, Osterreichischen Gesellschaft für Tropenmedizin und Parasitologie* 7, 85–91.

Jenkins, P., Dixon, J.B., Rakha, N.K. and Carter, S.D. (1990) Regulation of macrophage-mediated larvicidal activity in *Echinococcus granulosus* and *Mesocestoides corti* (Cestoda) infection in mice. *Parasitology* 100, 309–315.

Jenkins, P., Dixon, J.B., Haywood, S., Rakha, N.K. and Carter, S.D. (1991) Differential regulation of murine *Mesocestoides corti* infection by bacterial lipopolysaccharide and interferon-γ. *Parasitology* 102, 125–132.

Jenkins, P., Spiers, S., Dixon, J.B., Carter, S.D. and May, S. (1992) The effects of tumour necrosis factor on host-parasite relations in murine *Mesocestoides corti* (Cestoda) infection. *Parasitology* 105, 453–459.

Jones, J.T. and Kusel, J.R. (1989) Intraspecific variation in *Schistosoma mansoni*. *Parasitology Today* 5, 37–39.

Judson, D.G., Dixon, J.B., Clarkson, M.J. and Pritchard, J. (1985) Ovine hydatidosis: some immunological characteristics of the seronegative host. *Parasitology* 91, 349–357.

Judson, D.G., Dixon, J.B. and Skerritt, G.C. (1987) Occurrence and biochemical characteristics of cestode lymphocyte mitogens. *Parasitology* 94, 151–160.

Kadian, S.K., Dixon, J.B., Green, J.R., Carter, S.D. and Jenkins, P. (1994) Modification of macrophage-T cell interaction during infection of mice with *Mesocestoides corti* (Cestoda). *Parasitology* 109, 591–597.

Kamiya, H., Oku, Y., Kamiya, H. and Nomura, T. (1982) Characteristic responses of nude mice in angiostrongyliasis and echinococcosis. In: *Proceedings of the Third International Workshop on Nude Mice*, Montana State University, Bozeman, Montana, 6–9 September 1979, vol. 1. Gustav Fischer, New York, pp. 133–145.

Kizaki, T., Ishige, M., Bingyan, W., Kumagai, M., Day, N.K., Good, R.A. and Onoé, K. (1993a) Interleukin-1-dependent mitogenic responses induced by protoscoleces of *Echinococcus multilocularis* in murine lymphocytes. *Journal of Leukocyte Biology* 53, 233–239.

Kizaki, T., Ishige, M., Kobayashi, S., Bingyan, W., Kumagai, M., Day, N.K., Good, R.A. and Onoé, K. (1993b) Suppression of T-cell proliferation by $CD8^+$ T cells induced in the presence of protoscoleces of *Echinococcus multilocularis in vitro*. *Infection and Immunity* 61, 525–533.

Kizaki, T., Ishige, M., Bingyan, W., Day, N.K., Good, R.A., and Onoé, K. (1993c) Generation of $CD8^+$ suppressor T cells by protoscoleces of *Echinococcus multilocularis in vitro*. *Immunology* 79, 412–417.

Kroeze, W.K. and Tanner, C.E. (1987) *Echinococcus multilocularis*: variation among samples of cyst fluid in binding of parasite specific antibodies. *Annals of Tropical Medicine and Parasitology* 81, 393–403.

Liance, M., Bresson-Hadni, S., Vuitton, D.A., Lenys, D., Carbillet, J.P. and Houin, R. (1992) Effects of cyclosporin A on the course of murine alveolar echinococcosis and on specific cellular and humoral immune responses against *Echinococcus multilocularis*. *International Journal of Parasitology* 22, 23–28.

Mackenzie, C.D. (1984) Sequestration beneficial to both host and parasite. *Parasitology* 88, 593–595.

Osuna, A., Gamarro, F., Jansen, D., Carreras, M.A., Castanys, S., Adroher, F.J. and deRycke, P.H. (1987) Cytotoxicity of *Echinococcus granulosus* hydatid fluid: species specific activity and initial biochemical characterisation. *Revista Ibérica de Parasitología* 47, 47–52.

Pezzella, M., Galli, C., Delia, S., Vullo, V., Zennaro, F., Lillini, E. and Sorice, F. (1984) Fractionation and characterization of hydatid fluid antigens with identification of an antigen similar to human serum albumin. *Transactions of the Royal Society of Tropical Medicine and Hygiene* 78, 821–826.

Plancarte, A., Flisser, A. and Larralde, C. (1982) Fibronectin like properties of antigen B from cysticerci of *Taenia solium*. In: Flisser, A., Willms, K., Laclette, J.P., Laralde, C., Ridaura, C. and Beltran, R. (eds) *Cysticercosis: Present State of Knowledge and Perspectives*. Academic Press, London, pp. 453–463.

Plancarte, A., Flisser, A. and Larralde, C. (1983) Fibronectin-like properties in antigen-B from the cysticersus of *Taenia solium*. *Cytobios* 36, 83–93.

Playford, M.C., Ooi, H.-K., Oku, Y. and Kamiya, M. (1992) Secondary *Echinococcus multilocularis* infection in severe combined immunodeficient (scid) mice: biphasic growth of the larval cyst mass. *International Journal of Parasitology* 22, 975–982.

Quintieri, F., Siracusano, A., Rigano, R. and Pugliesi, O. (1992) Limit dilution analysis of autoreactive T cells in patients affected by hydatid disease. *Journal of Autoimmunity* 5, 733–744.

Rakha, N.K., Dixon, J.B., Skerritt, G.C., Carter, S.D., Jenkins, P. and Marshall-Clarke, S. (1991a) Lymphoreticular responses to metacestodes: *Taenia multiceps* (Cestoda) can modify interaction between accessory cells and responder cells during lymphocyte activation. *Parasitology* 102, 133–140.

Rakha, N.K., Dixon, J.B., Jenkins, P., Carter, S.D., Skerritt, G.C. and Marshall-Clarke, S. (1991b) Modification of cellular immunity by *Taenia multiceps* (Cestoda): accessory macrophages and $CD4^+$ lymphocytes are affected by two different coenurus factors. *Parasitology* 103, 139–147.

Rakha, N.K., Dixon, J.B., Carter, S.D., Craig, P.S., Jenkins, P. and Folkard, S. (1991c) *Echinococcus multilocularis* antigens modify accessory cell function of macrophages. *Immunology* 74, 652–656.

Rakha, N.K., Dixon, J.B., Skerritt, G.C., Carter, S.D. and Jenkins, P. (1992) Modification of accessory activity of sheep monocytes *in vitro* by a coenurus antigen from *Taenia multiceps*. *Veterinary Immunology and Immunopathology* 30, 293–304.

Rau, M.E. and Tanner, C.E. (1973) *Echinococcus multilocularis* in the cotton rat. The effect of preexisting subcutaneous cysts on the development of a subsequent intraperitoneal inoculum of protoscolices. *Canadian Journal of Zoology* 51, 55–59.

Rau, M.F. and Tanner, C.E. (1976) *Echinococcus multilocularis* in the cotton rat: growth of intrathoracic metastases following surgical removal of subcutaneous cysts. *International Journal of Parasitology* 6, 151–153.

Reuben, J.M. and Tanner, C.E. (1983) Protection against experimental echinococcosis by non-specifically stimulated peritoneal cells. *Parasite Immunology* 5, 61–66.

Rickard, M.D. and Williams, J.F. (1982) Hydatidosis/cysticercosis: immune mechanisms and immunization against infection. *Advances in Parasitology* 21, 229–296.

Riley, E.M., Dixon, J.B., Kelly, D.F. and Cox, D.A. (1985) The immune response to *Echinococcus granulosus*: sequential histological observations of lymphoreticular and connective tissues during early murine infection. *Journal of Comparative Pathology* 95, 93–104.

Riley, E.M., Dixon, J.B., Jenkins, P. and Ross, G. (1986) *Echinococcus granulosus* infection in mice: host responses during primary and secondary infection. *Parasitology* 92, 391–403.

Riley, E.M. and Dixon, J.B. (1987) Experimental *Echinococcus granulosus* infection in mice: immunocytochemical analysis of lymphocyte populations in local lymphoid organs during early infection. *Parasitology* 94, 523–532.

Rogers, W.P (1962) 2. The nature of parasitism; 3. Life cycles and infection. In: *The Nature of Parasitism*. Academic Press, New York, pp. 12–47.

Schantz, P.M. and Kagan, I.G. (1980) Echinococcosis (hydatidosis). Immunological investigation of tropical parasitic diseases: In: Houba, V. (ed.) *Practical Methods in Clinical Immunology*, vol. 2. Churchill Livingstone, Edinburgh, pp. 109–129.

Sealey, M., Ramos, C., Willms, K. and Ortiz-Ortiz, L. (1981) *Taenia solium*: mitogenic effect of larval extracts on murine B lymphocytes. *Parasite Immunology* 3, 299–307.

Shepherd, J.C. and McManus, D.P. (1987) Characterisation of *Echinoccocus granulosus* proteins and antigens from hydatid cyst fluid. In: Geerts, S., Kumar, V. and Brandt, J.J. (eds) *Helminth Zoonosis*. Martinus Nijhoff, Dordrecht, pp. 50–56.

Shepherd, J.C., Aitken, A. and McManus, D.P. (1991) A protein secreted *in vivo* by *Echinococcus granulosus* inhibits elastase activity and neutrophil chemotaxis. *Molecular and Biochemical Parasitology* 44, 81–90.

Siracusano, A., Teggi, A., Quintieri, F., Notargiacomo, S., DeRosa, F. and Vicari, G. (1988) Cellular immune responses of hydatid patients to *Echinococcus granulosus* antigens. *Clinical and Experimental Immunology* 72, 400–405.

Smithers, S.P. and Terry, R.J. (1976) The immunology of schistosomiasis. *Advances in Parasitology* 14, 399–422.

Smyth, J.D. (1969) Parasites as biological models. *Parasitology* 59, 73–91.

Thompson, R.C.A. (1991) *Echinococcus* and *Giardia*: variation on a theme. Australian Society for Parasitology presidential address. *International Journal for Parasitology* 21, 291–297.

Varela-Diaz, V.M., Williams, J.F., Coltorti, E.A. and Williams, C.S.F. (1974) Survival of cysts of *Echinococcus granulosus* after transplant into homologous and heterologous hosts. *Journal of Parasitology* 60, 608–612.

Vidor, E., Piens, M.A. and Garin, J.P. (1987) Host serum protein levels in cysts of human hydatidosis. *Transactions of the Royal Society of Tropical Medicine and Hygiene* 81, 669–671.

Vuitton, D.A., Bresson-Hadni, S., Laroche, L., Kaiserlian, D., Guerret-Stocker, S., Bresson, J.L. and Gillet, M. (1989) Cellular immune response in *Echinococcus multilocularis* infection in humans. 2. Natural killer cell activity and cell subpopulations in the blood and in the periparasitic granuloma of patients with alveolar echinococcosis. *Clinical and Experimental Immunology* 78, 67–74.

Willms, K. (1975) Caracteristicas biológicas de los antígenos de *Cysticercus cellulosae* de cerdo. [Biological characterization of porcine *Cysticercus cellulosae* antigen.] *Patologia* 13, 115–125.

Willms, K., Merchant, M.T., Arcos, L., Sealey, M., Diaz, S. and Diaz de Leon, L. (1980) Immunopathology of cysticerosis. In: Larralde, C., Willms, K., Ortiz-Ortiz, L., Sela, M. and Vogt, M.W. (eds) *Molecules, Cells and Parasites in Immunology*. Academic Press, New York, pp. 145–162.

Yakoleff-Greenhouse, V., Flisser, A., Sierra, A. and Larralde, C. (1982) Analysis of antigenic variation in cysticerci of *Taenia solium*. *Journal of Parasitology* 68, 39–47.

Yusuf, J.N., Frayha, G.J. and Malakian, A.H. (1975) *Echinococcus granulosus*: host lymphocyte transformation by parasite antigens. *Experimental Parasitology* 38, 30–37.

Zakroff, S.G.H., Beck, L., Platzer, E.G. and Spiegelberg, H.L. (1989) The IgE and IgG subclass responses of mice to four helminth parasites. *Cellular Immunology* 119, 193–201.

23

Epidemiology and Epidemiological Research Methods: Their Application in Parasitic Disease Investigation

K.C. SHEKHAR

Department of Parasitology, Faculty of Medicine, University of Malaya, Kuala Lumpur, Malaysia

Introduction

Epidemiology is the study of the causes and distribution of disease in a defined population and epidemiologic research methods are those tools that are applied in the investigation of diseases. There is a growing recognition of the important contribution that epidemiology and epidemiological research methods can make to the planning and management of parasitic disease control strategies in health services and ultimately to the improvement of health care in the general population.

Parasitic diseases remain widely distributed all over the world with high prevalence rates in many endemic regions. Though mortality to most parasitic diseases is relatively low, morbidity is very high, for example, opisthorchiaisis which affects 5.4 million people in northeast Thailand. In most endemic countries severe anaemia, diarrhoea, dysentery, impaired work capacity, malnutrition, blindness and reduced growth rate are some of the sequelae to infections with parasitic diseases and these pose important health and socioeconomic problems. With the advent of acquired immune deficiency syndrome (AIDS), new diseases or new clinical spectrums of old diseases such as microsporidiasis, isosporiasis, cryptosporidiasis and disseminated strongyloidiasis are on the increase. On the other hand, the incidence of food- and water-borne and vector-borne zoonotic infections are posing a medical challenge not only to clinicians but also to public health specialists.

With the increasing trend in the morbidity of parasitic diseases and in spite of the concerted efforts by various agencies to control and combat the diseases, the incidence pattern shows a rising trend. This could be attributed to flaws in epidemiologic research methods and improper surveillance which ultimately affects the implementation and execution of a control programme. Incorrect interpretations based on improper study designs have been detected in publications especially in relation to parasitic disease investigations or assessments. A literature search again shows that the utilization of epidemiological research methods in the investigation of parasitic diseases is somehow lacking except for classical studies on the impact of helminth infections on human nutrition. Epidemiological research recently has made substantial progress with the development of new techniques for diagnosis, isolation of parasites, identification and vector control. These have opened up new perspectives and introduced new simplified tools and to a certain extent standardized field studies.

The objectives of this chapter are:

1. to highlight the types of epidemiologic studies existing and basic principles needed to carry out these studies in relation to parasitic diseases; and
2. to impart sound epidemiological knowledge so that researchers are able to carry out good studies in their field of interest.

Epidemiological Research

In order to understand epidemiological research the basic outcomes need to be spelt out. They are:

1. to improve the health status of the population by minimizing morbidity or mortality; and
2. to decrease disability and increase the expectancy of life.

In this context the terms 'disease' and 'health' need to be defined. Disease will constitute pathological processes while health refers to the state of well being. In epidemiological research, the main target of study is the population and conclusions and inferences made are based on the study of many subjects.

Thus the main aims of epidemiologic research are:

1. to describe the health status of the population by enumerating occurrence of disease, the relative frequencies within groups and to observe any trends;
2. to explain the aetiology of diseases by determining the factors that are associated with the specific diseases, trends and modes of transmission;
3. to predict the number of disease occurrences and the distribution of health status within populations; and
4. to control the distributions of diseases in the population by preventing

new occurrences, eradication of existing cases, prolongation of life with the disease and ultimately improving the health status of the afflicted persons (Kleinbaum *et al.*, 1982).

Therefore, the aim of epidemiological research is to understand the natural history of the disease, its occurrence and to determine the aetiological cause so that prompt intervention and public health measures can then be instituted. At the intervention level the health practices, programmes and policies are from time to time evaluated to prevent disease occurrence and promote good health.

Four types of epidemiological research methods are available, each conducted by varied specialists in different ways utilizing different methodologies. They are:

1. laboratory-based research (carefully controlled animal studies);
2. epidemic investigation;
3. descriptive studies;
4. population research or analytic studies (community-based or field studies in humans – observational study).

Epidemiological Studies

There are certain concepts that need to be realized before an epidemiological study can be carried out:

1. a clear definition of the population at risk;
2. the definition of a case of the condition under study.

These two concepts should be of interest whether the investigator is carrying out a field epidemiological study or a clinical trial or analytical study.

Laboratory-based research (experimental epidemiology)

This is carried out in animals because the phenomenon and mechanisms of disease process are difficult to demonstrate in man. In these experiments, the investigator manipulates the allocation of an intervention measure or exposure to the study and control populations. The important condition is that the animal model of choice, the parasite and the pathological sequelae must be similar to the human–parasite system of interest. The parasite should occupy a similar anatomical site in both the animal models and in humans and provoke similar pathology, for example the utilization of mice in studying the pathogenesis of schistosomiasis. The variables to be considered include age, sex, duration of study, type of diet, inoculation techniques, route of administration of drugs, size and frequency of infective dose, the immunological status of host, etc. The animal age should correspond to the human age range in which the parasite is most likely to cause serious pathology (Stephenson, 1987). In other words, the experimental conditions should simulate as far as possible the natural conditions that humans face in an endemic situation. Research workers who conduct studies on parasitic disease should be aware of the differences that may exist in the laboratory and the field and be able to detect differences that might affect the results and be able to bridge the gap whenever possible.

Epidemic disease investigation

This deals with outbreaks of disease in specific populations. Epidemics can be classified as common source epidemics, point epidemics, propagative or progressive epidemics and mixed epidemics. In this study, the aetiologic agent of outbreak is investigated along with the mode of transmission. Preventive measures are then suggested after a thorough investigation.

All epidemics of parasitic origin are explosive, spread rapidly, have short latent periods, have an exposure period of short duration, are sudden in onset, have a point source and a large population is affected. Vector-borne epidemics are being reported with increasing frequency, e.g. malaria, leishmaniasis, African trypanosomiasis, onchocerciasis, which have cause considerable distress to individuals and families and disruption of local economies, particularly the wage earners.

Before carrying out an epidemiological investigation it should be established that an epidemic actually exists. In the investigation of the epidemic, the diagnosis should be verified. This may include not only clinical diagnosis confirmed by a physician or health personnel but also laboratory diagnosis. The reliability of the reporting source is of prime importance. The investigation includes search for additional cases within the defined community. The community to be investigated should be divided into three groups within the geographical area: definite, probable and suspect cases. Incidence of the disease should be compared with what is usual in the particular situation. Any significance of the disease should be noted.

The next step is to study the outbreak in terms of person (age, sex, ethnicity, race, occupation, residence), place (geographical area, availability of water supply, sanitation facilities) and time (hours, days, weeks, months and years). A preliminary statistical analysis should be made on the available data on the population based on these variables. The epidemic should be classified as to the mode of transmission (namely a common vector with single or multiple exposure or man-to-man transmission, by an arthropod vector or by animal reservoirs). The different modes of transmission tend to produce differing and characteristic distributions of new cases over time. An epidemic curve is then drawn and inferences can then be made on the likely mode of transmission. It is important to try to identify the target group and to discover the common source or sources to which members were exposed. For ethical reasons the sick are isolated in special wards to prevent spread of disease. It may be necessary to undertake special investigations using laboratory facilities or expert consultations in various fields to arrive at a conclusion.

Below are the basic steps in the investigation of an epidemic:

- Descriptive phase
  individual case investigation
  epidemic field investigations
  epidemiological surveys
  animal reservoir surveys
  demographic and environmental data.
- Formulation of hypotheses (as to type of epidemic, source, method of spread, etc.).
- Test the hypothesis.
- Make conclusions and formulate recommendations.
- Surveillance and prevention.

Preliminary assessment is usually based on existing clinical records of cases and past experience based on public health data on the disease in the community. Kelsey *et al.* (1986) have summarized very well steps to be taken for assessment of data collected within 24 h after an epidemic.

In water-borne outbreaks, the aetiologic agent or pathogen could be more than one. Epidemiological investigation requires a questionnaire to be distributed to all persons in the outbreak area within the week. A comparable control group should also be selected. Samples of faeces from patients should be collected for isolation of parasite. If possible blood samples from affected and asymptomatic patients should be collected for serology. After analysis of the questionnaire the following categories of people need to be interviewed to get further information: those people who were ill several weeks after the outbreak, those who were ill but who recovered, those people healthy throughout the period and the control group initially selected. Thus we have four groups of patients or subjects with data to be analysed. The prevalence differences between the groups should be tested statistically by use of two-sided tests and the conditional Poisson test.

Vector-borne parasitic disease epidemics are more complex and difficult to control. The epidemiologist will have to estimate qualitatively and quantitatively the parasite reservoir and its potential for transmission through specific vectorial capacity of the species. In addition, physical, environmental and biological factors related to the vector, the behaviour of man and the species of parasite involved will also have to be taken into account.

The mechanisms of vector-borne epidemics are more complicated. There are a number of publications which deal with outbreaks of vector-borne diseases, particulary malaria which has been well studied. There are methods for forecasting and the detection of malaria epidemics which could be applied for the control of other vector-borne disease epidemics with modification.

Among the diseases that have been studied well and where epidemiological methods of investigation and assessment are available are malaria, African trypanosomiasis, onchocerciasis, leishmaniasis and lymphatic filariasis. Epidemiological modelling, an important tool to study interrelationships between transmission, infection and disease have been formulated for carrying out epidemiological research in certain diseases.

Study designs

Most epidemic investigations are retrospective especially when the epidemic is over. If the epidemic has a common source and all exposed and susceptible persons have not developed the disease during the exposure period, then the follow-up of additional people known to be exposed but who are yet to be ill could be studied under a prospective cohort. Ill persons who are exposed could be followed prospectively and when exposure to contaminated food or water occurs in an occupational environment then a prospective cohort study can be undertaken.

Case control methods are commonly employed in epidemic investigations. This applies to common source food-borne outbreaks. One can determine the ratio of illness among those eating a specific food item relative to those not eating the food item among all those persons known to have eaten the meal.

Utilizing the case control method a study was carried out in Colorado on an epidemic of giardiasis (Wright *et al.*, 1977). The differences in the extent of exposure to a variety of activities was compared between cases and controls. Using the questionnaire method for cases and the control group, significant risk factors were found among campers and mountain climbers. The contamination of mountain water supplies by infected humans and/or beavers was identified as the ultimate source of infection.

When the source of the pathogen is known or strongly suspected a retrospective cohort study can be constructed and the attack rates among those exposed and unexposed can be determined.

Descriptive epidemiological studies

Descriptive studies provide information on patterns of disease occurrence in a population according to age, sex, race, marital status, social class, occupation, geographical area and time of occurrence. It is conducted when little is known about the occurrence, the natural history, or the determinants of a disease. The descriptive studies thus generated are used with clinical observations, laboratory studies and other sources of information to generate a hypothesis, for example that *Toxoplasma gondii* causes congenital defects or Burkitt's lymphoma and *Plasmodium falciparum* infections. Thus the aim is to estimate the disease frequency or time trend in a particular population and to generate more specific aetiological hypotheses.

Analytical studies

Analytical (or aetiological) studies are conducted when enough is known about the disease before investigation so that specific hypotheses can be tested. The objectives are then to identify risk factors for the disease, estimate their effects on the disease and suggest possible intervention strategies.

Analytical or observational research is often the most practical or feasible to conduct because there is no manipulation of the study factor. Besides, the studies are carried out in more natural settings so that the study population is more representative of the target population. These studies have important implications to health planners and policy makers who base their decisions on the results of these epidemiological investigations.

The limitations to analytical studies are that they afford the investigator the least control over the study situation. The results are more susceptible to distorting influences. In the data analysis, the investigator has to deal with potential sources of error that arise which threaten the validity of the findings.

Because the investigator achieves little control in the design stage, observational studies tend to be unique in a sense that they are difficult to replicate, therefore making scientific generalizations less secure (Jamison, 1980).

The three major types of analytical studies are:

1. Cross-sectional studies.
2. Cohort studies (prospective and retrospective).
3. Case control studies.

The most definite type of observational study is usually the prospective cohort study.

Cross-sectional survey (prevalence study)

This study indicates a single observation at one hypothetical point in time, i.e. the exposure and disease status are measured at one point in time or over a short period of time in the study subjects. The disease prevalence rates (number of cases of disease existing per population at risk in a given period of time) arising there with and without exposure are then determined and compared. Cross-sectional studies can be either horizontal surveys or longitudinal studies. The prevalence studies concerned with disease aetiology are most often carried out to learn about the risk factors for frequently occurring diseases.

The advantages of cross-sectional studies are: selection of a sample from the general population, carried out over a relatively short period of time, less expensive and easier to conduct. The main disadvantage of cross-sectional studies is the inability to separate cause and effect. If measurement of exposure and disease are made at one point in time it will be difficult to assess which came first. The second disadvantage is that the prevalent cases will have a higher proportion of cases with disease of long duration compared to the incident cases. Because of this, those who have died of the disease quickly or those who have recovered have less chance of being included in this disease group. This could lead to misinterpretations. An added problem is for diseases with periods of exacerbation and remission. A person in the remission stage may be falsely classified as not having the disease. The treated cases or those still being treated should be properly addressed so that they can be classified correctly in the final analysis.

Cross-sectional studies do not involve a follow-up period, hence it is usually used to generate new hypotheses regarding study factors and/or disease. It is not useful for studying rare diseases or diseases with short duration. Sampling of the dynamic population is done by random methods. Cross-sectional studies sometimes do not use random sampling. But without random sampling, a cross-sectional study has limited value for describing the frequency of a disease or other characteristics in a large population.

The diagnostic criteria for a particular disease must be clearly established. Confounding variables is another source of potential bias that can influence a cross-sectional study. This can be controlled by matching in the study design or adjustments or controlling in the data analysis.

Another cross-sectional study utilized by research workers is the use of serological techniques in the screening of certain populations to determine the presence or absence of antibodies, immune complexes or antigens (immuno-epidemiology). These studies have been employed extensively in the study of schistosomiasis, toxoplasmosis, invasive amoebiasis, malaria, leishmaniasis and Chagas' disease, to name but a few. Sero-epidemiology is carried out to determine past or current patterns of infection and disease. Most often the serological diagnostic results are difficult to interpret because the tests are often indirect and because cross-reactivity occurs. Furthermore, antibodies against an infection maybe found in the serum long after the parasite has been eliminated. Antigen detection then becomes more reliable as the living parasites and specific stages can be detected.

Most of the epidemiological studies in parasitic diseases has been by the use of cross-sectional prevalence studies. Horizontal and cross-sectional surveys have been applied for the study of epidemiology of many parasitic diseases, i.e. transmission dynamics, prevalence, morbidity and intensity, predisposing factors, seasonality and drug efficacy.

Observational cohort studies (prospective and retrospective)

In a prospective cohort study, individuals who are initially free of disease(s) under study are classified according to whether they are exposed or not exposed to the factors of interest. The cohort is then followed over a period of time and the incidence rates (number of new cases of diseases per population at risk per unit time) or mortality rates (number of deaths per population at risk per unit time) in the exposed and unexposed to possible risk factor(s) are compared. The relationship of the factors of interest to several different diseases can also be studied at the same time. The prospective study is generally preferred for making causal inferences.

The cohort study should initially include only persons free of the disease(s) of interest, but the disease may have begun in some cohort members, even though it cannot be detected by available methods. The set back of cohort study is the loss of subjects by migration, participation or death. Attrition of the study population during the follow-up period could lead to distortion of results which is difficult to rectify in the final analysis. The advantage of cohort studies is that it is relatively free of selection bias. The cohort can be either a fixed or dynamic cohort.

The outcome of the prospective cohort study will enumerate the incidence rates of the disease in those exposed and unexposed to a possible risk factor. Confounding variables should be taken into account when making estimates of the incidence or mortality rates associated with a given factor. Failure to consider confounding variables can cause the investigator to make wrong conclusions such as that the factor increases the risk for the disease when in fact it does not. The investigator may not be able to detect a causal association when in fact it exists.

In retrospective cohort studies (using secondary data) persons with exposures and appropriate comparison groups are identified retrospectively and followed from the time of exposure up to the present to determine disease incidence or mortality rates in the presence or absence of the suspected risk factor. The retrospective cohort study is more feasible for studying a rare disease or a disease with a long latent period.

The advantages of a retrospective cohort study is that the study can be completed within a specific period of time and is less expensive. In this case, the investigator selects the subjects on the basis of exposure status. The exposed cohort must be selected in such a manner that a comparable comparison group can be selected.

Case control studies

In this study, persons with a given disease (cases) and persons without disease (controls) are selected; the proportion of cases and control who have certain background characteristics or who have been exposed to possible risk factors are then determined and compared. Cases enter the study as they are diagnosed over time and controls enter as they are identified. Case control studies is the most frequently undertaken type of analytical epidemiological study. They require a short period of time, small sample size and are less expensive. Case control studies are good for the study of diseases with relatively rapid onset. The frequently used measure of association is the odds ratio. Sufficient detailed information on exposure, confounding variables and potential effect modifications should be collected to permit analysis.

Cases are selected from among persons seeking medical care for the disease(s) under study and usually newly diagnosed cases are preferable. The investigator selects these groups from separate populations of available cases and controls: therefore the study differs from cross-sectional design which involves selection from a single target population. Matching is often used in case control studies. Another difference is that case control studies, unlike cross-sectional studies may involve incident cases of disease. The occurrence of disease in a case control study are not usually limited to a single, well-defined population.

The case control design has been used by biomedical researchers. In recent years, the number of published case control studies in parasitic diseases has increased dramatically. For example, studies have been conducted to determine whether the presence of ascariasis contributes to poor growth of children under conditions where the infection is highly prevalent and risk factors of *Ascaris*-induced intestinal obstruction are high.

Case control studies are less time consuming and less expensive. The disadvantage of case control studies are that (i) the study information is obtained after the occurrence of the disease, and (ii) the compared groups of cases and controls are selected from two separate populations and so it is difficult to ensure cases and controls are comparable with respect to the risk factor and other sources of distortion. The introduction of recall bias may also distort results, making the observed association between study factor and disease stages greater than the actual association. Several studies using the case control method have been utilized in the investigation of onchocerciasis.

Experimental designs

Experimental designs can be either clinical trials (with patients as subjects), field studies (with subjects who are healthy) and interventional trials (with the intervention assigned to groups of healthy subjects).

Clinical trials

A clinical trial is an experiment with patients as subjects. The main aim is to evaluate one or more new treatments for a disease or condition. The subjects in a clinical trial must be diagnosed as having the disease in question and must be recruited to the study soon enough following the diagnosis to permit treatment assignment. Subjects who are very ill or who have a mild illness should be excluded from the study. Treatment assignment should be designed in such a way as to minimize variation of extraneous factors that might influence the comparison. This can be overcome by the use of random assignments. For example, the clinical trial 'The randomized clinical trial to compare the efficacy and safety of albendazole and mebendazole in the treatment of *Giardia lamblia* in children' is being carried in 13 countries worldwide.

Any type of study undertaken must be carefully designed to accommodate the objectives of the investigation, the nature of the disease being studied, the availability of human and financial resources and sociopolitical constraints. Failure to consider these factors in the early stages of design may result in problems that cannot be rectified.

Need for epidemiological investigation

The purpose of epidemiological investigation are thus for:

1. Planning purposes: planners need to know the frequency of disease, its severity, chronicity and socioeconomic effects such as manpower loss, crippling effects and natural history of the disease.
2. Surveillance: to keep track of disease occurrence with special attention to distribution in time and place. This will depend upon prevalence

surveys which will be based on screening techniques. These techniques have to be accurate, specific, sensitive and reproducible.

3. To establish the prevalence of incidence of a condition so as to determine the effects of a disease prevention or eradication programme.

Assessment methods

The harmful effects of parasitic diseases or infections to the health and welfare of individuals or communities at large depends much upon (i) the species or strains of parasite, (ii) the intensity and course of infection, (iii) the interaction between parasite species and other concurrent infections, (iv) nutritional and immunological status of the host, (v) socioeconomic factors, (vi) seasonal influences, (vii) diagnostic criteria variations, and (viii) diagnostic techniques utilized.

Parasitic diseases, on the whole, usually are asymptomatic, for example sarcocystosis or toxoplasmosis which sometimes go misdiagnosed or undiagnosed. In such circumstances, better quantitative techniques need to be introduced to evaluate the morbidity of the infections. Certain qualitative methods need to be sought to assess potential life lost, number of working days lost, incidence rates of diseases, prevalence and case fatality rates. The selection of potential indicators is of prime importance in determining the significance of these diseases on man. At the same time, one has to bear in mind the existence of interspecific and intraspecific variation in parasites that exist interregionally. In these cases, the consequences cannot be determined because of the low mortality. For other specific parasitic diseases, epidemiological assessment can be made of, for example, protein energy malnutrition, blindness, reduced productivity, reduced cognitive performance and loss of important nutrients.

In those cases where assessments cannot be made, accurate epidemiological, pathological and parasitological data need to be collected. These datas are then incorporated into the surveillance mechanism of a control programme. Morbidity studies that are carried out in different parts of the world have to be uniform, comparable and consistent. Case definitions also have to be comparable.

Recent developments in epidemiological research, particularly of intestinal parasitic infections, have led to improved understanding of the dynamics of infections and how the intensity of infections is regulated within hosts. Mathematical models for several parasitic diseases have been developed, e.g. ascariasis, malaria, schistosomiasis, filariasis, etc. These models can only be used as a control programme if accurate data such as age-related prevalence and incidence, and intensity of infection can be obtained. Whatever study design one intends to implement, the sociobehavioral and socioeconomic aspects are important factors to be considered. In order to study parasite epidemiology, the stratified epidemiological variables that need to be considered are the environment, man, vector and parasite. Once the epidemiological factors have been elucidated than the operational and socioeconomic factors need to be considered.

Research Needs

Some of the flaws in epidemiological research that need to be highlighted are:

1. An inability to estimate the number of cases attributed to parasitic diseases in each particular region of the world.
2. In many countries the total figure for the prevalence rate has been estimated by extrapolation from relatively small sample surveys and without accurate knowledge of the local distribution of endemicity. The figures derived this way will often underestimate the true level of prevalence.
3. The prevalence figures from different regions of the world are not standardized according to age, sex and so on. These are thus meaningless.
4. While specific prevalence rates are useful and may reflect important patterns and trends in a population, they must always be interpreted with caution. They may be influenced by differences in incidence, migration, case detection or survival.
5. The intensity and prevalence of an infection in a population should be relatively high if a meaningful relationship is to be derived or established. Some studies do not include this.
6. The number of newly detected or registered cases is frequently used as an estimate of incidence. It should be remembered that these figures may seriously misrepresent the actual incidence of the disease because many cases may not be recognized for some years after clinical onset.
7. Case definitions are not always clear or consistent and the enumeration of cases in many regions of the world is incomplete or irregular. Many cases are not recognized and some registered cases are in fact misdiagnosed. These may not conform to internationally accepted definitions.
8. The criteria for diagnosis, the population and the time may not be spelled out. If ascertainment of disease is not thorough then the number of cases recorded may be an underestimate of the true prevalence. If there is overdiagnosis in some populations the reported prevalence in fact will exceed the actual number of cases.

9. Because of lack of facilities in diagnosis and reporting systems baseline data on prevalence and incidence of many important diseases are lacking.

As such, research workers in endemic countries for parasitic diseases need to carry out proper studies using the designs discussed above. Application of new techniques (or molecular epidemiology), for example molecular biological and biochemical techniques, need to be expanded to gain precise information on infection rates in man and animal reservoirs. Any changes in prevalence and intensity of infection should be clearly reported in publications. Hospital data (which are a potential source of morbidity data though may not reflect morbidity in the population) should be carefully analysed. The frequency of hospitalization for any specific parasitic disease should be studied. Seroepidemiological studies for prevalence of important parasitic diseases should be carried out in different parts of the world. Thus epidemiological research has a wide range applications, including the study of routes of transmission, frequency of epidemics, significance of infections, life-threatening complications and, most important, the risk factors so that these can be incorporated into the surveillance mechanisms of a control programme.

References

Jamison, J.R. (1980) The use of inferential statistics in health and disease: a warning. *South African Medical Journal* 57, 783–785.

Kelsey, J.L., Thompson, W.D. and Evans, A.S. (1986) Epidemic investigation. In: *Methods in Observational Epidemiology*. Oxford University Press, London, pp. 212–253.

Kleinbaum, D.G., Kupper, L.L. and Morgenstern, H. (1982) *Epidemiologic Research. Principles and Quantitative Methods*. Van Nostrand Reinhold Co., New York.

Stephenson, L.S. (1987) The design of nutrition-parasite studies. In: Stephenson, L.S. (ed.) *The Impact of Helminthic Infections on Human Nutrition*. Taylor & Francis, London, pp. 21–46.

Wright, R.A., Spencer, H.C., Brodsky, R.E., and Vernum, T.M. (1977) Giardiasis in Colorado: an epidemiological study. *American Journal of Epidemiology* 105, 330–336.

24

Importance of Abdominal Angiostrongylosis in the Americas

P. MORERA

School of Medicine and Institute for Health Research, University of Costa Rica, Pathology Department, Hospital San Juan de Dios, P.O. Box 2177–1000, San José, Costa Rica

Introduction

Abdominal angiostrongylosis, caused by *Angiostrongylus costaricensis* Morera & Céspedes, 1971 (= *Morerastrongylus costaricensis* (Chabaud, 1972) is a parasitic disease characterized by a granulomatous inflammatory reaction with heavy eosinophilic infiltration of the intestinal wall, especially in the ileocaecal region. The inflammatory reaction, as well as vascular lesions, can cause sub-occlusion, occlusion and/or perforation of the intestine. Ectopic localizations of the nematode cause liver and testicle lesions.

Although the parasite was described in 1971 (Morera and Céspedes, 1971), the disease has been observed in Costa Rican children since 1952 (Cespedes *et al.*, 1967; Morera, 1967). Subsequently, the definitive (Morera, 1970) and intermediate (Morera and Ash, 1971) hosts were identified and the life cycle was elucidated (Morera, 1973). Since then, human cases of the disease have been reported from Mexico to Argentina, including some Caribbean Islands. More recently an autochthonous case of the disease has been reported from Africa (Zaire) (Baird *et al.*, 1987). In addition, naturally infected cotton rats (*Sigmodon hispidus*) have been found in the United States (Uberlaker and Hall, 1979).

Synonyms

Morerastrongylus costaricensis (Chabaud, 1972): in some Latin American countries this parasitosis is named Morera's disease.

Historical Background

Lesions caused by *A. costaricensis* were first observed in Costa Rica in the early 1950s, in surgical specimens excised with a clinical diagnosis of 'malignancy'. However, the histopathological study demonstrated an inflammatory reaction related to scattered structures similar to nematode eggs. In the early 1960s, in addition to the eggs, cross-sections of a nematode were observed within the vessels of the intestinal wall; however, since no similar parasite had been previously described, its identification proved to be impossible. In 1967, about 60 cm of a seriously damaged large intestine was excised from an 8-year-old girl (Fabiola). Many eggs, embryos and adult parasites were observed within the tissues. Thus, after washing and dissecting the surgical specimen, three complete worms, two males and one female, were obtained, and the identification of the worm was undertaken.

In 1967 two papers appeared simultaneously, one describing the lesions (Céspedes *et al.*, 1967), and one with a description of the nematode (Morera, 1967); however, this description did not lead to the identification of the new species.

The first description of the clinical symptoms appeared in 1968 (Robles *et al.*, 1968).

Meanwhile, field studies were carried out in order to identify the definitive and intermediate hosts. In 1970 the identification of the parasite was completed and the description of the new species, *Angiostrongylus costaricensis*, appeared in 1971. While this paper was in press, the definitive and intermediate hosts were found, and reported as 'research notes' that appeared in 1970. Later on, the life cycle of the parasite was elucidated, and since then, several studies on the epidemiology, experimental pathology, diagnosis and treatment have been carried out.

Aetiology and Life Cycle

Angiostrongylus costaricensis (Morera and Céspedes, 1971) is a Metastrongylid (super-family Metastrongyloidea, family Angiostrongylidae). It is a filiform nematode that normally lives within the mesenteric arteries of the definitive host. In both sexes the oral opening is surrounded by three small lips. The male is 20 mm long with a copulatory bursa and two spicules approximately 0.3 mm in length. The female is 33 mm long, and the anus and the vulva are located near the caudal end.

In the natural definitive host (often rodents) the adult worms live within the ileocaecal branches of the anterior mesenteric artery (Sierra and Morera, 1972). Here, the parasites mate, the female oviposits and the eggs are carried by the bloodstream into the intestinal wall, where they embryonate. The first stage larvae hatch in 4 days, migrate to the lumen through the intestinal wall, and reach the soil within the rat faeces.

The molluscan intermediate host, usually veronicellid slugs, become infected by eating the rat faeces (Sly *et al.* 1982). Two moults take place in the molluscan fibromuscular tissues, and after 18 days the infective third stage larva matures; these larvae may remain within the mollusc for several months or may be shed with its mucous secretion. The definitive mammal host becomes infected by eating the infected mollusc. The prepatent period is 24 days.

Although several molluscs are naturally infected, veronicellid slugs are considered the main intermediate hosts. There is no evidence that people intentionally eat slugs; however, small ones hidden in salad leaves could be finely chopped and inadvertently eaten raw. In addition, several cases are known of ingestion of these molluscs by infants. Nevertheless, most human infections are probably caused by ingestion of the infective larvae shed in the secretion of the molluscs. Slugs have been found in ripe fruits that have fallen to the ground. The characteristic mucous trails left by the molluscs can be observed throughout the endemic areas. The propensity for small children to put things in their mouths could explain why they show the highest infection rates.

Epidemiology

The first non-Costa Rican case of abdominal angiostrongylosis was found in Honduras (Sierra and Morera, 1972). During the 1970s, several cases were reported from Venezuela, Mexico and Brazil (Zambrano, 1973; Zavala *et al.*, 1974; Ziliotto *et al.*, 1975; Uberlaker and Hall, 1979).

Today, the disease is being reported from the USA to Argentina, including some Caribbean Islands. However, it has often been found as the result of the routine examination of surgical specimens in pathology departments. In addition, naturally infected cotton rats (*Sigmodon hispidus*) have been found in the United States.

In areas where clinical knowledge and the methodology for laboratory diagnosis have been improved, it is becoming clear that this is quite a prevalent parasitic disease. In 1993, more than 650 cases were diagnosed in Costa Rica (population 3 million), to give a probably underestimated rate of 21.6 cases per 100,000 persons per year. In that country, its distribution is universal, from sea level to an altitude of 2000 m.

We think that in areas where physicians are unaware of the problem, most clinical and subclinical cases probably go unrecognized or misdiagnosed. Although the cotton rat *S. hispidus* is considered to be the most important definitive host, 11 additional rodent species, one case of coati (*Nasua narica*) (Monge *et al.*, 1978) and dogs in Costa Rica (Arroyo *et al.*, 1988) and marmosets (*Saguinus mystax*) from Iquitos, Peru, were also found to be naturally infected (Sly *et al.*, 1982). Studies carried out in Costa Rica showed that the highest infection rate found for *S. hispidus* was 43.2%, similar to that found in Panama (35.0%) for the same species (Tesh *et al.*, 1973).

At least two species of veronicellid slugs in Costa Rica, one in Ecuador and one in Brazil have been found to be naturally infected with *A. costaricensis*; 50% of 6025 slugs from 20 Costa Rican localities, from sea level to an altitude of 2000 m, were found to be infected (Morera, 1985); more than 16,000 infective larvae were counted in a single specimen. In addition, some aquatic and terrestrial snails have also been found to be naturally infected.

Thus, the opportunities for humans, especially children, to become infected are high.

Pathogenesis

In most cases, the lesions are located in the ileocaecal region. They have also been observed in the hepatic flexure, descending colon, regional lymph nodes, liver and testicles.

Two major pathogenetic mechanisms are clearly distinguishable in the infections caused by *A. costaricensis*: first, the adult worms living within the mesenteric arteries damage the endothelium inducing thrombosis and, consequently, necrosis of the tissues formerly irrigated by the vessel; and second, eggs, embryos and larvae, as well as excretion/secretion products, cause inflammatory reaction. Combinations of these phenomena, as well as the patient's susceptibility, and the number and localization of parasites, determine the clinicopathological differences, ranging from cases in which only the appendix is damaged to those in which major surgery is required.

Sero-epidemiological studies are being done demonstrating that in endemic areas, many subclinical cases are present.

Clinical Features

Abdominal angiostrongylosis predominantly affects children. From 116 patients studied in a paediatric hospital in Costa Rica, 53% were of school age; 37% were of preschool age and 10% were infants. Of these patients, 64% were male and 34% female (Loría-Costes and Lobo-Sanahuja, 1980).

When the worms are located in the ileocaecal region, most patients complain of pain in the right iliac fossa and the right flank. Palpation in this area often causes pain. Rectal examination is also painful in about half of the cases, and most patients present with fever ranging from 38°C to 38.5°C (100.4F to 101.3F), but rarely accompanied by chills. In some children, intestinal bleeding can be confused with a Meckel diverticulum.

In chronic cases, a mild fever may persist for several weeks. Anorexia, vomiting and constipation are also present in about half of the patients. A very important finding is a tumour-like mass that, if present, can be palpated in the lower right quadrant and may be confused with a malignancy.

Although a few patients have no haematological abnormalities, leucocytosis and eosinophilia are usually present. White blood cell counts usually range between 15,000 and 50,000 mm^{-3} an eosinophilia from 20% to 50%. The leucocytosis has been as high as 169,000 mm^{-3}, with a 91% eosinophilia.

An X-ray examination is often helpful in making the clinical diagnosis. Radiologic changes are localized in the terminal ileum, caecum, appendix and ascending colon. The contrast medium shows incomplete filling and irritability of the involved areas. The lumen is reduced by the thickening of the intestinal wall.

Sometimes the patient complains of pain in the upper right quadrant; in these cases, the liver is almost always enlarged and tender to palpation. At laparoscopy, small yellowish spots are seen on the surface of the liver. Most patients have hepatic involvement along with intestinal angiostrongylosis (Morera *et al.*, 1982). When the testicle is involved, the most remarkable findings are acute pain, accompanied by redness that later changes to purple (Ruiz and Morera, 1983). Eosinophilia and leucocytosis are also conspicuous. All patients with testicular necrosis have been misdiagnosed as having testicular torsion, and the correct diagnosis was only made following surgery.

In non-surgical intestinal infections, the symptomatology could last from a few weeks to a few months.

Pathological Features

In humans, as well as in the rodent natural host, the adult parasites are located within the mesenteric arteries. In most cases, the lesions are located in the ileocaecal region. They have also been observed in the hepatic flexure, descending colon, regional lymph nodes, liver and testicles.

The gross examination of surgical specimens shows a hardened and thickened intestinal wall with yellowish foci on the serosal surface and in the mesentery.

The intestinal lumen is reduced, sometimes causing partial or complete obstruction. When necrotic areas are present, they could lead to perforation and peritonitis. In many cases, even though only an appendectomy is performed, the surgeon observes lesions on the caecum.

Histopathology demonstrates granulomatous inflammatory reaction with heavy eosinophilic infiltration, especially in the mucosa and submucosa, the serosa and muscular layers are often involved to a lesser degree. Eggs, embryos and larvae appear in small cavities lined by endothelium. Unfertilized eggs usually degenerate and are difficult to recognize. These structures as well as excretion/secretion antigens are easily identified by immunochemical techniques within microscopic areas of necrosis. Large necrotic areas are caused by arterial thrombosis. Eggs and embryos are also present in the mesenteric lymph nodes, which show reticuloendothelial hyperplasia and eosinophilic infiltration.

Hepatic lesions caused by *A. costaricensis* are similar to those caused by *Toxocara canis* (Morera *et al.*, 1982). However, the finding of eggs, embryos and even adult worms in the hepatic parenchyma establishes the diagnosis.

In excised necrotic testicles histologically showing extensive parenchymal haemorrhagic necrosis, worms could be observed obstructing the arteries of the spermatic cord (Ruiz and Morera, 1983).

Diagnosis

The clinical diagnosis is based on the above-mentioned features. The symptoms and other clinical findings can be confused with appendicitis. Sometimes the diagnosis is made after surgery. Barium enema examination can show filling defects of the colon that often resembles a malignancy.

In infected rats, the first-stage larvae can be easily identified in the stools; however, this is not true in humans. In man, an immunological test should therefore be used to confirm suspected cases (Sauerbrey, 1977). Immunoelectrophoresis and Ouchterlony immunodiffusion tests give good results. However, an inexpensive and rapid latex-beads agglutination test also gives high sensitivity and specificity rates.

During the last 17 years, more than 16,000 Costa Rican potential patients have been examined with this test. In more than 97% of the cases, no cross-reactions, nor false positive or negative results have been observed.

Treatment

Surgery, when necessary, is the treatment of choice for abdominal angiostrongylosis. However, as knowledge of this often self-limited disease increases, more non-surgical cases are being followed. Three drugs, diethylcarbamazine, thiabendazole and albendazole have been used with 'remission' of symptoms reported. However, there was no objective evidence to prove that cure was attributable to the drugs. In fact, *in vitro*, and *in vivo* trials in experimentally infected rats, demonstrate that the parasites are excited by the drugs instead of being killed, causing erratic migrations and worsening of the lesions (Morera and Bontempo, 1985). Thus, chemotherapy is not recommended until experimental studies demonstrate a more efficacious drug. Non-surgical patients should therefore be observed carefully and well-controlled palliative treatment could be used. Some patients may present with remission of the symptoms; others could eventually develop an acute syndrome that requires surgery.

References

Arroyo, R., Rodríguez, F. and Berrocal, A. (1988) Angiostrongilosis abdominal en *Canis familiaris*. *Parasitología al Día* 12, 181–185.

Baird, J.K., Neafie, R.C., Lanoie, L. and Connor, D.H. (1987) Abdominal angiostrongylosis in an African man: case study. *American Journal of Tropical Medicine and Hygiene* 37, 353–356.

Céspedes, R., Salas, J., Mekbel, S., Troper, L., Müllner, F. and Morera, P. (1967) Granulomas entéricos y linfáticos con intensa eosinofilia tisular producida por un estrongilideo (Strongylata). *Acta Medica Costarricense* 10, 235–255.

Chabaud, A. (1972) *Stefankostrongylus dubosti* n. sp. parasite du potamogales et essai de classification des Nématodes Angiostrongylinae. *Annales de Parasitologie Humaine et Compare* 47, 735–744.

Loría-Cortes, R. and Lobo-Sanahuja, J.F. (1980) Clinical abdominal angiostrongylosis. A study of 116 children with intestinal eosinophilic granuloma caused by *Angiostrongylus costaricensis*. *American Journal of Tropical Medicine and Hygiene* 29, 538–544.

Monge, E., Arroyo, R. and Solano, E.A. (1978) New definitive host of *Angiostrongylus costaricensis* Morera and Céspedes, 1971. *American Journal of Parasitology* 64, 34.

Morera, P. (1967) Granulomas entéricos y linfáticos con intensa eosinofilia tisular producidos por un Strongilideo (Strongylata Railliet y Henry, 1903). II Aspecto parasitológico. (Nota Previa). *Acta Médica Costarricense* 10, 257–263.

Morera, P. (1970) Investigación del huésped definitivo de *Angiostrongylus costaricensis* Morera y Céspedes, 1971. *Boletín Chileno de Parasitogía* 25, 133–134.

Morera, P. (1973) Life history and redescription of *Angiostrongylus costaricensis*

Morera and Céspedes, 1971. *American Journal of Tropical Medicine and Hygiene* 22, 613–621.

Morera, P. (1985) Angiostrongiliasis abdominal: transmisión y observaciones sobre su posible control. *OPS/OMS Serie de Co-publicaciones* 1. 230–235.

Morera, P. and Ash, L.R. (1970) Investigación del huésped intermediario de *Angiostrongylus costaricensis* Morera y Céspedes, 1971. *Boletín Chileno de Parasitología*, 25, 135.

Morera, P. and Bontempo, I. (1985) Acción de algunos antihelmínticos sobre *Angiostrongylus costaricensis*. *Revista Medica del Hopital Nacional de Niños* 20, 165–174.

Morera, P. and Céspedes, R. (1971) *Angiostrongylus costaricensis* n. sp. (Nematoda: Metastrongylidae) a new lungworm occurring in man in Costa Rica. *Revista de Biología Tropical* 18, 173–185.

Morera, P., Pérez, F., Mora, F. and Castro, L. (1982) Visceral larva migrans-like syndrome caused by *Angiostrongylus costaricensis*. *American Journal of Tropical Medicine and Hygiene* 31, 67–70.

Robles, G., Loría, R., Robles, A., Valle, S. and Cordero, C. (1968) Granuloma eosinofílico parasitario intestinal. *Revista Medica del Hopital Nacional de Niños*, 3, 67–80.

Ruiz, P.J. and Morera, P. (1983) Spermatic artery obstruction caused by *Angiostrongylus costaricensis* Morera and Céspedes, 1971. *American Journal of Tropical Medicine and Hygiene* 32, 1458–1459.

Sauerbrey, M. (1977) A precipitin test for the diagnosis of human abdominal angiostrongyliasis. *American Journal of Tropical Medicine and Hygiene* 26, 1156–1158.

Sierra, E. and Morera, P. (1972) Angiostrongilosis abdominal. Primer caso humano encontrado en Honduras (Hospital Evangélico de Siguatepeque) *Acta Medica Costarricense* 15, 95–99.

Sly, D.L., Toft, J.D., Gardiner, G.H. and London, W.T. (1982) Spontaneous occurrence of *Angiostrongylus costaricensis* in marmosets (*Saguinus mystax*). *Laboratory Animal Science* 32, 286–288.

Tesh, R., Ackerman, L., Dietz, W. and Williams, J. (1973) *Angiostrongylus costaricensis* in Panama. Prevalence and pathological findings in wild rodents infected with the parasite. *American Journal of Tropical Medicine and Hygiene* 22, 348–356.

Uberlaker, J.E. and Hall, N.M. (1979) First report of *Angiostrongylus costaricensis* Morera and Céspedes 1971 in the United States. *American Journal of Parasitology* 65, 307.

Zambrano, Z. (1973) Ileocolitis seudotumoral eosinofilica de origen parasitario. *Revista Latinoamericana de Patología* 12, 43–50.

Zavala, J., Ramírez, W., Reyes, A. and Bates, M. (1974) Angiostrongilosis costarricensis. Primeros casos Mexicanos. *Revista de Investigaciones Clinicas* 26, 389–394.

Ziliotto, A., Künzle, J.E., Ruas-Fernández, J.E., Prates-Campos, J.C. and Costa-Britto, R. (1975) Angiostrongiliase: a presentacão de um provavel caso. *Revista do Instituto de Medicina Tropical do São Paulo* 17, 312–318.

25

PCR-based Detection and Typing of Parasites

G.L. McLaughlin,[1] S.S. Ssenyonga,[2] E. Nanteza,[2] Rubaire-Akiiki,[2] O. Wafula,[2] R.D. Hansen,[3] M.H. Vodkin,[3] R.J. Novak,[3] V.R. Gordon,[3] S. Montenegro-James,[4] M. James,[4] H. Aviles,[5] R. Armijos,[5] C. Santrich,[6] K. Weigle,[7] N. Saravia,[7] E. Wozniak,[8] O. Gaye,[9] R. Mdachi,[10] S.Z. Shapiro,[11] K.-P. Chang[12] and I. Kakoma[3]

[1] Department of Pathology and Laboratory Medicine, Indiana University Medical School, Indianapolis, IN 46202, USA; [2] Department of Microbiology and Parasitology, Makerere University, Kampala, Uganda; [3] Department of Veterinary Pathobiology and Natural History Survey, University of Illinois, Urbana, IL 61801, USA; [4] Ochsner Medical Foundation and Department of Tropical Medicine, Tulane University, New Orleans, LA 70112, USA; [5] Laboratory of Immunology, Central University Medical School, Quito, Ecuador; [6] Biochemistry Department, Purdue, West Lafayette, IN 47906, USA; [7] CIDEIM, Cali, Colombia and University of North Carolina, Chapel Hill, NC, USA; [8] School of Veterinary Medicine, University of California at Davis, Davis, CA, USA; [9] Faculty of Medicine, Dakar-FANN, Dakar, Senegal; [10] Kenya Trypanosomiasis Research Institute, Muguga, Kenya; [11] Retrovirus Research Laboratory, CBER/FDA, Bethesda, MD 20892, USA; [12] Chicago Medical School, North Chicago, IL 60064, USA

Introduction

Broad-spectrum strategy and rationale

A limited number of taxonomic groups of pathogens cause important diseases among humans and animals. Although it is sometimes preferable to be able to identify pathogens at the species, strain or isolate level, decisions regarding treatment and control measures can often be made after identification at the genus or species level; e.g. similar drug regimens may be effective against an array of related pathogens. Since pathogens tend to occupy privileged sites such as the blood or central nervous system (CNS) in which the only other material is of host origin, detection of DNA of a parasitic family, genus, or species may suffice. Our goal is to define polymerase chain reaction (PCR)-based assays that can be used to achieve sensitive detection and sufficiently broad and narrow levels of specificity to be useful.

PCR provides the opportunity to detect infectious disease agents with high sensitivity and specificity in vertebrate and arthropod hosts (Saiki *et al.*, 1988; Erlich *et al.*, 1991; Howe *et al.*, 1991; McLaughlin *et al.*, 1991). Most of the PCR-based assays thus far evaluated have been designed to detect a single pathogen species, for instance, either *Babesia bigemina*, *Babesia bovis*, or *Theileria parva* (Fahrimal *et al.*, 1992; Figueroa *et al.*, 1992; Bishop *et al.*, 1994). A broad-spectrum approach to the detection of disease agents allows surveys for a variety of pathogens. Using primers with broad-spectrum specificities allows the use of fewer amplification reactions and, if designed properly, a single set of amplification conditions for all infectious disease agents.

Sequences within the ribosomal DNA (rDNA) cistron often are identifiable as 'signature' sequences, i.e. sequences that are characteristic of a particular taxonomic group of organisms (Olsen, 1988). Our goal is to define rDNA targets that allow the detection of most of the important infectious disease agents, using a single set of amplification conditions and a limited number of amplimers. We initially defined this strategy for *Acanthamoeba* amoeba (Vodkin *et al.*, 1992) and for *Neisseria* spp. and other eubacteria (McLaughlin *et al.*, 1993); the suggestion for use of the currently described strategy for additional eukaryotic parasites was discussed in a previous paper (McLaughlin *et al.*, 1993). We have continued to explore this strategy, and in this chapter, broad-spectrum rDNA primer sequences are presented along with preliminary estimates of thresholds of detection and sensitivity and specificity for those primers and groups tested thus far. More emphasis will be placed on apicomplexan and kinetoplastid pathogens, since results are encouraging and previously unpublished, except for a study involving snake haemogregarines (Wozniak *et al.*, 1994).

Apicomplexan species

Parasites of the evolutionarily very successful phylum Apicomplexa cause many chronic and severe diseases among humans and mammals throughout much of the world. Among cattle, *T. parva*, the aetiologic agent of East Coast fever in East and Central Africa, causes enormous losses in productivity. *B. bigemina* and *B. bovis* are also economically important diseases in cattle and *Babesia canis* is an important pathogen of dogs in much of the world. The malaria diseases caused by *Plasmodium* species are estimated to cause 1–2 million human deaths annually and significant morbidity. Considerable effort has been made to design

Table 25.1. Ribosomal DNA primers for Apicomplexa.

Name	Identifier*	Sequence (5′–3′)
A1	18AP 1950.FOR	GGG GAT TGA TTA TTG GAA TT
A2	18AP 1488.REV	CGG AAT TAA CAA GAC AAA TC
A3	5.8AP 27.REV	AAG ACA TCC ATC GTT AAG ACT G
A4	18AP 1059.REV	AAA TCT AAG AAT TTC ACC TCT GAC A
A5	18AP 853.FOR	GAG TAA ATT AGA GTG TTC CAA GCA
A6	18AP 1514.FOR	GAA CGA GAT CTT AAC CTG CTA A

* In the identifier, the prefix and number identify the location of the start of the primer, and the suffixes FOR and REV indicate forward and reverse directions, respectively; forward and reverse primers which lie close to each other are used in combination; an antiparallel primer from the same taxon-specific site would reverse orientation. Thus far, primers A2 and A5 have given the best sensitivity and specificity of *Babesia* and *Plasmodium* detection.

repetitive DNA or rRNA-targeted direct hybridization or amplification-based DNA probes (Jaureguiberry *et al.*, 1990; Kane and Lanar, 1991; Barker *et al.*, 1992; McLaughlin *et al.*, 1993a,b; Snounou *et al.*, 1993). Blood-borne apicomplexan diseases are characterized by anaemia and a number of other non-pathognomic signs. Diagnosis often depends upon the detection of piroplasm-like forms in blood using microscopy. Microscopic analysis is problematic for unskilled microscopists or when good-quality microscopes are not available, and when parasitaemias are low (McLaughlin *et al.*, 1993a). Hybridization assays in an *in situ* format with scoring by microscopy have been described, and are feasible with repetitive DNA or rRNA, but these may offer little advantage over stained blood smears (McLaughlin *et al.*, 1993a). Serology can supplement microscopy, but the distinction between carriers and cured individuals often remains problematic. We previously evaluated PCR-based detection and typing of *Babesia* and *Plasmodium* species using a primer pair designed from an analysis of *Plasmodium* and *Toxoplasma* rDNA; that set of primers did not robustly amplify variable levels of *Babesia* and *Plasmodium* DNA, and they amplified some heterologous bands even from host DNA (McLaughlin *et al.*, 1993a). Therefore, for the current study, alternative primers based on a narrower analysis of *Babesia* and *Plasmodium* sequences were chosen for blood-borne apicomplexa species (Table 25.1).

Kinetoplastid species

The family Trypanosomatidae is a well-defined taxonomic group of flagellates that is easily recognized by the characteristic intramitochondrial kinetoplast. Several *Trypanosoma* and *Leishmania* species, and many identifiable strains, are recognized as important pathogens.

For both genera, there may be swelling and a small lesion or chagoma near the vector site, which usually resolves. The *Trypanosoma brucei* complex contains several species that cause sleeping sickness in humans (*Trypanosoma rhodesiense* and *Trypanosoma gambiense*) and trypanosomiasis in domestic animals that makes some areas of Africa impractical for agriculture. *Trypanosoma cruzi*, which causes Chagas' disease, is a common cause of heart disease in Brazil, and is found from the southern United States to Argentina.

Leishmania species are usually classified by clinical characteristics and geographic origin, although isozyme analysis and other approaches have been described. Most species cause cutaneous diseases ranging from mild dry localized lesions, to wetter lesions, and some strains or species cause the necrotizing and/or metastasizing mucocutaneous leishmaniasis. *Leishmania major* and *Leishmania tropica* cause Old World cutaneous leishmaniasis, and *Leishmania mexicana* and *Leishmania braziliensis* species complexes in the New World cause cutaneous and mucocutaneous disease, respectively. Visceral leishmaniasis may be caused by *Leishmania donovani*, *Leishmania infantum*, or *Leishmania chagasi*. Strains that can cause visceral or mucocutaneous diseases are usually treated, but not milder *Leishmania* pathogens, because of the toxicity of the antimony-based treatment protocols.

Most earlier DNA probe research emphasized kinetoplast DNA detection, which provides abundant but highly polymorphic markers. However, several other groups have independently explored the potential for using the ribosomal RNA cistron to detect and type pathogens, using primer sequences different than those we have chosen (Persing *et al.*, 1992; Reddy and Dame, 1992; Bishop *et al.*, 1993).

Methods and Results

Primer selection

A schematic of the rDNA cistron is shown (Fig. 25.1); indicated are taxon-specific sites within genes for structural RNA to which forward and reverse primers hybridize, and similar sites between the 16/18S rDNA and the 5.8S rDNA sequence; the latter spans the more polymorphic internal transcribed spacer (ITS1) region. For each group of pathogens (Tables 25.1–25.3), taxon-specific primers were chosen with attention to achieving uniform melting temperatures. For each taxonomic group of pathogens, a database was constructed of the small rRNA sequences of important pathogens and a number of outgroups. For instance, for the Apicomplexa, sequences of *Plasmodium falciparum*, *Plasmodium vivax*, *B. bovis*, *Saccharomyces*, *Escherichia*, *Bos*, *Muris* and *Homo*

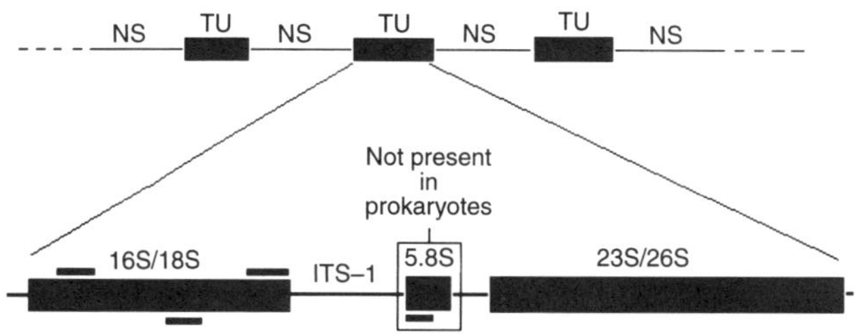

Fig. 25.1. Schematic of rDNA cistron and PCR strategy. Each rDNA transcriptional unit (TU) is usually in multiple copies, often (as shown here) in tandem arrays (top). Within the transcription unit (expanded bottom diagram), the small rDNA subunit gene (about 1500 bp in length) for the 16S (in prokaryotes) or 18S ribosomal subunit is separated from the 5.8/23/26S rDNA by non-structural DNA called the internal transcribed spacer (ITS); the ITS1 sequence is usually 300–800 bp in length. Forward primers homologous to rDNA sequences identified as characteristic of a given taxonomic group of pathogens are indicated on the top of the 16/18S gene, and taxon-specific reverse primers are shown underneath the corresponding genes (in prokaryotes, a separate 5.8S gene is absent and is part of the 5′ end of the 26/28S gene). Predicted sizes of amplified DNA fragments are determined by the distances spanned by the chosen forward and reverse primers. Abbreviations: NS, non-transcribed spacer; ■, primer location.

were assembled. These sequences were aligned and primers were selected from regions that were conserved in *Babesia* and *Plasmodium*, but not the other genera. Primer lengths were adjusted to achieve predicted melting temperatures of 54–58°C for each taxonomic group, and additional potential problems such as end homologies and lack of hairpin formation were also avoided when possible by using primer selection programmes (Devereux *et al.*, 1984); target lengths were chosen as less than 650 base pairs (bp) to improve sensitivity. Candidate primer sequences were then searched with the compiled rDNA database and with the complete database. For all taxonomic group primers, the few fortuitous high-level homologies (defined as over 60% sequence identity) were not from hosts or other species that would be found in the same habitats as the pathogens.

Amplification conditions

After comparative tests of primers using purified parasite DNAs and DNA-spiked samples, a uniform, multistage thermal cycler programme was chosen that progressed from an initially higher hybridization temperature to a lower temperature. We hypothesize that greater stringency of hybridization is necessary at earlier amplification steps, when primers are in excess and the number of amplification targets is low; and a large number of amplification cycles were chosen to achieve high sensitivity for clinical samples. A variety of brands of thermal cyclers, including Perkin-Elmer, MJ Research mini- and full-sized models and Hybaid thermal cyclers, gave similar results. Formulation of the PCR cocktail followed the default conditions included with the Cetus-Perkin Elmer amplification kit 1. After a 2-min denaturation step at 94°C, 15 cycles of amplification were performed with 1 min denaturation at 94°C, 1 min hybridization at 58°C, and 1 min elongation steps at 72°C. An additional 21 thermal cycles were run with hybridization at 55°C, and a 5-min elongation step at 72°C was used at the end of the amplification programme. Sizes of DNA products are estimated by agarose gel electrophoresis. Related species that show similar initial-sized DNA bands, are often distinguished after digestion with restriction endonucleases of four-base recognition. Typically, at least one amplification product is cloned, partially sequenced, and the structural rDNA regions are aligned to confirm the identity of the amplified DNA band; e.g. for the blood apicomplexans, a *B. bovis*-derived product showed the expected sequence.

Apicomplexa specificity test

Processed and amplified samples known to contain *Babesia*, *Theileria* or *Plasmodium* parasites, but not samples from mammalian hosts or other tested microbes, yielded one to several DNA band(s) of 600 to 650 bp in length (Fig. 25.2). Threshold sensitivity was 1–10 organisms. Samples from mammalian hosts and other tested microbes did not show amplified DNA of this size, although occasionally faint bands were seen at various sizes, especially when excess target DNA was tested. Some species evaluated for broad-spectrum reactivity are listed in the legend of Fig. 25.2 and include *P. falciparum*, *P. vivax*, *Plasmodium malarias*, *Plasmodium knowlesi*, *B. bovis*, *B. bigemina*, *B. canis*, *Babesia equi*, *Babesia microtos* and an unidentified *Haemoproteus* species from snakes (Ubzniak *et al.*, 1994). Surveys for non-reactivity included four *Leishmania* species, *Tr. cruzi*, *Tr. brucei*, *Saccharomyces cerevisiae*, *Anaplasma marginale*, *Escherichia coli*, *Homo sapiens*, *Bos taurus*, *Canis* and *Muris* in parallel amplification reactions. For several haemotropic Apicomplexa species (as for some kinetoplastids), a doublet was observed, but usually a single major band was observed, only from infected blood or from blood-fed ticks. Within the range of parasitaemias tested (0.1–40% infected erythrocytes), the degree of amplification was similar among the various species tested.

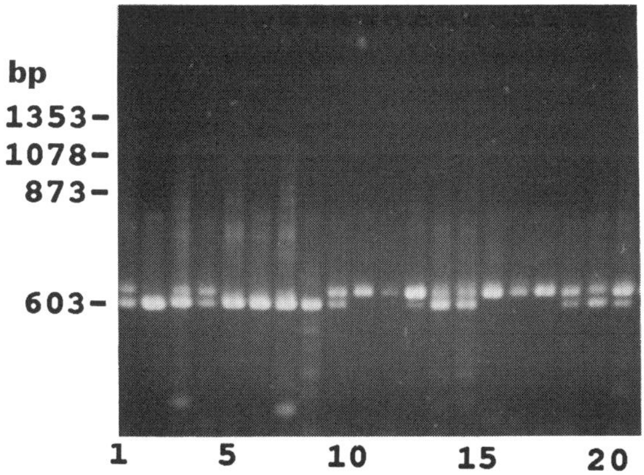

Fig. 25.2. Survey of apicomplexan primer specificity. Aliquots (1 μl) of infected blood from various *Babesia* and *Plasmodium* species were processed with 19 μl of Genereleaser matrix; 1 μl of the supernantant of the processed sample was amplified in 30 μl of reaction mix using primers A2 and A5 (Table 25.1), and 10 μl of the product was resolved on an 0.8% agarose gel and photographed after ethidium bromide staining. Amplifed DNA bands at 600 and/or 630 bp were observed using infected blood. Uninfected blood and blood containing 0.01–5% infected erythrocytes containing (left to right), lanes 1–9, *Babesia* species: 1, *B. bigemina*; 2, *B. bovis*; 3, *B. bigemina*; 4, *B. bigemina*; 5, *B. gibsoni*; 6, *B. canis*; 7, *B. microti*; 8, *B. equi*; 9, *B. microti*; lanes 10–19; 10, *Plasmodium* species; 11, *P. brasilianensis*; 12, *P. cyanomulgi*; 13, *P. falciparum*; 14, *P. malariae*; 15, *P. ovale*; 16, *P. vivax*; 17, *P. inui*; 18, *P. malariae*; 19, *P. vivax*; and lane 20, uninfected dog blood.

Apicomplexa field tests

When possible, these tests of candidate primers were done using blinded protocols. Tissues were successfully processed for PCR using many different protocols, but it was frequently observed that excess sample interfered with amplification; this seemed more of a problem than an inadequate target number. Therefore, even processed samples were typically diluted 10–1000-fold from the corresponding original tissue volume in the final PCR cocktail. The IsoQuick kit (Microprobe, Garden Grove, CA) or Geneclean II (La Jolla, CA) kits were convenient for isolating target DNA and/or to purify PCR products before analysis by restriction endonucleases. The Genereleaser II kit (Bioventures, Murfeesboro, TN) allowed rapid, convenient processing of samples. It uses only 1 μl of sample per reaction tube. In our hands, processed samples remained stable in Genereleaser at −20°C after processing; if samples were depleted or dried, mixing with 20 μl of water and a pulse-spin allowed reuse. Usually, 1 μl of the supernatant of pulse-spin Genereleaser-processed material was used in a 30 μl amplification reaction volume. It was often a problem to validate each candidate amplimer because large, well-characterized batteries of infected and uninfected samples were difficult to arrange or to transport to the United States. Fortunately, the equipment and supplies needed to perform the amplifications fit in several medium-sized suitcases, and on-site tests were feasible and performed. The multiple small batteries from different geographic regions available, gave some indications regarding the robustness of candidate assays.

In a test of apicomplexan primers, blood was collected in citrate-dextrose-glucose from cattle at a slaughterhouse in Uganda. From one animal with generalized lymphadenopathy, a single *Theileria* organism was found in 50 microscopic fields of a thick smear, and blood from four asymptomatic and non-parasitaemic animals was used for negative controls; these were correctly identified, both using the Genereleaser protocol and IsoQuick processing. Additional samples of known positive *Theileria* and negative infections were obtained from duplicate unfixed thick smears, methanol-fixed thin smears, and Giemsa-stained thin and thick smears from a survey conducted 11 months earlier 60 miles northeast of Kampala; these were processed by three repeated pipettings and removals of Genereleaser matrix (2 μl) directly over a *c.* 0.5-cm diameter spot of blood on the slide; this procedure dissolved the erythrocytes and left a clear circle on the glass microscope slide; four positive and two negative samples were also correctly identified after processing and amplification with A2 and A5 primers (Table 25.1). No positive amplification was obtained from even confirmed-positive blood smears that had been previously stained with Giemsa stain, but five positive and two negative samples that had been methanol-fixed were correctly identified, which suggests that retrospective studies of these types of samples may be feasible using PCR. Neither infected cattle nor characterized blood smears were available during this study for *B. bigemina*, and trials elsewhere were inconclusive. However, a dog was presented to the Makerere University Veterinary Clinic and was identified with lymphadenopathy and other clinical signs of babesiasis. By examination of Giemsa-stained blood smears, one *Babesia* organism was identified in 50 thick smear blood fields. Duplicate 1-μl samples of this dog's blood, and blood from several uninfected dogs presenting to the clinic, were processed using the Genereleaser protocol. The infected dog blood, and squashes of five blood-fed ticks were positive by PCR, but not three unfed ticks or uninfected dog blood (data not shown).

In a test of the A2 and A5 primers with a battery of 22 blood samples of *P. falciparum* malaria from Senegal, positive samples as identified by thick smear microscopy, the quantitative buffy coat (QBC) malaria analysis, and the ParaSight F® test (Becton Dickenson, Baltimore, MD), were also detected by PCR. The PCR detected 13 positive patients,

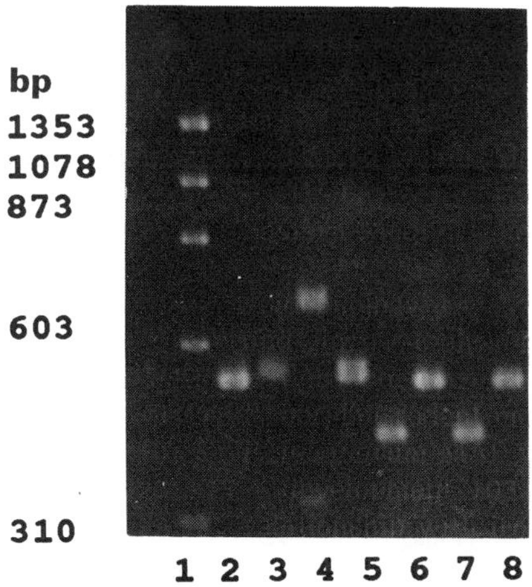

Fig. 25.3. Survey of kinetoplastid primer specificities. Aliquots (1 μl) of infected blood or suspended purified parasites were processed using Genereleaser II, and 1 μl of the supernatant was amplified in 30 μl of PCR cocktail with Kin 1 and Kin 2 primers (Table 25.2). Left to right: 1, molecular mass markers; 2, *Tr. brucei* MITAT 1.4; 3, *Tr. brucei* ILTB 117; 4, *Tr. cruzi* YG; 5, *Leptomonas seymouri*; 6, *L. donovani*; 7, *L. seymouri*; 8, *L. donovani*.

including one low-parasitaemic sample not detected by thick smear or the ParaSight F test which was also detected by the QBC, and two additional samples not detected by the other three methods. This indicates either that the PCR assay was more sensitive than the other assays, or resulted in two false positives.

Kinetoplastid specificity tests

In screens of kinetoplastid species using primers Kin 1 and Kin 2, a threshold detection of 1–10 organisms was obtained. Consistent size polymorphisms were identified among several genera; the largest fragment was from other *Trypanosoma* species; then *Leishmania* species; and *L. donovoni* showed the smallest amplified fragment (Fig. 25.3). The cutaneous *Leishmania* species showed a slightly larger Kin 1–Kin 2 product than did the mucocutaneous species, and the visceral species yielded a significantly smaller product (Fig. 25.3). Additional *Leishmania* species were resolved using a variety of four-base-specific endonucleases; some *Hae*III or *Sau*3A digestions of the ITS1 target are shown (Fig. 25.4). Most recognized species of *Leishmania* have been resolved using this strategy, but *Leishmania guyanensis* and *Leishmania panamensis* have thus far not been distinguished.

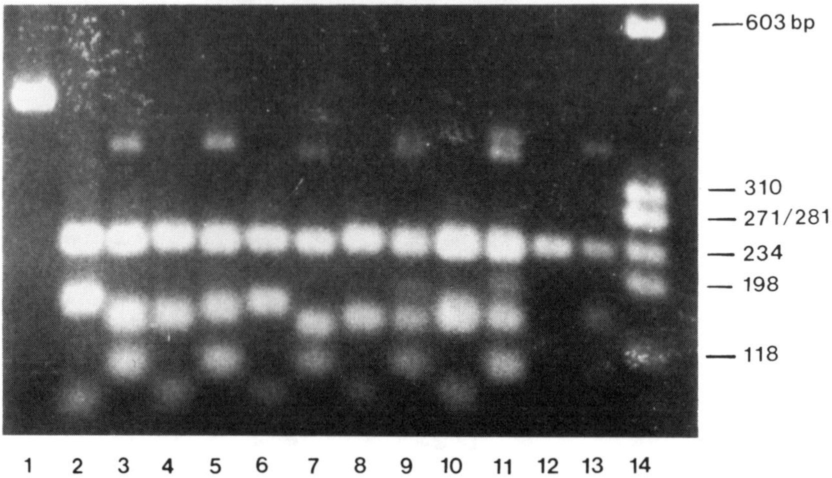

Fig. 25.4. Restriction endonuclease analysis of amplified *Leishmania* ITS with *Hae*III or *Sau*3A using Kin 1 and Kin 2 primer-amplified DNA. 1, Undigested *L. donovani* amplified ITS; 2, digestion of the same DNA with *Hae*III; and 3, with *Sau*3A. Lanes 4 and 5, second isolate *L. donovani*, digested with *Hae*III and *Sau*3A, respectively. Lanes 6 and 7, *L. infantum*. Lanes 8 and 9, *L. tropica*. Lanes 10 and 11, *L. major*. Lanes 12 and 13, *L. mexicana amazonensis*.

Kinetoplastid field tests

A single-blind screen of a characterized battery of 20 lesion aspirates from Colombia with Kin 1 and Kin 2 primers showed only a single apparent false positive (a clinically plausible positive) in a battery of 20 specimens, and the 16 positive *Leishmania* samples were correctly identified (Table 25.2). Cultured *Leishmania* species and rodent-passed *Trypanosoma* species from Ecuador, Colombia and elsewhere were analysed and yielded similar banding patterns for corresponding species; one laboratory contaminant was confirmed using the PCR. From all sources, *Tr. cruzi* strains showed the largest amplified band, but this band was typically weak and variable relative to DNA bands generated from the other *Trypanosoma* or *Leishmania* species. This may reflect the typically lower numbers of *Tr. cruzi* both in cultures and parasitaemias among patients, or there may be other technical or genetic explanations.

No field tests were attempted for *Trypanosoma* species. However, 12 clones and four uncloned isolates from the *Tr. brucei* complex showed uniform-sized amplification products using Kin 1 and Kin 2. In an attempt to distinguish the isolates and species, ten different endonucleases were screened. Only *Tr. gambiense* was resolved from the other isolates, using *Hae*III digestion (data not shown).

Table 25.2. Ribosomal DNA primers for Trypanosomatidae.

Name	Identifier*	Sequence (5′–3′)
Kin 1	ITb58S106r	GCG TTC AAA GAT TGG GCA AT
Kin 2	2Lers2118s	CGC CCG AAA GTT CACC
Kin 3	3Lers1726s	CAA ACT CAT CGG CGG G
Kin 4	4Lers1726s	TCT TGT TCT CAC TGA CAT TGT
Kin 5	5T28s4511s	TAC CCT CAT TTT CAC AAG ATC T
Kin 6	58Tbr4220s	TGT GTA TCT CTA TAT GGC TC
Kin 7	6Ler18s20r	CTA GAC CAA CTA AGA CGG TC

* Nearby 's' and 'r' primers are used in combination for PCR, but not all combinations have been tested. Synthetic DNA primers Kin 1 from the 5.8S gene and Kin 2 near the end of the 18s gene, yield abundant, characteristic DNA products (largely ITS1) from all tested *Leishmania* and *Trypanosoma* species, and product sizes allow identification of some species; *HaeIII* digestion of the amplified products resolves additional *Leishmania* species, and distinguished *Tr. gambiense* from other members of the *Tr. brucei* group.

Table 25.3. Ribosomal DNA primers for bacteria.

Name*	Identifier	Sequence (5′–3′)
B1	NEG 16S 1294.REV	TCG GTT TTG TGA GAT TGG C
B2	EUB 16S 1354.REV	AGC GAT TCC GAC TTC ATG
B3	NEG 16S 1424.SEQ	GGA TAC CAG AAG TAG GGT
B4	EUB 16S 919.SEQ	CCG CAC AAG CGG TGG A
B5	NEG 16S 999.SEQ	GTG CGG AAT CCT CCG G
B6	EUB 5.8.REV	CCA GGG CAT CCA CCG T

* Most combinations have not yet been tested, but the primer combinations: B6 with B4 or B5; B1 and B4; and B3 and B5, were able to distinguish the 14 tested genera of pathogenic eubacteria.

Bacterial typing

As for the other taxonomic groups, most of the bacterial primer combinations of the sequences listed in Table 25.3 have not been tested; but in general, more combinations of primers amplified more polymorphic-sized targets among the bacteria than were observed for eukaryotic pathogens (Fig. 25.5). Those combinations tested thus far have been specific for bacteria, and have amplified DNA bands that vary between bacterial species in length at either structural rDNA or ITS loci. Tentative identification appears to be feasible at the species level even without restriction endonuclease analysis, and strain identification has been demonstrated for one group using RFLP analysis of amplified products (McLaughlin *et al.*, 1993c). Among the genera that gave characteristic patterns using primers B3 and B6, are *Neisseria*, *Pneumococcus*,

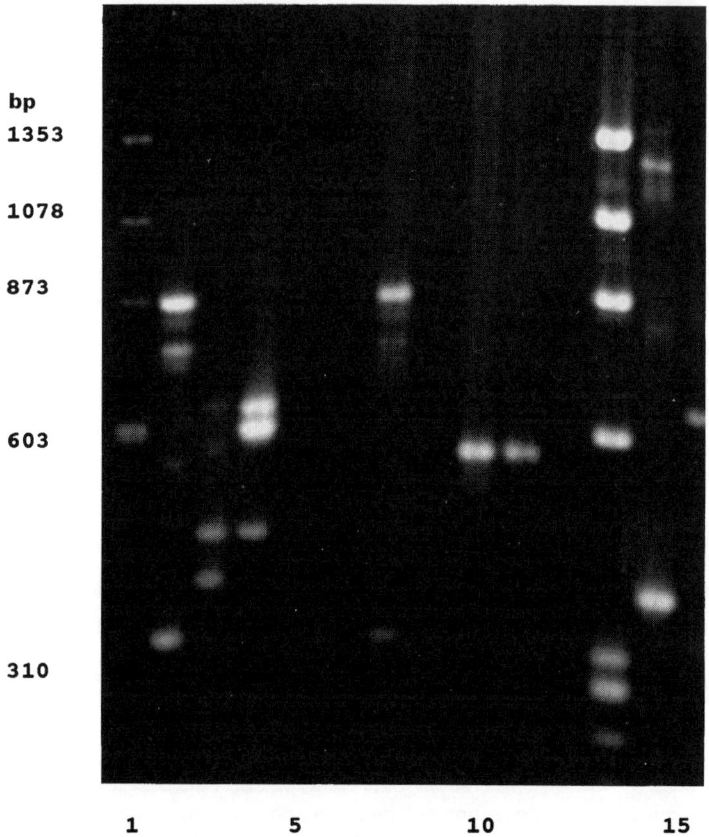

Fig. 25.5. Amplification of selected eubacteria using primers B3 and B6 (Table 25.4). Lanes 1 and 13, molecular mass markers. Lane 2, *Staphylococcus aureus*; lane 3, *Staphylococcus epidermidis*; lane 4, *Staphylococcus helicus*; lanes 5–7, empty lanes; lane 8, *Streptococcus equi*; lane 9, empty lane; lanes 10 and 11, *Corynebacterium renale*; lane 12, empty lane; lane 14, *Actinomyces pyogenes*; lane 15, *Listeria monocytogenes*.

Afipia, *Rochalimaeae*, *Pasterella*, *Anaplasma*, and *Mycobacterium* (Fig. 25.5 and data not shown).

Fungi and nematodes

We have designed similar sets of primers based on fungal-specific sequences, that distinguished *Pneumocystis* and *Histoplasma*, two common fungal pathogens of the lung (data not shown). However, we found that primers from the rDNA cistron of nematodes did not resolve the common nematode parasite species. Somewhat better resolution of nematode species and strains was obtained from primers targeting polymorphic mitochondrial DNA markers.

Discussion and Conclusions

This study indicates that a broad-spectrum PCR approach is feasible for the detection and typing of many members of many groups of infectious disease pathogens. Using selected rDNA amplimers, all of the tested *Babesia*, *Theileria*, and *Plasmodium* parasites were detected; both low- and high-parasitaemic samples were robustly detected; and host DNA was not detected. All of the species of pathogenic kinetoplastids were successfully detected, and most of the broad categories and species of recognized kinetoplastids were successfully resolved. Within the *Leishmania* complex, most common species that cause cutaneous, mucocutaneous, and visceral disease were resolved.

As a relatively slow-changing locus, hybridization targets in the rDNA tend to vary less than would targets such as repetitive DNA, e.g. telomer-linked repetitive DNA or kinetoplast minicircle DNA. Consistent with this expectation, our results indicate that genera, species and isolates from several continents are identifiable using relatively simple protocols, i.e. these markers are stable enough for widespread use. Although earlier research on DNA-based detection of parasites focused on repetitive DNA such as telomere-linked nuclear DNA or kinetoplast DNA to achieve the necessary sensitivity required for direct-hybridization detection, the sensitivity provided by PCR may allow other strategies.

Specific hybridization probes and primers have been designed for the identification of many of the pathogens analysed in the current study. PCR-based detection has been described for *Plasmodium* spp (Jaureguiberry *et al.*, 1990; Kain and Lanar, 1991; Barker *et al.*, 1992; Snounou *et al.*, 1993), *B. bigemina* (Figueroa *et al.*, 1992), and *B. bovis* (Fahrimal *et al.*, 1992), and other blood-borne apicomplexan species (Dalrymple *et al.*, 1992; Ellis *et al.*, 1992; Persing *et al.*, 1992; Reddy and Dame, 1992; Bishop *et al.*, 1993). Research on the kinetoplastids has also been extensive, with (usually species-specific) probes described for *Leishmania*, *Tr. cruzi* and *Tr. brucei* (Dalrymple *et al.*, 1992; de Bruijn and Barker, 1992; Ellis *et al.*, 1992; Reddy and Dame, 1992; Smyth *et al.*, 1992; Bishop *et al.*, 1993; de Bruijn *et al.*, 1993; Gebre-Michael *et al.*, 1993; Hassan *et al.*, 1993; Lopez *et al.*, 1993). The broad-spectrum strategy for apicomplexans and kinetoplastids should allow any of these organisms to be detected in only two reactions. Theoretically, assays that amplify highly repetitive and polymorphic DNA loci will achieve greater sensitivity and specificity for a single species than our suggested strategy which targets the moderately repetitive rDNA. However, relative convenience and cost may be significant considerations for PCR-based detection and typing, and the strategy we have outlined may have advantages in this regard, and in regard to the stability of the genetic markers among geographic isolates. A uniform-target strategy also has the advantage that the copy number of rDNA genes is believed to be similar or identical throughout a taxonomic group; this helps to standardize conditions and may become more significant if quantitative PCR is developed to estimate parasitaemias. However, the major likely advantage of the described strategy is efficiency; one reaction each is necessary to determine the presence of any haemotropic Apicomplexa parasite; another reaction could detect any kinetoplastid; and similar single reactions could detect either fungi or bacteria. This provides an efficient strategy for diagnosis and screening efforts, relative to the dilemma posed by using the many optimal primers specific for all possible pathogen species in a given habitat.

As expected (Olsen, 1988), the degree of polymorphism of rDNA markers was related to the taxonomic 'depth' (time since divergence) of the particular disease agents. Structural rDNA sequences are under constraints to maintain secondary structure and to associate with a variety of proteins and other nucleic acids. Structural rRNA changes are relatively slow and mark major divergence between taxonomic groups; the members of the taxonomic group are therefore all detected, and the size and/or sequence of the ITS1 should allow finer levels of taxonomic resolution. However, few polymorphisms were detected using nematode-specific primers from the rDNA locus; metazoan species have diverged relatively recently and show few polymorphisms. In contrast, most species of bacteria showed marked length and sequence polymorphisms both within the structural rDNA regions and in the ITS. Amplification of the ITS gave potentially useful length polymorphisms for kinetoplastid parasites, but it was necessary to digest the amplified ITS1 DNA with endonucleases to resolve most species of *Leishmania*. Digestion of the amplified ITS with four different endonucleases did not resolve 12 of 13 *Tr. brucei* isolates, which probably diverged recently. Nor was this strategy effective at distinguishing the common nematode parasites. It seems likely that more polymorphic loci will be necessary for the metazoa, and to identify isolates or strains in most groups, although the hemogregarines do show RFLP heterogeneity (Wozniak *et al.*, 1994); *Babesia* and *Theileria* species have not been analysed for RFLP-based polymorphisms.

Although the predominant amplification DNA band(s) were usually detectable for a given species using long-term cultured, long-term stored, and freshly obtained organisms, it should be cautioned that attention to the quality of samples was necessary to achieve reliable amplification and identification. The use of separate sample preparation and amplification/analysis rooms helped to prevent high-copy product contamination. Few of the theoretical primer combinations from Tables 25.1–25.3 (and the corresponding taxon-specific antiparallel primers) have thus far been evaluated, and only a few combinations have been tested either with purified parasites or with clinical batteries. Less amplification product is typically obtained from clinical batteries, and coupling the amplification with a more sensitive and/or convenient detection method may be desirable.

Our goal in these studies has been to identify correlative rather than causative genetic markers for pathogenicity. However, the recent suggestions that some rDNA sequences encode antigens that include virulence factors in organisms as diverse as *Giardia*, *Entamoeba histolytica*, and bacteria (Upcroft *et al.*, 1990; Ralphand McClelland, 1993; Jansson *et al.*, 1994) is intriguing. Haemolysis bioassays similar to those described for *Entamoeba histolytica* indicate that a similar level of consitutive haemolytic activity is present in cloned *Acanthamoeba polyphaga* rDNA (G.L. McLaughlin, unpublished observations). We do not as yet understand the replication or regulation of such putative rDNA-encoded virulence factors, but it would be interesting to survey additonal parasites for such genes.

The correct identification of aetiologic agents is the basis for rational treatment and control policies for infectious diseases. The design of improved diagnostic methodologies for parasites is a low priority for most basic researchers and private companies. However, the PCR is becoming available for use in the detection and speciation of parasites throughout the world. Hopefully, laboratories with ready access to fresh clinical material and training in use of the PCR, will help to define optimal PCR-based diagnosis and typing assays.

Acknowledgements

This study was supported by USAID PSTC projects 9.229 and 9.312 to I.K. and G.L.M., by NIH EY82905 to G.L.M., by the Pragma Corporation and Fullbright programmes, and by Pittsburg Supercomputer Center grant PSCB DMB890077P to M.V. We acknowledge the technical help of Unoba, Buga, Mrs Onapito and Mr Musisi and the assistance and warm hospitality of Dean Situma of Makerere University in this study. We thank Roberto DoCampo and Fernanda Gadehla of the University for supplying additional kinetoplastid isolates, Terry Bowersock and Della Richer of Purdue University for supplying bacterial strains, and William Collins of the Centers for Disease Control for supplying blood infected with *Plasmodium* species.

References

Barker, R.H. Jr, Banchongaksorn, T., Courval, J.M., Suwonkerd, W., Rimwungtragoon, K. and Wirth, D.F. (1922) A simple method to detect *Plasmodium falciparum* directly from blood samples using the polymerase chain reaction. *American Journal of Tropical Medicine and Hygiene* 46(4), 416–426.

Bishop, R.P., Sohanpal, B.K., Allsopp, B.A., Spooner, P.R., Dolan, T.T. and Morzaria, S.P. (1993) Detection of polymorphisms among *Theileria parva* stocks using repetitive, telomeric and ribosomal DNA probes and anti-schizont monoclonal antibodies. *Parasitology* 107, 19–31.

Bishop, R., Sohanpal, B., Kariuki, D.P., Young, A.S., Nene, V., Baylis, H., Allsopp, B.A., Spooner, P.R., Dolan, T.T. and Morzaria, S.P. (1994) Detection of a carrier state in *Theileria parva*-infected cattle by the polymerase chain reaction. *Parasitology* 104, 215–232.

Dalrymple, B.P., Dimmock, C.M., Parrodi, F. and Wright, I.G. (1992) *Babesia bovis, Babesia bigemina, Babesia canis, Babesia microti* and *Babesia rodhaini*: comparison of ribosomal RNA gene organization. *International Journal of Parasitology* 22(6), 851–855.

de Bruijn, M.H. and Barker, D.K. (1992) Diagnosis of New World leishmaniasis: specific detection of species of the *Leishmania brasiliensis* complex by amplification of kinetoplast DNA. *Acta Tropica* 52(1), 45–58.

de Bruijn, M.H., Labrada, L.A., Smyth, A.J., Santrich, C. and Barker, D.C. (1993) A comparative study of diagnosis by the polymerase chain reaction and by current clinical methods using biopsies from Colombian patients with suspected leishmaniasis. *Tropical Medicine and Parasitology* 44(3), 201–207.

Devereux, J., Haeberli, P. and Smithies, O. (1984) A comprehensive set of sequence analysis programs for the VAX. *Nucleic Acids Research* 12, 387–395.

Ellis, J., Hefford, C., Baverstock, P.R., Dalrymple, B.P. and Johnson, A.M. (1992) Ribosomal DNA sequence comparison of *Babesia* and *Theileria*. *Molecular and Biochemical Parasitology* 54, 87–96.

Erlich, H.A., Gelfand, D. and Sninsky, J.J. (1991) Recent advances in the polymerase chain reaction. *Science* 252, 1643–1651.

Fahrimal, Y., Goff, W.L. and Jasmer, D.P. (1992) Detection of *Babesia bovis* carrier cattle by using polymerase chain reaction amplification of parasite DNA. *Journal of Clinical Microbiology* 30(6), 1374–1379.

Figueroa, J.V., Chieves, L.P., Johnson, G.S. and Buening, G.M. (1992) Detection of *Babesia bigemina*-infected carriers by polymerase chain reaction amplification. *Journal of Clinical Microbiology* 30(10), 2576–2582.

Gebre-Michael, T., Lane, R.P., Frame, I.A. and Miles, M.A. (1993) *Leishmania donovani* infections in phlebotomine sandflies from the kala-azar focus at Aba Roba in Ethiopia: DNA probe compared with conventional detection methods. *Medical and Veterinary Entomology* 7, 294–296.

Hassan, M.Q., Ghosh, A., Ghosh, S.S., Gupta, M., Basu, D., Mallik, K.K. and Adhya, S. (1993) Enzymatic amplification of mini-exon-derived RNA gene spacers of *Leishmania donovani*: primers and probes for DNA diagnosis. *Parasitology* 107(5), 509–517.

Howe, D.K., Vodkin, M.H., Novak, R.J., Shope, R.E. and McLaughlin, G.L. (1991) Use of the polymerase chain reaction (PCR) to detect St. Louis encephalitis viral RNA. *Journal Virological Methods* 36, 101–110.

Jansson, A., Gillin, F., Kagardt, U. and Hagblom, P. (1994) Coding of hemolysins within the ribosomal RNA repeat on a plasmid in *Entamoeba histolytica*. *Science* 263, 1440–1443.

Jaureguiberry, G., Hatin, I., D'Auriol, H. and Galibert, H. (1990) PCR detection of *Plasmodium falciparum* by oligonucleotide probes. *Molecular and Cellular Probes* 4, 409–414.

Kain, K.C. and Lanar, D.E. (1991) Determination of genetic variation within *Plasmodium falciparum* by using enzymatically amplified DNA from filter paper discs impregnated with whole blood. *Journal of Clinical Microbiology* 29(6), 1171–1174.

Lopez, M., Inga, R., Cangalaya, M., Echevarria, J., Llanos-Cuentas, A., Orrego, C. and Arevalo, J. (1993) Diagnosis of *Leishmania* using the polymerase chain reaction: a simplified procedure for field work. *American Journal of Tropical Medicine and Hygiene* 49(3), 348–356.

McLaughlin, G.L., Vodkin, M.H. and Huizings, H.W. (1991) Amplification of repetitive DNA for the specific detection of *Naegleria fowleri*. *Journal of Clinical Microbiology* 29(2), 227–230.

McLaughlin, G.L., Montenegro-James, S., Vodkin, M.H., Howe, D.K., Toro, M., Leon, E., Armijos, R., Kakoma, I., Greenwood, B.M., Hassan-King, M., Marich, J., Ruth, J. and James, M.A. (1993a) Molecular approaches to malaria and babesiosis detection. *Memoranda of the Osvaldo Cruz Foundation* 87, Suppl. III, 57–68.

McLaughlin, G.L., Subramanian, S., Lepers, J.P., Raharimalala, L. and Deloron, P. (1993b) Evaluation of a nonisotopic DNA assay kit for diagnosing *Plasmodium falciparum* malaria in Madagascar. *American Journal of Tropical Medicine and Hygiene* 48(2), 211–215.

McLaughlin, G.L., Howe, D.K., Biggs, D.R., Smith, A.R., Ludwinski, P., Fox, B.C., Tripathy, D.N., Frasch, C.E., Wenger, J.D., Carey, R.B., Hassan-King, M. and Vodkin, M.H. (1993c) Amplification of rDNA loci to detect and type *Neisseria meningitidis* and other eubacteria. *Molecular and Cellular Probes* 7, 7–17.

Olsen, G.J. (1988) Phylogenetic analysis using ribosomal RNA. *Methods in Enzymology* 164, 793–812.

Persing, D.H., Mathiesen, D., Marshall, W.F., Telford, S.R., Spielman, A., Thomford, J.W. and Conrad, P.A. (1992) Detection of *Babesia microti* by polymerase chain reaction. *Journal of Clinical Microbiology* 30(2), 2097–2103.

Ralph, D. and McClelland, M. (1993) Intervening sequence with conserved open reading frame in eubacterial rRNA genes. *Proceedings of the National Academy of Sciences of the United States of America* 90, 6864–6869.

Reddy, G.R. and Dame, J.B. (1992) rRNA-based method for sensitive detection of *Babesia bigemina* in bovine blood. *Journal of Clinical Microbiology* 30(7), 1811–1814.

Saiki, R.K., Gelfand, D.H., Stoffel, S., Sharf, S.J., Higuchi, R., Horn, G.T., Mullis, K.B. and Erlich, H.E. (1988) Primer-directed enzymatic amplification of DNA with a thermostable DNA polymerase. *Science* 239, 487–491.

Smyth, A.J., Ghosh, A., Hassan, M.Q., Basu, D., de Bruijn, M.H., Adhya, S., Mallik, K.K. and Barker, D.C. (1992) Rapid and sensitive detection of *Leishmania* kinetoplastid DNA from spleen and blood samples of kala-azar patients. *Parasitology* 105(2), 183–192.

Snounou, G., Viriyakosol, S., Jarra, W., Thiathong, S. and Brown, K.N. (1993) Identification of the four human malaria parasite species in field samples by the polymerase chain reaction and detection of a high prevalence of mixed infections. *Molecular and Biochemical Parasitology* 58, 283–292.

Upcroft, J.A., Healey, A., Mitchell, R., Boreham, P.F.L. and Upcroft, P. (1990) Antigen expression from the ribosomal DNA repeat unit of *Giardia intestinalis. Nucleic Acid Research* 18(23), 7077–7081.

Vodkin, M.H., Howe, D.K., Visvesvara, G.S. and McLaughlin G.L. (1992) Identification of *Acanthamoeba* at the generic and specific levels using the polymerase chain reaction. *Journal of Protozoology* 39(3), 378–385.

Vodkin, M.H., Streit, T., Mitchell, C.J., McLaughlin, G.L. and Novak, R.J. (1994) PCR-based detection of arboviral RNA from mosquitoes homogenized in detergent. *Biotechniques* 17, 114–116.

Wozniak, E.J., Telford, S.R. and McLaughlin, G.L. (1994) Employment of the polymerase chain reaction in the molecular differentiation of reptilian hemogregarines and its application to preventative zoological medicine. *Journal of Zoo and Wildlife Medicine* 25(4), 538–547.

Index